Contributions to Management Science

More information about this series at http://www.springer.com/series/1505

Clemens Dachs

Viable Project Business

A Bionic Management System for Large Enterprises

Springer

Clemens Dachs
Dormitz, Germany

Dissertation University of Erlangen-Nuremberg 2020, accepted with the original title "Viable Project Business - A strategic success factor for mastering the complexity in large technical plant manufacturing".

ISSN 1431-1941 ISSN 2197-716X (electronic)
Contributions to Management Science
ISBN 978-3-030-62903-8 ISBN 978-3-030-62904-5 (eBook)
https://doi.org/10.1007/978-3-030-62904-5

This Springer imprint is published by the registered company Springer Nature Switzerland AG
The registered company address is: Gewerbestrasse 11, 6330 Cham, Switzerland

Dedication

For Cornelia, Felix, Regina and Valentin

Abstract

Large plant manufacturers face severe challenges caused by stagnating markets and new competitors. High pressure on project cost and duration lead to a demand for improvements in high speed.

Unfortunately, many organizations have large problems to learn fast enough. They have a long backlog of improvements that should be done, but their improvement machine which implements them is too slow.

The deeper reasons for this slow learning process come from inefficient work in teams and poor collaboration. This paper wants to find the best practices for collaboration between teams. It wants to identify how it can be managed that every team has a perfect environment and every business process is very efficient.

For this purpose, this paper uses an unconventional benchmarking partner: LIFE. In multicellular organisms trillions of cells work perfectly together. Every cell has a perfect environment and is highly efficient. Every chemical reaction of a cell is highly accelerated by catalysts.

This works analyzes the system dynamics of living organisms and translates them to a business environment. This translation is based on Maturana's and Varela's theory of autopoietic systems. However, it does not follow the known approaches of the Stafford Beer or Niklas Luhmann. Instead, it translates directly from the state-of-the-art literature of molecular biology to the world of business with the target to develop a system that can be practically implemented. This new translation comes to different results than the other approaches that translate the system dynamics of life.

All the translations are used to define requirements for a Viable Business Management System that is cross-checked with definitions of ISO 9000. The work also designs a specialized version Viable Project Business that can be used by industrial plant manufacturers.

Four case studies show that such a biologically inspired system can be implemented and has a significant business impact.

The work also contains several proposals for additional research.

Keywords:

Learning organization, Autopoiesis, Autocatalysis, System Dynamics, Life. Cell, Organism, Organization, Self-organization, Project Business, Lean, Agile, Cybernetics, System Theory

Preface

Dear reader,

The first time I thought deeper about "System Thinking" was around 2008 when I read the book "The 5th discipline" written by Peter Senge (Senge, 2006). He described an interview with a manager of Toyota who talked about the visitors that wanted to understand how Toyota works and what the reason for their success is. The Toyota manager said in the interview something like "Most of the visitors want to see all the methods and how we work. But they do not understand that everything is part of a system."

This statement really impressed me. In the following months and years, I understood more and more that all my expert knowledge was only about puzzle pieces and fragments. And all the books I read about Lean and Agile and every conference and discussion only added more and more puzzle pieces. But I could not see the big picture for an entire business.

When I started to implement Lean in a project business organization in 2012, I changed the perspectives away from lean methods back to teams. This was the first time I saw an analogy between cells in an organism and teams in an organization. Why do billion of cells work together so perfectly? They have no thinking in silos. Can we utilize that?

Since I had no idea about biology (school is now 30 years ago), I started to learn organic chemistry, biochemistry and molecular biology in detail, to understand living systems in detail. It was difficult to identify the core concepts and to translate them into my world of complex project business. In parallel we tried to implement the concepts in a management system. As expected, this showed great results.

But back to the book: This book is not another collection of success stories, experiences and principles or new methods. Every detail in this book is knowledge that already exists.

Instead the book describes how life manages to be so successful. It explains the structures and dynamics of life. And it translates it to social systems in a way, that it shall be possible to create social systems like business that have the same dynamics and with that the same success.

The subject matter might be difficult, but I hope I can take you on a very interesting journey which goes through nature from physics, chemistry to biology and applies it to social systems and business.

Maybe you can share my point of view that also in business there are deeper patterns at work. And these patterns are purely mathematical and proven in 4 billion years of evolution.

This is the first edition of this book. It focuses mainly on the core concepts and the levels of translation. This makes it possible to discuss the translation with various experts and extend it in later editions. Thus, there is only one thing left to say:

Go for a coffee and enjoy reading!

Clemens Dachs

Acknowledgements

I would never have had the energy to write such a complex story if there were not some colleagues supporting me by motivating me or contributing with ideas and feedback:

Researchers from FAU Erlangen for trusting in the idea of Viable Business and encouraging me to publish it as a thesis.
Prof. Dr. Eberhard Schlücker, Prof. Dr. Markus Beckmann, Dr. Michael Krenz, Prof. Dr. Kathrin Castiglione

Moritz Hornung for reading and discussing the countless versions of this book.

Amy Schol for helping me translate this work from denglisch into English.

The Viable Business Team developing the methods for large project business:
Moritz Hornung, Ralf Spateneder, Mona Karl, Englbert Westermeier, Christoph Fuchs, Harald Gebert, Rene Bernhard, Klaus-Peter Galuschki, Tobias Müller-Zielke

Senior Managers of Energy Transmission trusting in the idea and leading the change to a viable organization:
Hermann Kirchberger, Frank Grunert, Hauke Juergensen, Armin Moosburger

The Authors of "Mastering Disruption and Innovation in Product Management" who wrote their excellent book in parallel and motivated me in the darkest hours to continue writing:
Christoph Fuchs and Franziska Golenhofen

My friends and colleagues that supported me with their knowledge and experience and by motivating me (in alphabetic order):
Roland Berner, Katharina Beumelburg, Klaus Biehl, Rainer Blessing, Stefan Brem, Angelika Denk, Max Fleischer, Markus Forthaus, Thomas Fröhlich, Volker Ganz, Franziska Golenhofen, Jürgen Grabenhofer, Thomas Hahn, Bahram Hamraz, Martin Heider, Bodo Ikinger, Jürgen Kirsch, Wolfgang Korn, Stefan Lauter, Johannes Merkel, Daniel Rauter, Frank Rexer, Michael Seidler, Uwe Splettstoesser, Britta Stutzmann.

And my family: My wife Cornelia and our children Felix, Regina and Valentin for all their patience and their love

Contents

Figures

Tables

1 Introduction

1.1 Challenges in Project Business

Many large plant manufacturers face severe challenges for the future. The markets do not grow like in the 80s and 90s and new competitors want to get their share of the market. The competition is getting harder, which leads to a high pressure on quality, cost and duration.[1]

Solutions for these challenges can be new technical developments. Modular Product Platforms allow the configuration of complex plants based on defined components. Modern tools can present the plant in 3D in virtual reality or augmented reality. Additive Manufacturing can simplify production of components and Artificial Intelligence can improve plant control systems. All these ideas are presented at many conferences.[2]

All these new technical developments can only be used if the large plant manufacturer is able to introduce them. This requires change: change of products, processes, tools and roles. Competition means the demand for change is getting bigger and bigger. The ability to change is prerequisite for all other improvements.

Plant manufacturers that are too slow in implementing the necessary changes have problems to keep their competitive advantage. This leads to an even higher cost pressure and to projects with very low profit margins, increased risks to keep the margin and finally to crisis projects with high non-conformance costs. In such a situation it is even more difficult to improve the organization. In other words: A vicious circle!

Thus, every large plant manufacturer must have a focus on its capability to introduce complex changes in the organization. This capability itself does not come for free. The capability to improve also has to be improved itself. The benefit of this capability is the higher speed for all other improvements.

In reality this capability for change is very often insufficient. Large plant manufacturers do not just have problems to implement large changes fast. Even small improvements or lessons learned from projects can be problematic.[3]

1.2 Learning from Project to Project

Let us take an example for learning from project to project.

A Project Business executes two projects A and B and works on a new proposal C. It has a Top Management and several line management teams that are responsible for processes (Figure 1-1).

---

[1] See (VDMA, 2018)
[2] See (VDMA, 2016)
[3] See (Wiewiora, et al., 2009)

C. Dachs, *Viable Project Business*, Contributions to Management Science, https://doi.org/10.1007/978-3-030-62904-5_1

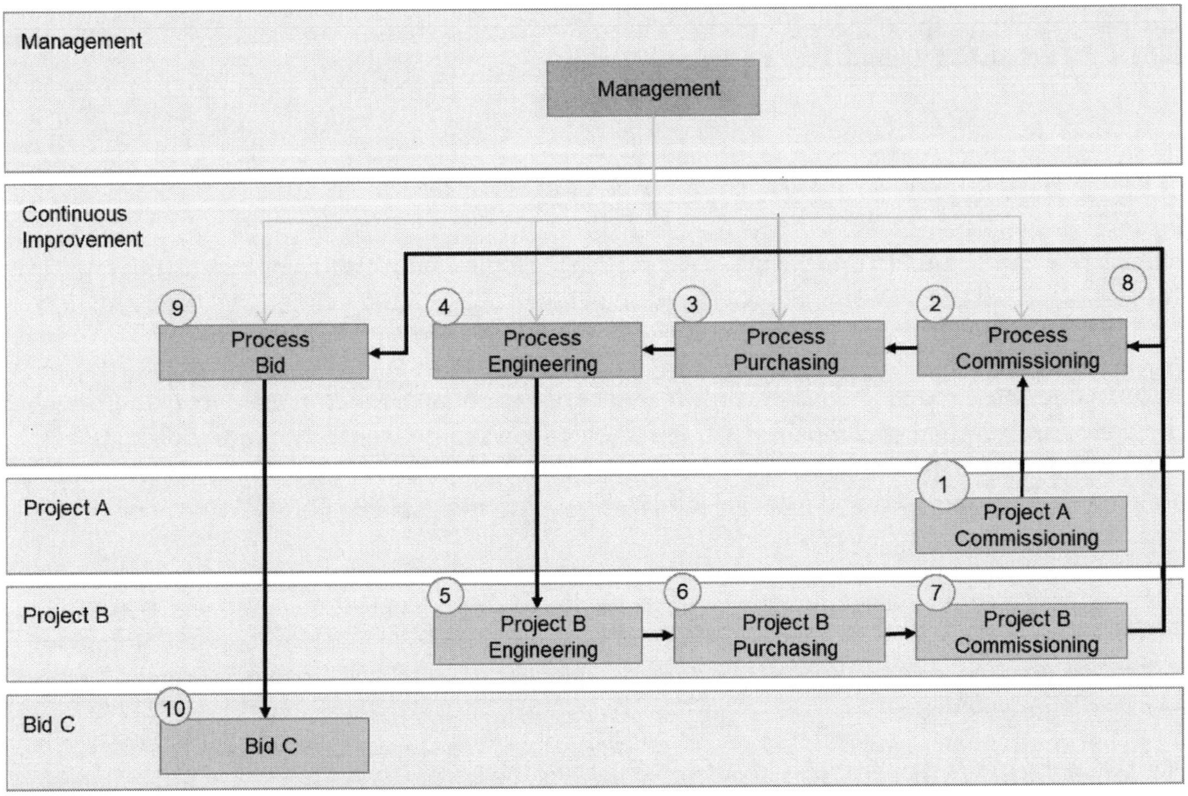

Figure 1-1: Learning in Project Business. Project A has a problem during commissioning. It needs a strong collaboration of many teams to solve the root cause.

(1) Project A has a problem during the commissioning phase: A cable is missing on the construction site. The reason for the missing cable happened during engineering. For the commissioning manager of project A it is no longer possible to change the root cause, instead he has to work on the symptom and try to get a cable as fast as possible. He also has to report the problem to his line manager, who is responsible for the process commissioning.

(2) The commissioning team is responsible to get the problem solved at the root cause. Like in many cases the root cause is in a different team, which has to be identified. Here the commissioning team has to contact logistics and procurement to find out why the cable is missing

(3) The procurement team has not ordered the cable, but it was due to a lack of information from engineering.

(4) The engineering team can detect the problem and change the standard process for cable engineering. Then it has to ensure that every project has to use the changed processes in the future. This happens by training, coaching and support for the next projects.

(5) Engineering subproject of project B now uses the new standard process that helps to avoid the problem

(6) Procurement in project B gets the right list

(7) ... and project B has all the cables.

(8) Since the standard calculations and standard schedules cover some of these problems, the solving of the problems can also reduce cost and duration. So, every process improvement team has to check, whether cost and duration can be reduced.

(9) Now the organization uses new standard values
(10) And applies them to new proposals

If all the steps work properly and at a high velocity, then a project business is able to learn fast and solve root causes of problems in projects. But it not only applies to the analysis of problems. The same mechanisms can also be used to collect ideas from projects and implement them or to make continuous improvement according to business targets.

Unfortunately, it is not so easy to implement this chain of 10 processes. The probability that it works is the product of the probabilities of every step. In reality there are many obstacles which slow down the improvement and thus avoid fast learning processes (Figure 1-2).

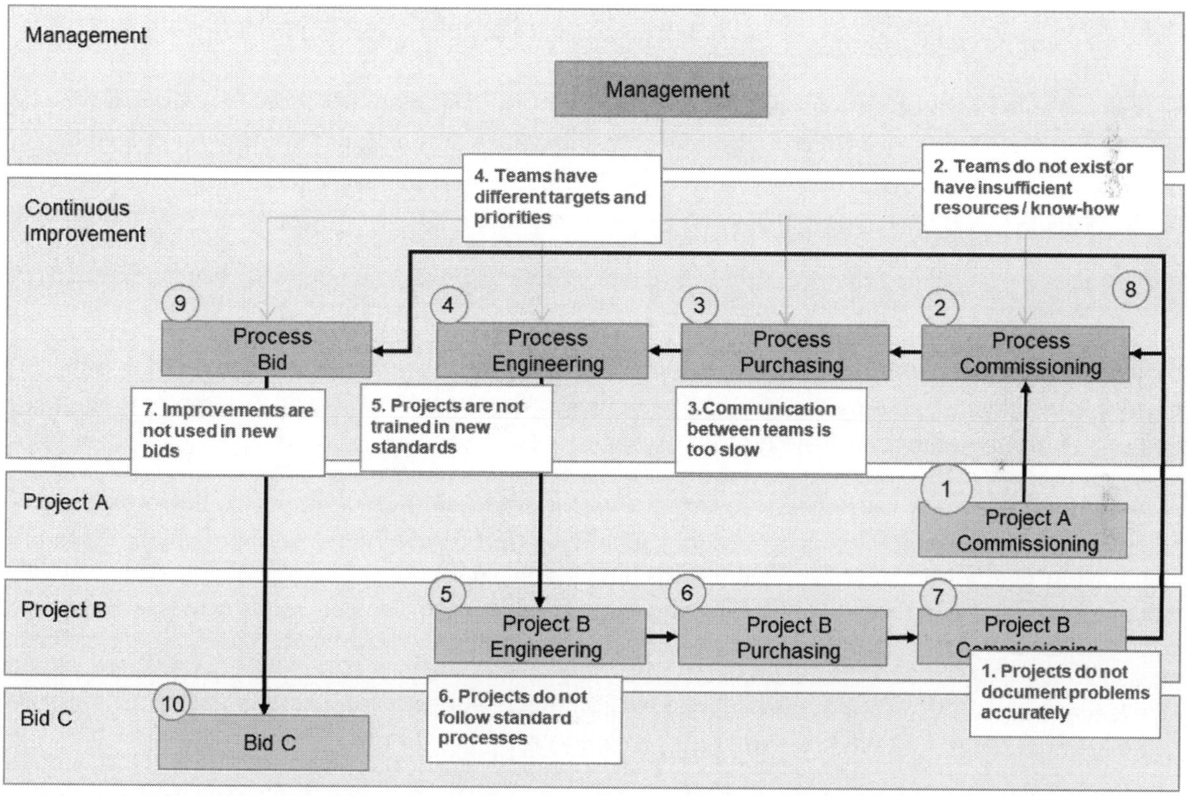

Figure 1-2: Problems with improvement. Each of the many teams involved in improvement introduces specific collaboration problems. Improvement is only possible if the entire chain of collaboration works.

Here are only some examples which are well-known in project business:

(1) Project Teams do not describe the problems accurately enough in non-conformance-reports.
(2) There are no process improvement teams at all or they have insufficient resources or know-how.
(3) The communication between the teams is slow (monthly meetings only). Escalation processes are not used.
(4) The different improvement teams have different targets and priorities (Engineering does not care what happens onsite), because of an insufficient target setting process.

(5) Even if a process is changed, the new projects are not trained, coached or not informed at all.
(6) New projects are informed, but the project manager wants to do it his way, instead of using a standard. This very often comes from unclear responsibilities in the matrix, where the project manager decides about processes.
(7) Improvements are not considered in new proposals.

This list contains three major categories of problems:

1. Teams do not exist or have insufficient resources. This problem cannot be solved by the team itself. It leads to the question which other team has created the team and who has to assign resources.
2. Teams exist, but they do not follow best practices. This problem has to be solved by the team itself. Very often this is a problem with missing discipline or know-how.
3. Teams do not collaborate rapidly and precisely with their neighbors.

All these problems reduce the speed of improvements. This results in the loss of competitiveness and even higher challenges, which cannot be solved because of the lack of improvement speed.

1.3 Improvement Machine

Let us use the term **improvement machine** for everything that is needed to implement improvements very fast. This includes all structures, processes, tools and roles that are needed to implement improvements. A good improvement machine allows the implementation of lessons learned from project and ensures that problems do not occur again. It also allows implementing all kind of new improvement ideas generated by employees.

Even more important, it is a necessary prerequisite for all larger improvements like introducing modular platforms, digitalizing processes or exchanging tool chains. All these major changes require such an improvement machine as well. Teams have to be aligned to the same targets, standard processes have to be defined and trainings have to be conducted. Projects really have to use the new standards.

An excellent organization needs an excellent improvement machine. What is the best organizational structure? What are the best processes for it? What is the benchmark?

1.4 Benchmarking

The improvement machine leads us to three questions:

1. How can we organize project business in a way that every single process has the best working conditions?
2. How can teams be highly effective and efficient?
3. How can we organize a perfect collaboration between a large number of teams, where every team contributes to the overall success in the best way?

What would be the benchmark for such an excellent team-play?[4]

Perfect team-play is needed in every project business. A first approach can be internal benchmarking with neighbor departments or benchmarking with other plant manufacturers. Perfect team-play is also required in many other industries. Maybe it is a better idea to make a best-of-best benchmark with other industries.

One possibility could be the benchmarking with automotive manufacturers like Toyota. They use lean methods to provide the best working conditions for people in production. Modern workers in production are trained in every detail of a process, they have all the tools they need, and they have all the material in high quality. Many teams have to work perfectly together to deliver cars just-in-time.

Another possibility are software companies. They focus on innovation. They have other processes in which they are good. Innovative organizations need to have best working conditions for innovation processes. They have excellent, creative people, who have all the information they need. They have a culture of trust, and they have enough time and budget. The organization has created a network of experts that help the innovators to develop ideas and to maintain motivation. This could be another benchmark for best working conditions and collaboration.

Both types of organizations are good at the same thing. They are able to provide the best working conditions for their core processes. In innovative organizations there seems to be the best working condition for innovators. In lean production companies there is a best working condition for production. In all cases the best working conditions make the important processes much more efficient and effective.

Which organization is the best? And if we could find out who the best is, then it is not so easy to get all the information. We would have to analyze their management system in detail and understand the full system dynamic. If we have understood all the details, then we have to re-contextualize this knowledge to plant manufacturing.

Fortunately, there is a much better way: we do not focus on companies. We even do not focus on social systems.

Because there is a benchmark which outperforms all of them: LIFE

Life is a type of system which creates perfect conditions for chemical reactions. The important chemical reactions needed by cells are accelerated up to 100 billion times. Cells can not only build themselves. Bacteria can duplicate themselves every 30 minutes. One bacterium is able to produce billions of descendants in a couple of days. [5]

The human body contains trillions of cells that work together perfectly. Every cell in the human body has a clear contribution to the human as a whole.[6] Life is also very stable in the long-term. After 4 billion years we can consider it a well-proven system.

---

[4] See (Gerberich, 1998) for the method of benchmarking

[5] See (Alberts, et al., 2015)

[6] See (Sadava, et al., 2011)

All these considerations are not new: Even large consulting companies think that the organizations of the future will look like biological organisms. McKinsey writes about "the new paradigm - organizations as living organisms" that an agile organization is a "network of teams [...] that co-creates value for stakeholders."[7] This already shows the idea to compare teams with cells and large organizations with large multi-cellular organisms. However, this paper refers to agile methods only, and doesn't compare them with biological organisms.

The consulting company Accenture writes about "Living Business" that organisms are a benchmark for sustained growth.[8] They also don't give deeper biological insights.

Both papers show that the idea is not new. However, it seems to be a trend, to understand organizations more and more as an organism. What is missing, is a detailed translation that can be used in real organizations.

And this leads to the research idea.

1.5 Research Idea

The last sentences showed that life forms have excellent solutions for our questions. Thus, the idea is:

Develop a Management System that utilizes the System Dynamics of Life.

We need a model that provides a biological perspective on organizational design with a clear structure, where every concept has a clear logical explanation. It should be possible to understand this model without any biological knowledge. And even more important: It must be possible to implement it in a real environment to make organizations successful.

1.6 Research Questions

For the realization of the research idea we need the answer to the following questions:

1. **Can the understanding of biological organisms be used to design the system dynamics of business?**

The first question is about the benchmarking of organizations with life forms. Benchmarking only makes sense if it is able to produce new insights. It shall give insights how to do something. Thus, it would not be sufficient to identify the analogy that both organisms and organizations have to adapt to a changing environment. Benchmarking shall transfer the solution. It shall show how biology masters this challenge, why the biological approach makes sense and how it can be translated to business.

2. **Can the system dynamics of life be used to design a management system that helps to significantly improve project business?**

[7] See (McKinsey & Co, 2018 S. 6)
[8] See (Accenture, 2018)

This second question is about a model for viable organizations that includes all translations. Such a model has the benefit that it has a clear system dynamic that is well-proven in 4 billion years of evolution. It can be used as a reference to understand the system dynamics of specific methods or method frameworks.

Besides the deeper understanding of organizations, it has a second target. It must help organizations to become more successful. It shall deliver the improvement machine that has been discussed in the introduction and thus allow organizations to improve at high speed.

3. **Can typical methods of Lean and Agile be explained by the system dynamics of life?**

A validation of the model is only possible if all the findings can be practically tested in a large organization. Thus, it would be impossible to design completely new structures and methods. Maybe this is not necessary. Lean production contains many well-documented methods which work perfectly together. Maybe the reason for the success of Lean and Agile is that they already implement some of the system dynamics of life.

This third question tries to understand the existing methods in the context of the system dynamics of life. Our model shall not be another puzzle piece alongside existing methods; instead, it should be the template that can be used to locate and connect the already existing puzzle pieces.

2 State of the Art

In this section we want to review the current literature to understand whether system dynamics of life can help to improve the learning in project business.

We will start this review with learning in projects and project organizations, before we focus on the learning organization in general. The learning organization has been proposed by Peter Senge and is based on a thinking in systems. We are especially interested in those systems that utilize the system dynamics of life. We will follow two directions: Management cybernetics focus on the control of complex systems, whereas autopoiesis focuses on the self-creation of systems. Finally, we will discuss the current knowledge of autocatalytic systems, which realize this autopoiesis.

2.1 Projects

Two international organizations work on the standards for project business and give first answers: The Project Management Institute (PMI)[9] and the International Project Management Association (IPMA)[10].

Project Management Institute

In the PMBOK the PMI describes a project as "a temporary endeavor to create a unique product, service or result" and project management as "the application of knowledge, skills, tools, and techniques to project activities to meet the project requirements."[11]

These characteristics already show the basic reasons for the problems with learning. Projects have a unique purpose, unique results and a unique set of processes. These characteristics make it difficult to learn inside the project. If many processes happen only one time in the project, there is no repetition and thus no chance to do it better a second time.

The PMBOK describes that lessons learned can be created at any time in a project and be added to a lessons-learned knowledge base.[12] Lesson learned are especially collected in process *perform quality control* [13] or in *closing procurement*[14] or *project closure*[15]. The collected lessons learned become part of the organizational process assets, which contain processes and procedures and the corporate knowledge base.[16]

PMBOK refers to lessons learned or organizational assets on several pages. However, there is no description, how the lessons learned are used to make process improvements. This happens somewhere outside the project.

---

[9] See (PMI, 2018)
[10] See (IPMA, 2018)
[11] See (Project Management Institute, 2008 pp. 5-6)
[12] See (Project Management Institute, 2008 p. 437)
[13] See (Project Management Institute, 2008 p. 214)
[14] See (Project Management Institute, 2008 p. 344)
[15] See (Project Management Institute, 2008 pp. 101-102)
[16] See (Project Management Institute, 2008 p. 32)

C. Dachs, *Viable Project Business*, Contributions to Management Science, https://doi.org/10.1007/978-3-030-62904-5_2

International Project Management Association

The IPMA has three main publications: The Project Excellence Baseline, the Individual Competence Baseline and the Organizational Competence Baseline.[17]

In the Project Excellence Baseline (IPMA-PEB) the IPMA refers to a project as "a unique set of processes consisting of coordinated and controlled activities with a start and end date, performed to achieve project objectives"[18]. This definition can also be found in ISO 9000 with minor differences[19].

In its executive summary the IPMA-PEB emphasizes that continuous improvement of project management methods is one of the goals of its project excellence model.[20] Continuous Improvement is based on the PDCA-cycle and can be done inside the project.[21] It can be especially used for the repetitive tasks of project management. The transfer of knowledge to other projects is not explicitly described.

Summary

Project management literature does not describe the details of systematic improvement, because it happens outside projects. Thus, we have to understand the entire organization.

What are typical organizational structures for large plant manufacturers?

2.2 Organizations

Robbins describes an organization as "a consciously coordinated social unit, composed of two or more people that functions on a relatively continuous basis to achieve a common goal or set of goals"[22] .

If there are more persons involved, the organization has an organizational structure. Robbins and Judge describe the following classical structures[23]:

- Simple structure / Flat structure without hierarchy
- Bureaucracy
- Functional structure (structuring by functions like marketing, production, sales)
- Divisional structure (structuring by products, services or structuring by markets)
- Matrix structure (combination of functional and divisional structures)

Krajewski et al compare these typical organizational structures for projects[24]:

- Projects in a functional structure have problems to ensure that every project team member works with high priority on the project

[17] See (IPMA, 2018)
[18] See (IPMA-PEB, 2018 S. 38)
[19] See (ISO 9000, 2015)
[20] See (IPMA-PEB, 2018 S. 7)
[21] See (IPMA-PEB, 2018 S. 45)
[22] See (Robbins, et al., 2017 S. 44) and (Robbins, 2001)
[23] See (Robbins, et al., 2017 S. 538-542)
[24] See (Krajewski, et al., 2016 S. 260)

- Projects in a divisional structure have all people on board. The project manager has the authority to align everybody. Typically, these projects have problems in exchanging knowledge between projects
- The matrix organization is a mixture that wants to eliminate the disadvantages. The challenge here is to develop clear split of responsibilities between the project manager and the functional manager.

The matrix is very typical for project business. The project manager drives the project, whereas the line managers take care about competences, processes and tools for their trade.

There are a few other organizations to be mentioned, which became popular in the last years:

Laloux describes new forms of organizations in "Reinventing Organizations". Additionally, he describes the new approaches in a historical context. Every new management style was the answer for the problems of its time. This historical perspective also shows that not everything in conventional models is bad. New models have to ensure that they solve not only new challenges but also the old owns.[25]

Holacracy is a specific model that implements this new thinking. Robertson describes how to design an organization based on roles and circles. Roles are the elements of the design. Circles can contain roles or other circles. The top management is called the Anchor circle. The design has the same structure as a file structure of directories (circles) and files (roles). It is hierarchical. The important difference is, that every circle is allowed to choose its own representative, the rep-link. Thus, each team selects its own manager[26]. Holacracy has described all the rules in a constitution, that can be downloaded at the website [27] .

Exner et al compare organizations with living organisms. The work describes that the organization consists of parts that could be autonomous (e.g. business units) and some infrastructure parts (e.g. corporate units) that keep everything together. The main focus of the text is on the control of such an organization. The most interesting aspect at this work is the comparison with life forms.[28]

Summary

Divisional structures are focused on projects and have problems with lessons learned. Functional structures allow to learn fast, but they slow down work in projects. The matrix organization combines both approaches and allows fast work in projects and fast learning, but they introduce more complexity and role conflicts.

How can lean and agile methods be used in such a matrix organization?

2.3 Lean and Agile

Most large industrial plants have a software and a hardware part. In the last years software development more and more follows agile methods rather than the waterfall approach.

---

[25] See (Laloux, 2017).
[26] See (Robertson, 2014)
[27] See (Holacracy , 2018)
[28] See (Exner, et al., 2009)

Manufacturing of hardware components very often follow lean methods. This introduced the concepts of lean production and agile development to large plant manufacturers.

Agile

Schwaber described Scrum as a project management method that helps to work in a team when the future is uncertain. Scrum uses several standardized roles, like the Scrum master and the product owner, and some structures like the daily standup meeting, the sprint planning and review meeting and the retrospective. The team collaborates to develop software features that are defined in the product backlog. The focus of Scrum is to manage teamwork in software development.[29]

Larman and Vodde showed with LeSS (Large-Scale Scrum) that Scrum can also be used with a larger set of teams. They propose a specific teamplay between the single Scrum teams. LeSS Huge can be used for development with more than 8 teams.[30]

Butterhof analyzed whether agile methods can be used in large plant manufacturing projects. The work shows the specific challenges in plant manufacturing processes, which are different from SW-development. These differences include

- Project scope is fixed, high penalties in case of delay
- High influence of material cost
- Long lead-times for main components
- Dependency from other organizations (suppliers, consorts,...)
- High technical dependencies

Some of the agile methods, like visualization and retrospectives, can be used despite these challenges. Butterhof proposes to define different classes of agility, from pure agility, via hybrid models to pure waterfall models and proposes a project classification to select the best type.[31]

Lean

Lean production was developed by Toyota. The former production manager Taiichi Ohno described the evolution of the Toyota Production System. His work includes key concepts like standardized work, improvement, kanban and just-in-time. He already writes that the organization is similar to the human body and needs an autonomous nervous system. Obviously, the thinking in systems and the biological metaphor is not new to Toyota.[32]

In the 1990s, lean production became popular in western manufacturing companies. Womack et al described the differences between Toyota and the American automotive industry to explain the differences in profitability.[33]

Later, Liker developed 14 principles that describe the various aspects of Lean. These principles cover areas like the long-term philosophy, process flow, pull systems, standardized work,

---

[29] See (Schwaber, 2004)
[30] See (Larman, et al., 2017)
[31] See (Butterhof, 2017)
[32] See (Ohno, 1988 p. 73)
[33] See (Womack, et al., 2007)

continuous improvement, development of people and suppliers and other principles. The list of principles helped to see that Lean is not only about one-piece flow and kanban. Instead, it requires implementing a different culture.[34]

Other books are more specialized on continuous improvement. Sobek describes the A3 problem solving process that is used for continuous improvement. A3 is named by the paper-size of the reports. The method is based on the PDCA-cycle of Deming. In his work, Sobek pointed out how systematical and logical thinking helps to identify root causes of problems and allows eliminating them.[35]

Rother also describes continuous improvement with A3, but he emphasizes that they have to be aligned to a future state. The organization has a current state that describes how it works today, and it needs to have an agreed future state that is better. Improvements with A3 contribute to reach this agreed future state.[36]

Jackson describes how this future state, and the main improvements can be defined during a process called Hoshin Kanri or lean policy deployment. Hoshin Kanri helps to define a clear future state for the entire organization and helps to align all improvements to that direction.[37]

Summary

In the matrix organization we can use agile methods to design agile project teams. Projects are endeavors and therefore not completely predictable. Agile methods can help to continuously adapt to the situation. Lean methods can be applied by the functional managers. The functional managers have to provide best working conditions for the projects and continuously improve the processes towards a future state. Both together help the organization to learn and to adapt.

The challenge is only to combine all the lean and agile methods in the right way. Which of the many hundreds of methods and principles shall be used? Having a big picture of the organization, or in lean terms an overall long-term future state, would help to understand the details. Unfortunately, Hoshin Kanri describes how to develop a future state, but it does not describe a reference model for future states. Such a reference model needs another perspective.

2.4 Learning Organization

All lean and agile methods can be seen as puzzle pieces of a larger picture: The learning organization. The learning organization aligns the activities of all employees to one vision and it allows continuously improving the organization.

Learning Organization

In his bestseller "The 5th Discipline – The Art and Practice of the Learning Organization" Peter Senge described five disciplines that have to be mastered to achieve a learning organization:[38]

---

[34] See (Liker, 2006)
[35] See (Sobek, et al., 2008)
[36] See (Rother, et al., 2009)
[37] See (Jackson, 2006)
[38] See (Senge, 2006 pp. 6-10)

1. Personal Mastery
2. Mental Models
3. Shared Vision
4. Team Learning
5. Systems Thinking

Personal mastery is about the individual striving for excellence. It includes the individual's lifelong learning and seeking for perfection. This individual energy to reach excellence drives the other four disciplines.[39]

Mental models are about the unconscious assumptions that influence how individuals perceive the world and how they act. Senge points out how important it is to improve these mental models. Otherwise, all energy to reach excellence can be wasted, even if the target is correct.[40]

A **shared vision** aligns many individuals to the same target. Senge highlights that the shared vision is not just a management statement. The shared vision is that what individuals really want to achieve. It gives them the direction for their ambition to achieve personal mastery[41]

Team learning is the discipline that goes beyond individual learning. Senge proposes the dialogue between team members to find solutions that are better than the initial ideas. He contrasts it with a discussion, where team members only want to convince the others, instead of learning from others.[42]

Finally, Senge sees the fifth discipline "**Systems Thinking** "as the most important link between the other disciplines. Systems Thinking creates a logical model of the organization that shows how individual contributions are linked to the shared vision. Such a model can be created via team learning by bringing all information together. This joint system model also helps to change the mental models. Each individual begins to understand the big picture and can align its individual contributions[43]

The system model Senge proposes with System Thinking is similar to the Current and Future States that are used in the lean policy deployment process Hoshin Kanri. Both create a logical picture of the organization that is aligned to the shared vision. Both use the models to align to individual contributions.

However, one aspect is different between Peter Senge's system model and most Hoshin Kanri implementations. Peter Senge explicitly sets the focus on System Dynamics.

2.5 Systems Dynamics and Theory of Constraints

System Dynamics

Forrester describes a language to express the time behavior of complex systems. This language is used to simulate complex systems, e.g. the works of the Club of Rome.[44]

---

[39] See (Senge, 2006 p. 7)
[40] See (Senge, 2006 p. 8)
[41] See (Senge, 2006 p. 9)
[42] See (Senge, 2006 p. 9)
[43] See (Senge, 2006 p. 11)
[44] See (Randers, 2012)

Sterman adapts this to a business environment. The language of System Dynamics allows to describe logical dependencies between observables. The observables are written as nominal phrases and the dependencies are drawn as unidirectional arrows. A plus sign indicates that the effect increases whenever the cause increases. A minus sign indicates that the effect decreases whenever the cause increases. [45]

System Dynamics is about the graph as a whole that consists of these elements. This graph can have causal loops, where one observable directly or indirectly influences itself. There are two main types of causal loops.

The reinforcing loop is a positive feedback loop (Figure 2-1). A change of the observable creates an additional change to the same direction. In the case of positive changes this leads to growth. Reinforcing loops are marked with an "R" in large diagrams.[46]

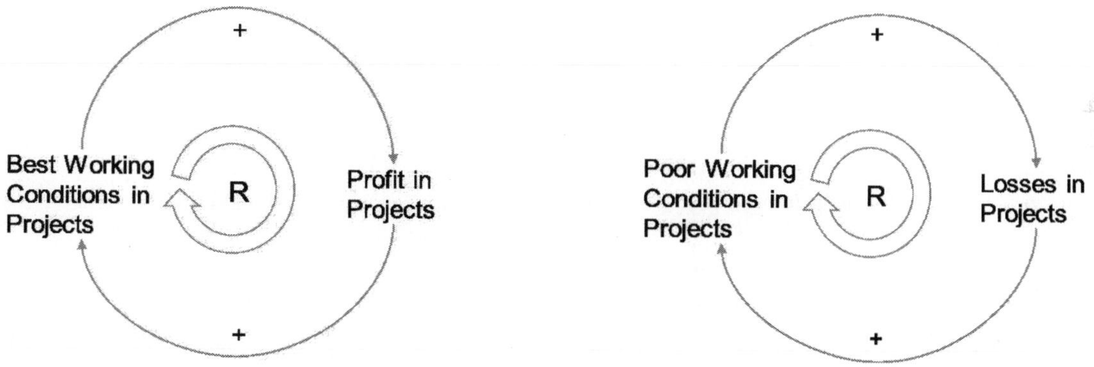

Figure 2-1: Reinforcing loops. Best Working Conditions lead to better projects lead to higher profit that can be reinvested. The plus sign indicates that the effect will be increased when the cause is increased. Poor working conditions increase losses and make the working conditions even worse. This is also a reinforcing loop or a vicious circle.

Balancing loops have a negative feedback (Figure 2-2). These loops do not result in growth; they stabilize themselves. A homeostat is a balancing loop. Balancing loops are marked with a "B" in large diagrams.[47]

---

[45] See (Sterman, 2010)
[46] See (Sterman, 2010)
[47] See (Sterman, 2010)

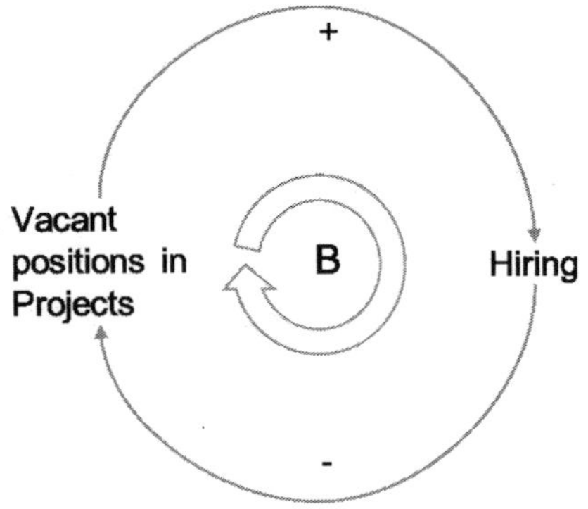

Figure 2-2: Balancing loops. The more positions are vacant the more people can be hired. The more people are hired the less positions are vacant. The loop balances itself. At the end no positions are open and nobody has to be hired.

Both types of loops appear in complex systems. Additionally, all the logical dependencies can have some time delay. This language already allows to model the System Dynamics of complex systems. More sophisticated models use additional symbols for stocks and flows.[48]

System Archetypes

Senge shows that the System Dynamics of organizations consists of typical patterns and presents a set of so-called System Archetypes. These are patterns that occur and have a specific behavior.[49]

Assumed, the organization has created such a detailed system model as its future state: How can such a complex system by optimized to achieve sustained growth?

Theory of Constraints (TOC)

In his work "Theory of Constraints" Goldratt described the concept in 5 points[50]

1. Identify the system's constraints
2. Decide How to Exploit the Systems Constraints
3. Subordinate Everything else to the Above Decisions
4. Elevate the System's Constraint
5. If in the previous steps a constraint has been broken, go back to step 1

This concept takes a systemic perspective. It can use the System Dynamics model to find the bottlenecks. The five steps can be understood as the team learning process of Senge that follows the PDCA logic. The team models the System Dynamics in order to identify and eliminate the bottlenecks.

On the other hand, the identification and elimination of bottlenecks are also processes of the organization. They form a balancing loop that tries to keep a system in balance.

---

48 See (Sterman, 2010)
49 See (Senge, 2006).
50 See (Goldratt, 1990)

Summary

System Dynamics allows creating a model of the organization with the focus on causal loops. Reinforcing loops are positive feedback loops and grow with exponential speed, whereas balancing loops are negative feedback loops that try to reach a balance.

A large model consists of many of these loops and can contain System Archetypes as patterns. Its growth is limited at its systemic bottleneck. The Theory of Constraints shows how to identify and eliminate it, in order to maximize growth. This identification and elimination of bottlenecks is itself a balancing loop that has to be part of the system.

Obviously, there are many system dynamics possible. For our purpose we want to have the best one: The system dynamics of life. This is the system dynamics that let living organisms live long and prosper. It allows E-coli bacteria to duplicate every 30 minutes. The system dynamics of life can then be used in the sense of Peter Senge's 5th discipline to provide a shared model that links individuals to the shares target. In the more practical world of Lean and Agile this can be understood as an excellent long-term future state that is used for continuous improvement. This helps to align improvements also in project business.

Thus, the question is: What are the system dynamics of life?

2.6 Autopoiesis

Autopoiesis as a Holistic Perspective on Living Organisms

Maturana and Varela introduced the term "Autopoiesis", which emphasize that a system produces ("poiese") itself ("auto").

Maturana and Varela showed that Autopoiesis is one of the main properties of living systems. Living systems are not based on other substances of other physical forces than non-living systems. It is not the components that make a system viable; it is its system dynamics.[51,52]

They compare living organisms with machines. Both produce something, but there is an important difference. A machine is allopoietic; it creates a product that is different from itself. The living organism is autopoietic; the product is its own components (Figure 2-3). The authors point out that the purpose of the machines is always what they actually do. It is the purpose of a machine to produce some kind of product. If a living organism creates itself, then it is its purpose. The living organism has an internal driving force to reach this purpose that is created by the organism itself.[53]

The product of any machine has more to do with the processes than with the exact design of the components. All parts of the machine that do not influence the process or the relations between the components could be removed or changed.[54] For a living organism this means that the processes in the cell and the relations between the components are more important than the exact design of the components. This perspective makes viability independent from the world of chemical

---

[51] See (Maturana, et al., 1980 S. 77)

[52] This explains why the value of a company differs from the value of its tangible assets. The value of a company is in its system dynamics not in its components. This will be clearer after our detail translation.

[53] See (Maturana, et al., 1980 S. 77-78)

[54] This is also the base for design to value. See Fuchs and Golenhofen for an excellent introduction (Fuchs, et al., 2019)

substances and allows comparing living organisms with other system that consist of very different components.[55]

Maturana and Varela define this autopoietic machine as "*a machine organized (defined as a unity) as a network of processes of production (transformation and destruction) of components that produces the components which: (i) through their interactions and transformations continuously regenerate and realize the network of processes (relations) that produced them; and (ii) constitute it (the machine) as a concrete unity in the space in which they (the components) exist by specifying the topological domain of its realization as such a network.*"[56]

This abstract definition makes a cell to an autopoietic machine, because its metabolic network (of chemical reactions) create the DNA and RNA and the proteins, which (i) catalyze all the chemical reactions, and (ii) which (together with DNA, phospholipids and other molecules) constitute the cell as a unity, and (iii) where the DNA specifies the RNA and proteins and with that the reactions.

The abstract definition of Maturana and Varela has now the advantage that it can be used for other systems that are not based on DNA, RNA and proteins or even not based on physical components at all.

Maturana and Varela also stated that autopoiesis makes living systems to systems independently from any observer. All other systems, like cars, have a system boundary that is defined by an observer. The autopoietic system defines its own boundary. The system is that what is produced by the system itself.[57]

Maturana and Varela claimed that "**autopoiesis is necessary and sufficient to characterize living systems**".[58]

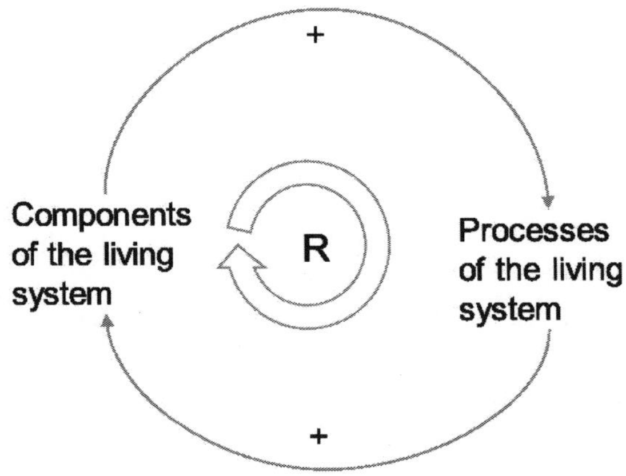

Figure 2-3: Autopoiesis as the main reinforcing loop of living systems

---

[55] See (Maturana, et al., 1980 S. 77-78)

[56] See (Maturana, et al., 1980 S. 79). Comment: This definition of the autopoietic machine already shows two problems. The human language has problems to express a phenomenon like self-creation in one simple sentence. Additionally, such sentences are even harder to understand if they are written in abstract terms rather than for a specific example. Both facts make the concepts difficult to understand and difficult to apply for a specific situation.

[57] See (Maturana, et al., 1980 S. 81)

[58] See (Maturana, et al., 1980 S. 82)

Maturana and Varela also describe the Homeostasis of living organisms as a homeostat that tries to maintain the ideal organization (which includes the process network and the relations between the components). The homeostat tries the minimize the difference between the actual state and the ideal state of the organization (Figure 2-4).[59]

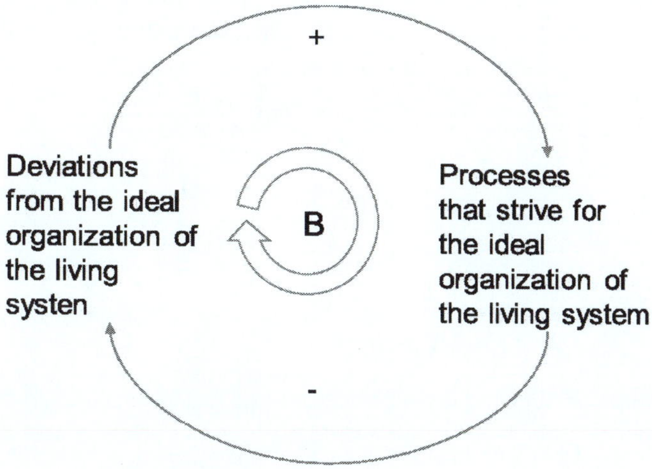

Figure 2-4: Homeostasis as the balancing loop of living systems

Today the mechanisms of self-creation or well-understood and documented in biology textbooks.[60]

What Maturana and Varela do not describe is how the components manage to create the network of processes. They already speak about enzymes and catalysis[61], but they don't connect the growth of the organism with the increased speed of catalyzed reactions. As a consequence, they do not explain how the hen-egg problem can be solved: It is impossible to understand the components, if we do not understand which processes and vice versa. This hen-egg problem leads to enormous difficulties to construct such autopoietic systems in a different context than biology.

Stafford Beer, who wrote the preface to the book, was especially excited in the observation, that living organisms are homeostats.[62] This links autopoiesis, the theory of living systems, with cybernetics, the theory of control.

2.7 Cybernetics and the Viable System Model

Bio-cybernetics and neuro-cybernetics are maybe the best-known examples for biologically inspired system theories which can be practically used in a business environment.

Cybernetics

Wiener can be seen as the founder of the science of cybernetics. Cybernetics is about the control of complex systems. Wieners book "Cybernetics: Or Control and Communication in the Animal and the

---

[59] See (Maturana, et al., 1980 S. 78)

[60] See (Alberts, et al., 2015) or (Sadava, et al., 2011) for the description of self-creation of cells.

[61] See (Maturana, et al., 1980 S. 90-92)

[62] See (Maturana, et al., 1980 S. 66)

Machine" from 1948 made a link from Biology to technical systems. This linkage can be seen as the starting point for a comparison of the human nervous system with technical systems.[63]

Viable System Model

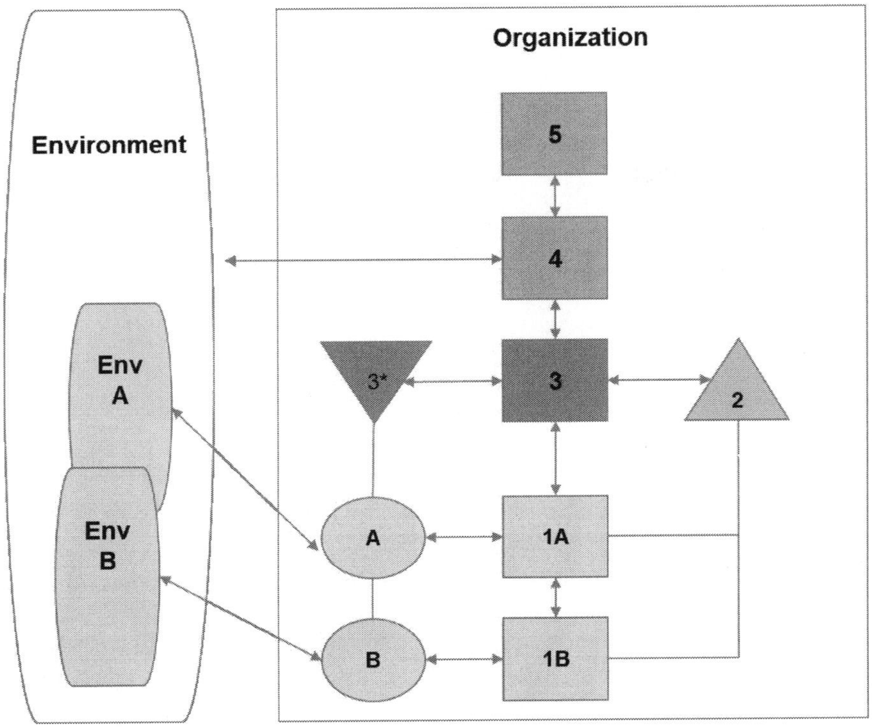

Figure 2-5: Simplified version of the Viable System Model. An organization interacts with the environment with two divisions a and b. Both divisions A and B have operations, represented by circles, and management, represented by a box. On a higher level the organization coordinates the work of the divisions with systems 2-5[64]

A more specific model gives Beer with his Viable System Model (Figure 2-5). It explains how multi-level systems can coordinate activities and adapt to the outside world. Beer compares management of an organization with the central nervous system of humans that is able to control the activities of many organs. Each organ has its own purpose and contributes to the viability of the entire organism. Beer points out that the brain has not the task to control every activity. Some activities like digestion, respiration or the management of blood pressure are managed decentralized without the need to control them consciously. However, sometimes these activities have to be coordinated by the autonomous nervous system (sympathetic and parasympathetic system) or by the brain itself. On a higher level the brain integrates various perceptions of eyes and ears and internal perceptions to make decisions that can be executed by the organs.[65]

Beer translated these functions in the Viable System Model (VSM). VSM is a recursive model where the entire diagram focuses on one defined level, for example the corporate level. Some parts of the diagram refer to the divisions of the organization. On a lower level each division can be modelled with the same techniques. The graphical elements (boxes, circles and triangles) represent

[63] see (Wiener, 1949)
[64] See (Beer, 1990 S. 136) or (Beer, 1995 S. 130-131) for the more detailed picture. Other presentations can be found at (Malik, 2008 S. 458)
[65] See (Beer, 1995 S. 73-149)

subsystems that have to be realized. The arrows represent the channels that transfer variety (the measure for complexity) between the subsystems. [66]

System 1 represents the divisions of the organization. It consists of operations that interact with the environment and the management of these operations. These divisions earn the resources for the organization. The arrows indicate that operations have a different complexity than the complexity of the market in the environment. For example, the divisions can produce only a variety of products, which does not meet every customer need. The complexity of operations is on the other hand not fully known to management. Management needs a simplified picture of operations[67]. Beer compares system 1 with the organs that fulfill viable functions and feed the nervous system.[68]

System 2 contains all mechanisms that reduces conflicts between the divisions of system 1 without the need of control by a senior management. Beer uses the example of a simple timetable in a school that helps to reduce conflicts. The work with a shared timetable can be done decentralized or by support functions without the need of a higher authority.[69] Beer compares the system 2 with the sympathetic system.[70]

System 3 contains the here and now activities of the senior management. This function has the authority to control all divisions (System 1). Senior management sets the legal and corporate rules and negotiates resources to narrow down the variety in the divisions. By giving authority to the divisions it can reduce its own variety. System 3 is also connected with System 2, which keeps the systems 1 stable, and with a System 3*, the audit system, that provides additional information about operations that is not filtered by the division management of System 1.[71,72] Beer compares the system 3 with pons and medulla, and system 3* with the parasympathetic system.[73]

System 4 adds a perception of the environment as a whole and a perception of the operative management and System 4 itself (!). System 4 ensures that the entire organization can adapt to the outside world.[74] Beer compares system 4 with the diencephalon and base ganglia, which integrate the perceptions from eyes and ears.[75,76]

System 5 is the highest management level which can make decisions based on the perceptions of System 4, especially if internal and external perception give no indications which decision is best. Thus, system 5 reduces variety in the other systems by making clear decisions.[77] Beer compares

---

[66] See (Beer, 1990 S. 1-18) for the discussion of recursion and (Beer, 1990 S. 19-35) for the discussion of variety and its attenuation and amplification.
[67] see (Beer, 1990)
[68] See (Beer, 1995 S. 131)
[69] See (Beer, 1995 S. 66-70)
[70] See (Beer, 1995 S. 131)
[71] See (Beer, 1990 S. 37-53)
[72] In lean management this is called genchi gembutsu (Go-and-see). The senior manager observes reality directly and thus avoids to be fully dependent on the reports of others.
[73] See (Beer, 1995 S. 131)
[74] See (Beer, 1990 S. 107-121)
[75] See (Beer, 1995 S. 131)
[76] Eyes and ears are connected to the thalamus
[77] See (Beer, 1990 S. 123-135)

system 5 with the cortex that has associations based on the perceptions of system 4 and experience.[78]

Even if the Viable System Model looks like a kind of organizational chart, it is not. The systems 2-5 can be implemented by a single person who creates structures that simplify autonomous coordination (System2) of the divisions, who does the operative management (System3), the strategic management (System 4) and leads the organization as whole towards a purpose (System 5).

Even more important: The arrows between the systems are not about resources and not about information. Instead, they are about variety. They describe how the complexity of the organization can be systematically managed. The activities of the systems are about amplification of attenuation of variety. Managers who decide between many alternatives, reduce the variety for others. Managers can also decide to amplify variety, for example by adding new product variants.

In today's VUCA world, VSM helps to manage the complexity for multi-level organizations. VSM explains that variety has to be managed explicitly to avoid a complexity overload or a inflexibility in a part of the organization.

Management of Complexity

Malik creates a bridge from the theoretical works of Stafford Beer to management reality. He explains the principles and the design of each of the 5 Systems and compares it with real situations. Corporate managers need to decide how much autonomy can be granted to their divisions. If all divisions run autonomously, they need some autonomous coordination. If this coordination does not work, the corporate manager is constantly asked to decide. [79]

He emphasizes that the cybernetic point-of-view can help to identify bottlenecks in an existing organization. Examples for typical bottlenecks are very often in System 2 and System 4. A missing or inefficient system 2 leads to a lack of coordination that has to be closed by system 3 (which cannot do it because a lack of variety). An insufficient system 4 leads to a blind organization that has too less information about the environment. Such organizations can be disrupted by competitors that could not be seen by the divisions (system 1). [80]

Holistic Complexity Management

Whereas Malik focuses on the strategic management in complex situations, Kirchhoff describes the connection of structural and dynamical complexity. Structural complexity can come from exogenous factors like markets which influence the product structure and increase variety. This variety leads to more complex processes. This work connects the management of variety of products and the management of the organization itself.[81]

---

[78] See (Beer, 1995 S. 131)
[79] See (Malik, 2016 S. 107-143)
[80] See (Malik, 2016 S. 107-143)
[81] See (Kirchhof, 2003)

Summary

Cybernetics and the Viable System Model focus on the question how to manage complexity in a multi-level system.

This question is already important in a small company that consists of only one team. The CEO has to decide about the product variety that is required to meet the market demands. The CEO also has no full picture what each employee is doing in operations. Larger organizations have to manage variety on a senior management level, and they need a strategic management that reacts to changes in the world.

However, the Viable System Model does not claim to answer the question, how the organization earns the resources to feed all the management activities. The Viable System Model does not translate an entire organism; it translates its (central) nervous system only. Thus, it can give no answer to growth, because it does not contain the reinforcing loops with the self-creation processes. For the Viable System Model the self-creation is located in the operations of System 1, which is seen from the perspective of homeostasis.

Which other translations of autopoietic systems exist? The most famous one is the translation of the German sociologist Niklas Luhmann, who interpreted the term autopoiesis for social systems and especially for organizations.

2.8 Autopoietic Organizations

Autopoiesis in Sociology

Biological ideas are well-known in sociology. The idea of translating concepts from biology to sociology is much older than the detail understanding of cell biology. The similarity of the words organism and organization inspired many researchers since Aristotle. Both words are based on the Greek word “organon” which means “tool”. Here it is good to understand which biological discoveries influenced sociology at which time.

Charles Darwin and his theory of evolution had a big influence not only on biology but also on sociology. The biologist and sociologist Herbert Spencer-Brown, who is known for the term “survival of the fittest” introduced the term “social organism” to indicate, that societies also follow the laws of evolution. This evolutionary perspective could explain, why societies have different structures.[82]

Luhmann

A milestone was the introduction of autopoiesis by Varela and Maturana in “Autopoiesis and Cognition” in 1972. This concept heavily influenced Luhmann’s system theory, it is also known as „Autopoietische Wende“. [83].

Luhmann starts his system theory with the difference that separates the system from its environment. This difference has to be created by the system itself. Like in autopoiesis the difference has to be created by operations that constitute the system.[84]

---

[82] See (Vester, 2009)
[83] See (Vester, 2009), (Vester, 2010) and (Berghaus, 2011)
[84] See (Luhmann, et al., 2017)

Luhmann compares social systems with biological systems. Where biological systems have chemical reactions as operations, social systems have communication as operations. However, he differs slightly from the notion of Maturana (compare the definition in the recent chapter): Luhmann speaks about operations that create operations: "Autpopoiesis heißt in der Definition von Maturana, dass ein System seine eigene Operationen nur durch das Netzwerk der eigenen Operationen erzeugen kann.".[85] This definition omits the components in the original definition and focuses on the network of processes, which indirectly ensures these processes.

A second important aspect in Luhmann's theory is the reentry: The system itself contains a representation about itself and the environment. Simplified: It understands the difference of itself from the environment and thus knows its identity.[86]

Luhmann also uses the concept of structural coupling from Maturana and Varela. Autopoietic systems are always operationally closed, but they are interactionally open. Maturana and Varela already pointed out that the psychic system and the physical system of a human being are operationally independent from each other. The psychical system consists of thought not of chemical reactions. Thoughts create new thoughts. However, the psychical system is structurally coupled with the nervous system in the physical world. Chemical reactions in the brain are structurally coupled with subjective impressions.

In the same way, Luhmann describes a third autopoietic system that exists in the social world and is structurally coupled with the psychical world. He described two candidates for the operation type: Actions or communications. Luhmann decides to use communication and explains that communication always includes the perception by a recipient, whereas actions can be done without a recipient. The perception of a recipient is necessary to trigger new communication. This triggering of communication creates a continuous stream of communication which is needed to keep the social system alive.

In later works Luhmann specialized this autopoietic system of communication for organizations. Here he focuses especially on the communication of decisions. The organization can be seen as an autopoietic system that is created by the operation of decision making. Decisions in organizations lead to other decisions.

Practically, decisions lead to a reduction of uncertainty. Here we can see, how Luhmann come to a similar conclusion as the cybernetic researchers: Decision reduce uncertainty or, as Malik or Beer would say, they attenuate variety.

In his lectures 1991/1992 Luhmann stated „Ich denke, dass der Begriff der Autopoiesis und die Theorie autopoietischer Systeme mit diesem Begriff zugleich unterschätzt und überschätzt werden. Sie werden in der Radikalität des Ansatzes unterschätzt." and some sentences later "Andererseits ist der Erkenntnisgewinn außerordentlich gering. Das muss man gerade im soziologischen Kontext betonen. Eigentlich kann man mit Autopoiesis nichts erklären." [87]

In other words: The theory of autopoiesis explains everything ("the system exists, because it produces itself") and nothing ("what can we conclude from this theory"). Especially it is not helpful for a practical purpose, because it does not contain enough details to get real insights.

---

[85] See (Luhmann, et al., 2017 S. 116)
[86] See (Luhmann, et al., 2017)
[87] See (Luhmann, et al., 2017 S. 110).

Martens and Ortmann describe the reception of Luhmann's work as follows: Some social theorists argue that Luhmann has created an inhuman system. It focuses on an abstract system and does not include humans and their needs. For Luhmann humans are in the environment of the organization. The critic of other social theorists was not that he describes a system without people. It was more that he didn't show interested in this question at all.[88]

Summary

Niklas Luhmann uses autopoiesis in a social context. It supports the idea that a system creates itself from inside. The definition of communication as the operation that creates the system omits the duality of form and function. Whereas in the cell the components drive the reactions and the reactions create the components, in organizations everything is about communication of decisions. This results in a loss of information and a reduction of usability of the concept.

Undoubtly, the connection of these authors are a strong base for an understanding of living systems. Autopoiesis is connected with cybernetics and can be translated to communications. A missing part seems to be a more detailed understanding of the relations between processes and components.

We do not understand components if we do not understand the processes they drive, and we do not understand these processes, because we do not know what components they produce. How can this hen-egg-problem be solved?

Here we have to step back from the work of Maturana and Varela and enter molecular biology and chemistry again.

2.9 Autocatalysis

Autocatalysis as the Reason for Autopoiesis

Autocatalysis is a chemical phenomenon where a chemical reaction produces its own catalysts. This leads to an acceleration of the chemical reactions. The acceleration makes these reactions faster than all alternative reactions that are not catalyzed. This difference of speeds result in the growth of a system.[89]

Autocatalytic Sets

Hordijk named such a set of catalysts an autocatalytic set, which has auto-poietic properties. Such an autocatalytic set grows, as long as sufficient input is available. He also proposed to translate this idea to a business environment, without creating a detailed translation himself [90].

Autocatalysis in Social Systems

Padgett and Powell have given a detailed description about autocatalysis in social systems in their work "The emergence of organizations and markets". The book contains three chapters which describe the emergence of autocatalytic systems and several historical cases that contain the

---

[88] See (Kieser, et al., Stuttgart S. 413-448)

[89] See (Alberts, et al., 2015) for a discussion of enzymes and catalysis

[90] See (Hordijk, et al., 2012)

abstract patterns. Padgett starts with the description of "Autocatalysis in Chemistry"[91]. He describes four aspects of living systems: Thermodynamic throughput of energy, autocatalysis, cellular enclosure and evolution. Padgett compares two competing theories how autocatalytic structures have originated in nature: the RNA-world hypothesis, which assumes that RNA molecules existed first, and the metabolism-first theory, which assumes that networks of proteins existed first. Both theories have different explanations how the autocatalysis in cells originated. Padgett also analyzes the term autopoiesis and its usage by Maturana and Luhmann. He mentions that the definition of autopoiesis is mainly inspired by control theory and emphasizes too much on autonomy and controls, and provides no empirical evidence.[92]

In chapter "Economic Production as Chemistry II", Padgett, McMahan and Zhong describe a multi-agent-software that can be used to understand the evolution of autocatalysis. The software implements a model consisting of products, production skills and firms. Each firm has a set of production skills that convert inputs to outputs. Whenever the produced output can be used by another firm, the creating firm (or the receiving firm) gets a reward: The production skills are duplicated. The authors are interested in the question, whether several cycles of the game, starting with a random configuration, can lead to reinforcing loops of autocatalysis. They have identified several influencing factors that have an influence on the emergence of autocatalysis. One of them is altruistic behavior, another has to do with the number of neighbors of each firm. This model is the base for the system theorical understanding of the historical cases in the later chapters.[93]

Padgett describes in "From Chemical to Social Networks", how the model in the last chapter can be refined. Whereas the original model only creates new production skills, when a product has been used by another firm, the new model strengthens the relationships between the firms. Each skill learns, where the consumer of the product is. This concept is similar to neural networks: Fire together – wire together.[94]

All later chapters of the book use these concepts to explain how organizational inventions could emerge in history. One example is "the birth of partnership systems in renaissance Florence". Padgett shows that the connection of the political system, the economic systems, and the kinship between families led to a coevolution. Rich merchants got more political power and could change the environment for economy in their sense. Marriages between families influenced the accumulation of capital. All this led to an autocatalytic system with reinforcing loops[95]

Their publication already showed that the basic idea of autocatalysis can be transferred to social systems. But it is still very abstract. It does not translate the many biological concepts on a detail level. Padgett and Powell are more interested in macro-economy than in a practical implementation in a single company. [96]

---

[91] See (Padgett, et al., 2012 S. 33)
[92] See (Padgett, et al., 2012 S. 33-70)
[93] See (Padgett, et al., 2012 S. 70-91)
[94] See (Padgett, et al., 2012 S. 92)
[95] See (Padgett, et al., 2012)
[96] See (Padgett, et al., 2012 S. 92)

Summary

Autocatalysis explains the construction of auto-poietic systems. Autocatalytic systems can be modelled mathematically as autocatalytic sets. They can be used to describe phenomena in economy.

2.10 Intermediate Summary

The literature review in the recent sections showed many aspects that are horizontally connected. Let us make an intermediate summary before we continue:

1. Learning between projects has aspects in the projects themselves, but it requires activities in the functional management of the organization. This is necessary to ensure that improvements can be generated and used in all projects
2. Lean and agile methods help to make improvements. Lean contains several improvement methods like kaizen or A3. Good improvement requires to have a defined future state that explains how the organization shall be in the future.
3. Senge shows that the future state should not be a collection of isolated ideas about the future. Instead, the future state should be a mutually agreed picture of the future system dynamics, which is based on the well-understood current system dynamics.
4. Sterman describes a language for this system dynamics that consists of causal dependencies, causal loops. Goldratt's Theory of Constraints helps understanding, how a future system dynamic can be realized by eliminating the bottlenecks.
5. Maturana and Varela describe the benchmark of such a system dynamic. The system dynamics of life. Autopoietic systems are able to produce themselves and realize growth.
6. In their cybernetic models Beer and Malik detail the controls aspect of living organism and create the Viable System Model, which can practically be used in organization. The focus is on distributed control of complex systems.
7. Luhmann creates an autopoietic system of decisions that is structurally coupled with the psychic system of thoughts. These decisions trigger other decisions. The organization is the result of this stream of decisions.
8. Autocatalysis focuses more on the self-creation aspect than on the control-aspect. Padgett and Powell give examples how such autocatalytic systems evolve in social systems.

The system dynamics of autopoietic systems are the ideal long-term future state for organizations and ideal to align all improvements. These improvements would pull the organization towards an outstanding system dynamic.

What is missing? There is currently no business interpretation of autopoietic systems available that contains the self-creation and the control aspects in one model and is practically usable in large organizations. Practical usability means that an organization is able to implement such a system in reality.

To be more specific: Large industrial plant manufacturers, whose challenges were explained in the introduction, require a theory that can be practically applied. Of course, all these theories are excellent. No doubt about that! However, we need an business interpretation of autopoiesis that can be practically used to solve urgent problems in organizations.

2.11 Decision for Further Research

At this point we are at a crossing point and have to decide for one direction:

1. Focus on Maturana and Varela and get a deeper understanding of autopoiesis. This requires a deeper understanding of organic chemistry molecular biology
2. Focus on Beer and Malik can give a deeper understanding of cybernetics and the Viable System Model. This leads to a rich literature about the distributed control of systems.
3. Focus on Luhmann and get a deeper understanding of the theories of social systems.
4. Focus on Padgett and Powell and their translations of autocatalytic systems.
5. Try a mixture of all directions.

All directions have their advantages and disadvantages. Deciding for one of the alternatives always leads to an ignorance of the other directions. However, it is practically impossible to go all directions at the same time. Alternative 1 requires studying molecular biology on a detail level, alternative 3 requires studying sociology.

My decision here is to go back to the roots, to follow Maturana and Varela and try to translate the autopoietic systems from scratch. This direction allows using the state-of-the-art knowledge of molecular biology and creating a more detailed picture of autopoiesis than both authors could create in the 1960s and 1970s. This deeper understanding of biology allows also a more detailed translation to business. The source of the translation is more detailed. The target of the translation to business organizations is also more specific than social systems in general.

Thus, the first part of the literature research ends here, and sets the starting point for a new and more detailed translation of living organisms to business. This second part of the literature review is based on books of Alberts et al about molecular biology and Sadava et al about biology in general[97] and translates it in the world of lean and agile methods and the world of management systems as defined by ISO 9000.[98]

2.12 Summary

Before we start with the translation, let us summarize the findings of the literature research and map them to our questions.

Question 1: Can the understanding of biological organisms be used to design the system dynamics of business?

The literature review revealed that there have been many approaches to use the knowledge of biology in a management context. All of them focus on specific aspects of living organisms. Approaches like neuro-cybernetics are excellent to explain how large systems can be controlled in several levels but they don't explain why a system grows by creating catalysts. The focus of cybernetics is on information, communication and the balancing loops. Autopoiesis is a correct answer, how growth can be managed, but it is quite abstract, and it focuses on the reinforcing loops only. Other mechanism like the protection mechanisms are not explained by these

[97] See (Alberts, et al., 2015) and (Sadava, et al., 2011)
[98] See (ISO 9000, 2015)

approaches. Several isolated translations exist, but they are not sufficient to create a complete management system. Thus, there is a need to systematically translate the current knowledge of biology to business again. This translation has to be based on the current knowledge of modern molecular biology and has to try to translate as many aspects as possible und connect them logically. Of course, the full translation of a cell is not possible. But it should be possible to go through the main aspects on a very coarse level and connect the translations. The translation should be able to explain many different aspects of management in one integrated model.

Question 2: Can the system dynamics of life be used to design a management system that helps to significantly improve project business?

This question is not answered by the scientific literature yet. There is no attempt known to translate biology to business and map it to ISO 9000. And, as expected, all the excellent researchers who worked on this topic have not focused on project business or on the needs of large industrial plant manufacturers. In the case of large plant manufacturing the application in reality is especially difficult. Plant manufacturers are quite conservative, because customer projects have high risks. No CEO wants to play the guinea pig for novel autopoietic experiments of a scientist. Even if the CEO agrees, a new management system changes teamplay for several hundred people in parallel. Such a change takes several years.

Question 3: Can typical methods of Lean and Agile be explained by the system dynamics of life?

The literature review also revealed that the existing translations are too abstract and theoretical for a practical use. Autopoiesis is a fascinating concept, which explains everything in theory and helps nothing in practice.

Here is a research gap that needs to be closed. This work wants to close this research gap with the following methodic approach:

3 Methodical Approach

3.1 Benchmarking

3.1.1 Basic Idea

The basic idea is to identify patterns in the structure of biological organisms that can be translated and utilized by organizations in project business (Figure 3-1).

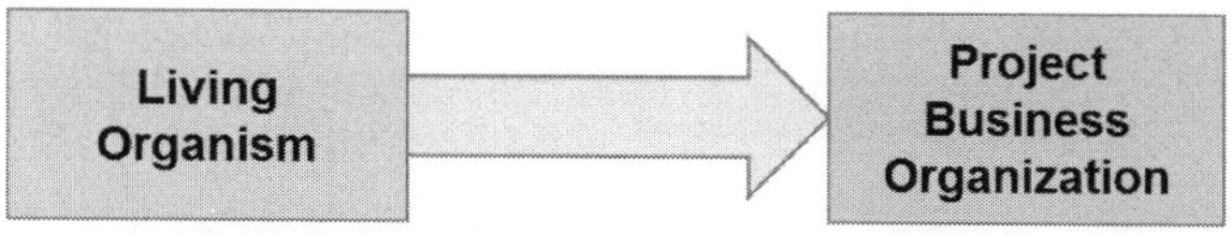

Figure 3-1: Translation from living organisms to project business organizations.

At the end of the translation process we expect some rules or principles of living systems. For an implementation this is not sufficient. The results of the translation have to be connected with well-established processes and methods like project management, agile development or lean management methods. This is important to bring all the rules and principles to life and to test them in a reality.

This linkage has one additional advantage. It makes it possible to see all existing methods in a new perspective: the system dynamics of living systems. The network of patterns coming from living organisms act like the background where the puzzle pieces of methods can be aligned and connected to the big picture of a Viable Business Management System.

3.1.2 Benchmarking Approach

The translation of the system dynamics of an organism to business organizations can be understood as benchmarking.

Benchmarking can be done with different partners. Internal benchmarking is the simplest one. External benchmarking compares the organization with others. This can be the best competitor or a completely different organization which has the best practice of a certain process. This is called best-of-best benchmarking.

Learning from living organisms is such a best-of-best benchmarking. Many bacteria can duplicate their mass every 20 minutes. Chemical reactions are highly accelerated by enzymes. And large animals are able to have 100 trillion cells working perfectly together. The science of biology understands many details of these highlights and the knowledge is available for free. The problem here is only the re-contextualization. It is difficult to translate these best practices of nature to a business world and to make them useable. Exactly this is the purpose of the work.

Gerberich differentiates three different forms of benchmarking: process benchmarking wants to find out whether the organization has the right processes, whereas strategy benchmarking wants to

C. Dachs, *Viable Project Business*, Contributions to Management Science, https://doi.org/10.1007/978-3-030-62904-5_3

find out whether the organization is doing the right things, and structure benchmarking is about the structure of the organization.[99]

Benchmarking with living organisms is not limited to one of these three aspects. We will see that chemical reactions of the cell, survival strategies and the structure of cells are strongly connected. We can expect that the benchmarking with life will lead to processual, structural and strategical implications.

3.1.3 Benchmarking Question

Let us reconsider the benchmarking question. The collaboration of cells in the human body, the perfect acceleration of every chemical reaction and other aspects are extremely interesting, but they show only single aspects of life.

What we are really interested in is the big picture:

"How is it possible to create sustained growth in a changing environment?"

The answer is already given in many biology books which describe how cells work. For that reason, it is not our research question. The answer is already given.

Until now, the answer has not been translated well enough to apply them in the business context. Our three research questions from the introduction section focus on the translation and the development of a model that can be practically used.

To give this translation, we first have to understand the existing answer in the biological language and find a way to connect the two worlds.

3.1.4 Benchmarking Partner

Our benchmarking partners are living organisms on two levels: Cells and Multicellular Organisms.

On the cell level there are two organisms that are used as model organisms in biology: [100]

1. Bacteria: *Escherichia coli (E. coli)*
2. Eukaryotic cell: *Saccharomyces cerevisiae* (Brewer's yeast)

Both organisms are well-understood and documented in databases. Bacteria can be used to understand the basic principles of life. Eukaryotic cells introduce the concept of organelles, which seem to be necessary for all complex multicellular organisms like animals or plants.

The smallest multi-cellular organism that is used as model organism is *the Caenorhabditis elegans (C. elegans)* worm. It consists of a fixed number of cells and thus is used to study the origin of organs and tissues. It has a small neural network of 302 neurons that can already be simulated by artificial neural networks.[101].

---

[99] See (Gerberich, 1998 S. 18-19)
[100] See (Alberts, et al., 2015)
[101] See (Achacoso B. Theodore, 1991)

The Viable Teams we want to have are the translation of eukaryotic cells. Since we are only interested in the basic concepts, we use the work presented in the book “Molecular Biology of the Cell”.[102]

For multicellular organisms there is a different reason for not using the *C. elegans* worm: The structure of business organizations can be as varied as like the variety of animals or plants. Thus, the specific blueprint of *C. elegans* is not needed, and for a general explanation of tissues and organs, the blueprint of humans is easier to understand.

Sometimes it is important to highlight that the human body is only one possible blueprint of multicellular organisms. In such a case we can use a tree as a counterexample. It is obvious that a tree is a successful life form, but it has very different organs.

Thus, the benchmarking partner for multi-cellular organisms is the *Homo sapiens* that is presented in many biology books [103] or in books about inner medicine.[104] Trees as alternative benchmarks are taken from botanic books.[105]

This selection will be discussed again in the literature review.

3.1.5 Identification of Deviations

One challenge of a benchmarking approach is to compare the findings at the benchmarking partner with the own organization. In the case of a best-of-best benchmarking it can be very difficult to create comparability.[106]

The benchmarking with Toyota to learn about lean management requires a re-contextualization of the findings from product business to project business. The differences of the business types are huge. The slow adaptation of lean methods in project business indicates that already this re-contextualization is difficult.

When benchmarking with life this re-contextualization is even more complex. Thus, the focus of this work is on a good translation from biology to business and the provision of a model.

If we can consolidate all our translation in the Viable Business Management System, it can be understood without any biological knowledge. If an organization wants to benchmark itself with life forms, then it can use the Viable Business Management System as a first approximation.

In Chapter 1 we will show how the Viable Business Management System can be used as a future state of the organization. It contains functions that describe how to identify the bottlenecks in the organization and to eliminate them. In other words: The model contains the methods to identify deviations to a viable system and to define countermeasures systematically.

---

[102] See (Alberts, et al., 2015)
[103] See (Taylor, 2017)
[104] See (Renz-Polster, et al., 2012)
[105] See (Weiler, et al., 2008)
[106] See (Gerberich, 1998)

3.1.6 *Definition and Implementation of Countermeasures*

Of course, the real benchmark has to be done by the organization. Viable Business can help to create a clear system dynamic, but definition and implementation have to be done by the organization.

Viable Business already has been successfully implemented in four large organizations by the author and a consulting team.

3.2 *Six-Step Approach*

The translation from biology to business and the creation and piloting of a model is quite complex. It has to be done carefully to clearly separate requirements of the model from design decisions. We will create not only one model. Instead we want to separate the more general Viable Business from the more specific Viable Project Business.

The overall approach uses the following six steps (Figure 3-2):

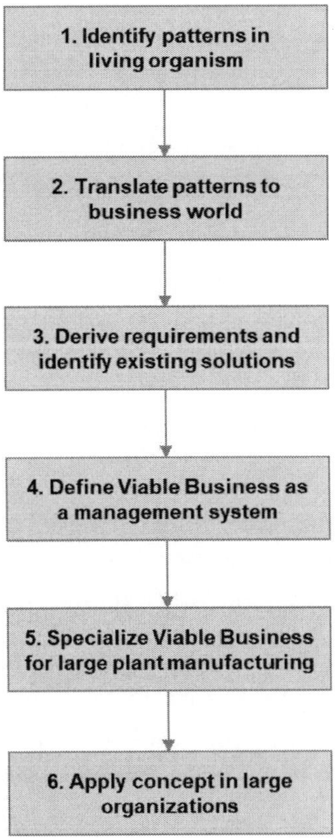

Figure 3-2: Six-step approach.

3.2.1 *Identify Patterns in Living Organisms*

If we want to understand the patterns of sustainable growth of life, we have to start from the basics in physics and chemistry and follow the evolutionary process.

Every step from the RNA world through the invention of DNA and proteins, the cell membranes, organelles and cytoskeleton introduced an evolutionary advantage for an organism in its environment. Some of these advantages (e.g. existence of DNA, cell membranes) are so important that they have not been changed again.

For that reason, we follow the evolution from bacteria, via protists and multicellular organisms to the human society.

The heart of the benchmarking is the pattern detection process. For each concept or mechanism of nature we need to understand the benefits. One example is the cell membrane. One of its benefits is to increase concentration of enzymes inside and this leads to an acceleration of chemical reactions and thus to faster growth.

These causal chains show the importance of some core concepts like "concentration" and their linkage:

"Existence of Cell Boundary" increases "Concentration of Enzymes" increases "Speed of Reactions" increases "Speed of Growth" which leads to an evolutionary advantage.

The second step is now to understand the abstract terms in a business environment. "Concentration" means number of items per volume in space. This requires understanding the term "Space" in a social system like business.

The identified terms can then be compared with methods of well-known management systems like Lean and Agile or Holocracy. These management systems have some methods that contribute to the found concepts. Agile Software Development wants to have collocated teams and tries to avoid split-heads. Lean management wants to optimize workplaces with 5S and ensures that everything is placed at the best possible location ("best point").

Such examples show that boundary-concentration-speed seems to be a valid translation from biology to business.

The pattern-finding process itself is iterative incremental. Every new pattern that can be linked to lean and agile methods creates associations on the business side. These associations give new inspiration. By reading the same sections in biology books again, there is a chance to identify additional patterns that have not been found the first time.

3.2.2 Translate Patterns to Business World

In biology it is not possible to sequence the sections in such a way that everything can be explained logically step by step, because a cell has many mechanisms that evolve in parallel.

One important example is the understanding of catalysts like ribozymes and enzymes. In today's cells there is no strict sequence to explain both. The creation of ribozymes (RNA molecules) is catalyzed by enzymes (proteins), and vice versa. It is like a chicken-egg problem. In the early stages of evolution, it most probably started with RNA only[107], so some of the translations require explaining the historical context.

---

[107] Most researchers follow the RNA hypothesis. See (Alberts, et al., 2015 pp. 362-366)

Additionally, it is difficult to present the business translation in the same logical sequence. Business is based on humans that collaborate in a society. All the motivations of people and all their collaborations are important to understand business. Typically, we associate business with value creation for customers and making profit for investors. We will see that the biological equivalent of value creation is the symbiosis, which was not the first concept in evolution.

So, it has been decided to choose a sequence of sections which makes the translations easy to follow, even if it does not follow the evolutional sequence, and even it does not follow the expectations in the business context.

At the end it is important that the translations in business language are clear enough to be collected and consolidated.

Chapter 5 contains the translation of single-cellular organisms. Chapter 6 focuses on multicellular organisms.

Every translation is described on two pages, where biology is always on the left-hand side and business on the right-hand side. Both biology and business are described in the same structure. This allows comparing biology and business very easily.

This strict layout of translations leads to the fact that not every page is 100% utilized. It also requires that the introduction and the transformation at the end have the right length and do not change the left/right sequence. But these disadvantages should be more than compensated by the chance to see a translation of biology to business on one double-page including the key messages and the illustrating figures (Figure 3-3).

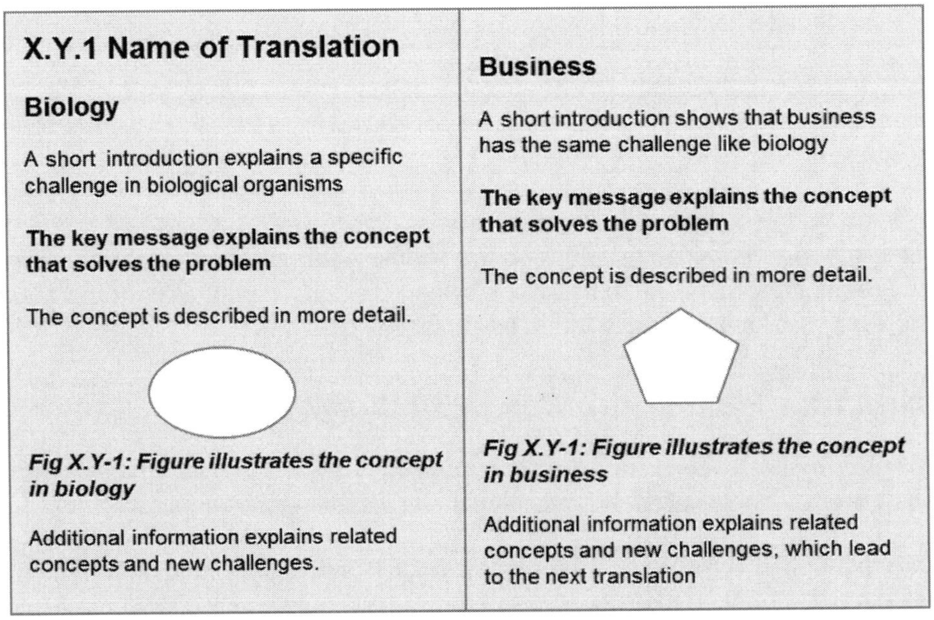

Figure 3-3: Structure of Translations. Selected mechanisms of biological organisms are analyzed and translated to business.

3.2.3 Derive Requirements and Identify Existing Solutions

Every section summarizes and discusses the results of a group of related translations. It has the focus to get the big picture again. Then it defines requirements that have to be implemented by a Viable Organization. These requirements are the essence of the translations and can be used for the design of organizations. The subsection also contains the most important lean and agile methods that can be used to realize the requirements. The annotations contain additional information about the translation itself. This includes special problems with translations or untranslated concepts.

3.2.4 Define Viable Business as a Management System

At the end a model for a Viable Business is presented that covers the requirements in a functional overview diagram. The functional overview diagram captures the system dynamics of a living organism. The functions of the Viable Business can then be implemented by known lean and agile methods.

Chapter 7 describes the result of the integration to the Viable Business Management System. This section is written in business language only.[108] This is necessary to show that the system dynamics can be understood and explained without referring to biology.

The Viable Business Management System is a purely functional description. This is necessary to allow many potential realizations. It can be used by a single team or by a huge organization.

3.2.5 Specialize Viable Business for Large Plant Manufacturing

One of the problems of a generic model is that it is not specific enough. In biology there is also not a unique blueprint for multicellular organisms. Plants have different organs than animals. To get a specific blueprint that can be used, the Viable Project Business Management System has been derived, which is especially designed for organizations that execute many large projects.

The model for Viable Project Business can be more specific and can refer to project management and specific methods. Like a specific animal consists of organs and cell types it is possible to define sub-organizations and team types.

The Viable Project Business Management System proposed in Chapter 8 focuses instead on organizations that run many projects of a similar type and want to enhance their ability to improve. Here we can see the design of the improvement machine which we discussed in the introduction.

3.2.6 Apply Concept in Large Organizations

Four case studies show the implementation of Viable Project Business in large project businesses. The main challenge of such case studies is that it takes several years to establish such a system. Thus, it is only possible to implement those parts of the model with the highest business relevance.

In the four case studies the focus was on the ability of work preparation and improvement. The implementation of these functions was mainly done with lean management methods. The Viable

---

[108] Footnotes will add the links to biology

Business Management System can show how the known lean methods have been connected to realize the viable functions of catalysis, auto-catalysis and self-reference.

The rollout in one of the case studies followed the replication model, where the implementation starts with one Viable Team only. The Viable Team then grows and splits itself until its descendants have covered the entire organization.

The case studies describe how the models could be used in reality and what the results have been. The exact design cannot be published here, but the models should provide enough information to understand how it works.

3.3 Remarks

3.3.1 Humans

This work describes the system dynamics of the organization. Humans, especially as employees, interact with organizations. In section 5.7 we will compare the relationship between employee and organization with the symbiosis between living systems.

This mutually beneficial symbiosis contains a give and take. The organization provides a great workplace and pays the salaries. The employee gives knowledge and skills, motivation, willingness to learn and to collaborate in teams, and other factors that are needed by the organization.

Today, many aspects of lean and agile, address this relationship. There is a rich literature about the behavior of humans in organizations. Books about organizational behavior, like Robbins and Judge, cover topics like leadership, communication, teamwork, group behavior, motivation, emotions and mood.[109] All this knowledge is centered around individuals and groups, in other words: the relationship to employees.

This human factor is extremely important. However, the purpose of this work is not to detail this existing knowledge. Instead, it creates a picture of the entire organization, where the employee relationship is one of several symbiosis.

3.3.2 Evolution

Other works, like those of Padgett and Powell, focus on the evolution of autocatalytic system.[110] In our case, we focus on the mechanisms of living organisms, not on their evolution. This allows understanding their form and function. Future works can then focus on evolution and explain, how more complex systems evolve from simpler ones. The decision to omit evolutionary aspects is reflected by the choice of literature. Alberts et al "Molecular biology of the cell" describe all details of eukaryotic cells in detail, but they have no special focus on their evolution. Future work that translates evolutionary principles could by based on "Evolution" of Futuyma.[111]

---

[109] See (Robbins, et al., 2017) for a broad discussion of organizational behavior.

[110] See (Padgett, et al., 2012)

[111] See (Alberts, et al., 2015) and (Futuyma, 2013)

4 Viability

This section describes the basics for our translation from biology to business. These basics are necessary for the translation of single-cellular organisms to Viable Teams in Chapter 5 and multi-cellular organisms to Large Viable Organizations in Chapter 6.

What are these basics?

There are many characterizations of life that can be found in biological literature. These characterizations describe what can be seen (e.g. "Every organism consists of cells") and they describe what living organisms do (e.g. "Every organism is able to replicate").[112]

So, there are two main perspectives on life: form and function. Anatomy describes the normal form of organisms. Physiology describes the normal function of organisms, and it describes how all the functions are realized.[113]

In this section we separate these two perspectives and discuss form and function independently before we focus on the question how form and function are related.

The discussion of form allows us to identify the first objects we want to translate: the cell and the multi-cellular organism. Their correlates, the Viable Team and the Large Viable Organization, will be the carriers of requirements.

The discussion of function allows us to get a first understanding what forms do. We will start with the overall goal of organisms, which can be seen as the root function in the function tree, and then find some main functions that have to be realized by all organisms. Since these functions are needed by every organism, we can see them as requirements that have to be implemented by the forms.

Finally, we will go into physics and chemistry to find how forms and functions are related in detail. The understanding of these deep relations is necessary because we need to discuss the foundation in business as well, before we can translate lifeforms.

Let us start with the first group of translations. Every group will be described in the same way. It starts with an introduction, before every single translation is presented. Biology is always on the left-hand side; business is on the right-hand side to allow an easy comparison. At the end, we summarize the translations and collect important requirements.

The first group of translations is about form.

---

[112] See (Alberts, et al., 2015 pp. 1-10)
[113] See (Renz-Polster, et al., 2012)

C. Dachs, *Viable Project Business*, Contributions to Management Science, https://doi.org/10.1007/978-3-030-62904-5_4

4.1 Forms of Life

Business

In business we can see that a wide range of organizations would like to achieve sustained growth.

It starts from small organizations with only one or few employees. Such an organization can be a small bakery or a software startup. It does not need to have complex structures. The small organization is only one legal entity. All the employees work closely together as one team.

If the organization grows and becomes larger, it is no longer possible to work as one team. Instead the organization creates specialized teams (sales, production, HR) that cooperate. Still everything is at one location and in one legal entity. The range of such an organization can be 10 employees to 1000 employees or more.

Organizations that are much larger can have regional companies in other locations or even other countries. These are other legal entities. They also can have business units and corporate units. So, the entire organization is more like a cooperation between mid-size organizations.

How can organizations of such different sizes be compared with organisms?

Biology

These different levels can also be seen in biology.

Escherichia coli bacteria (*E. coli*) and brewer's yeast live as independent cells. One cell is enough to implement all functions of viability.[114]

There are also multicellular organisms. Many cells cooperate. They contribute in a specific way to the success of the multicellular organism, which in return provides the best environment for each of its cells.[115] A well-known example is the human body, where every human cell has a clear purpose for the overall organism, and the human body ensures that every cell gets food and is protected.

Multicellular organisms do not exist by pure chance. They have a benefit for all cells that are included. And every cell plays its specific role to achieve that.

On an even higher-level multicellular organisms cooperate in swarms, flocks or herds. The human society is one example. Humans in a society have specialized roles. Some organize food, some are responsible for protection and some lead the group. Every human contributes to the success of the society and in return gets a more comfortable environment.

There is a clear dependency between the different levels. If we want to understand the beehive, we need to know the bee. If we want to understand the bee, we have to learn about the cell.

---

[114] See (Alberts, et al., 2015 pp. 22-32)
[115] See (Alberts, et al., 2015 pp. 32-39)

Translations

On the next pages we discuss the different levels of forms (Table 1).

On the lowest level there are simple viable systems like cells or teams. On the higher levels there are multi-level viable systems like multicellular organisms or large organizations that consist of other viable systems.

	Translation	Biology	Business
1	Simple Viable Systems	Prokaryotes: i.e. *E. coli*-bacteria	One-man show
2	Simple Viable Systems with substructures	Eukaryotes: i.e. brewer's yeast	Viable Team
3	Multi-level Viable Systems	Multicellular organisms: animals, plants, fungi	Large Viable Organization

Table 1: Translation of forms

At the end of the translation we will be able to define the various types of Viable Organizations.

All of them want to realize sustained success by utilizing the system dynamics of life. With our translations from biology to business we will be able to describe requirements that have to be implemented by them.

Let us start with the smallest Viable System in nature: the bacterium.

4.1.1 *Simple Viable Systems*

Biology

All living organisms consist of cells. The cell is the smallest unit that is able to create, protect, adapt and replicate itself[116]. The most basic version is the prokaryotic cell.

Prokaryotic cells are the most basic versions of cells. Prokaryotes include all bacteria and archaea.[117]

Biology focuses on selected several organisms which are studied in detail: so called model organisms. *E. coli* bacteria are model organisms for prokaryotic cells.[118]

What is the form and function of the prokaryotic cell? The form is that which we can see, whereas the function is that which the cell does. Let us have a look at all the components a prokaryotic cell has (Figure 4-1). Every component contributes to the overall function: "Live"

Cells have a cell membrane that separates the cell from its environment. Inside the cell there is the cytosol, a fluid which mainly consists of water and organic molecules like proteins, amino acids, nucleic acids and their precursors. A DNA molecule stores genetic information, which is used in the Ribosomes. Here all the catalysts are created that accelerate each chemical reaction. Some prokaryotic cells can move with their flagella.[119]

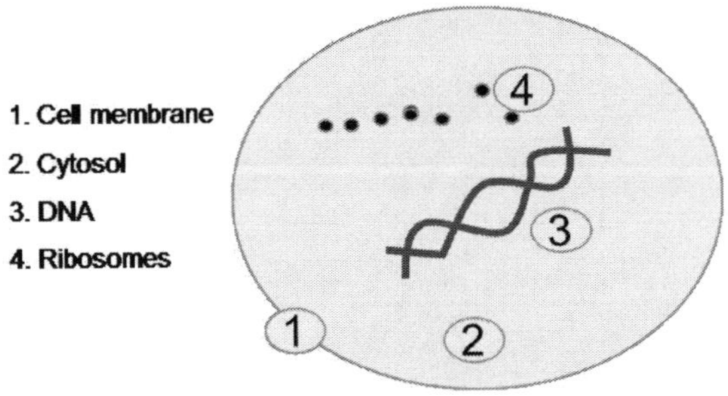

Figure 4-1: Prokaryotic cells. They are the simplest life forms. Prokaryotic cells have a cell membrane that encloses the cytosol. Inside the cell is the DNA with the genetic material and ribosomes.

Prokaryotic cells have only one compartment, the cytosol. All catalysts and material are located there. This limits the ability to increase the number of different chemical reactions, because the relative concentration of the needed catalysts would go down when the number of different molecules increase. We will see in the next section how eukaryotic cells solve this problem by introducing organelles.

---

[116] A virus contains DNA, but is not able to replicate on its own.
[117] See (Alberts, et al., 2015 pp. 12-14)
[118] See (Alberts, et al., 2015 p. 22)
[119] See (Alberts, et al., 2015 pp. 2-9) for the universal features of cells

Business

Every organization consists of groups or teams. In the simplest case it has only one team, maybe only a single person[120]. If this organization wants to have sustained success, it has to realize some basic functions: It has to create its own work environment, protect itself, and adapt it to the changing world.

Small teams can realize all processes that are needed to have sustained success.[121]

Typical examples for small businesses are startups (Figure 4-2). Startups have to create their work environment on their own and they have to protect their resources. They have to perceive the outside world and react flexibly. Their products have to be best aligned to customer needs. This allows them to earn the money that is needed to pay employees, tools and material. We will come back to all these things when we speak about function.

For the moment it is sufficient to understand that the team as an entity can realize that. Thus, it makes sense to collect all the mechanisms of life, translate them and assign them as requirements to the small team.[122]

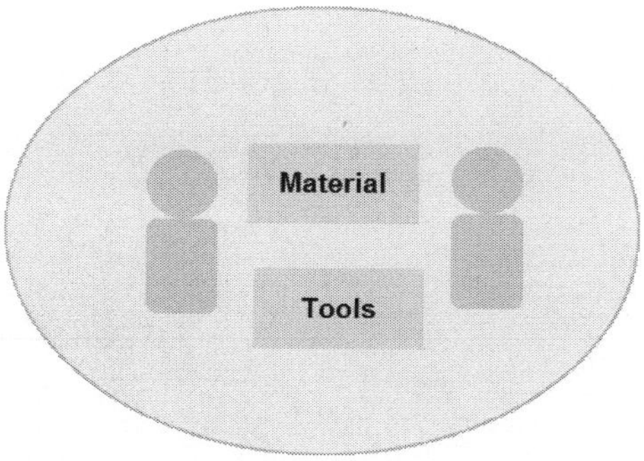

Figure 4-2: Small teams. Small teams can be successful without having complex substructures like roles.

If an organization consists of only one person, it does not need complex substructures like roles or meetings. This business-style is very limited and not able to build large organizations. So, we focus on teams that have more structures and are able to produce a more complex behavior.

---

[120] Organizational behavior theories define an organization as "a consciously coordinated social unit, composed of two or more people, that functions on a relatively continuous basis to achieve a common goal or set of goals" (Robbins, et al., 2017 S. 44) , whereas the ISO 9000 can start an organization with one person.

[121] Even organizations with only one person can have sustained success.

[122] Section 5 will show in detail how teams realize the system dynamics of life.

4.1.2 Simple Viable Systems with Substructures

Biology

Eukaryotic cells are much bigger cells. They have organelles as compartments that structure the cell interior. All the organelles are connected via the cytoskeleton.

Eukaryotic cells have several compartments which are used as specialized reaction spaces. Every compartment provides the best environment for a set of chemical reactions. This allows them to run a larger number of chemical reactions.[123]

Brewer's yeast is the model organism for eukaryotic cells.[124] Every eukaryotic cell has a basic set of organelles (Figure 4-3). The nucleus is the core that protects the DNA. It contains all reactions that deal with reading the genetic information from the DNA and transporting it to the cytosol. The nucleolus is the location where the parts of the ribosomes are created. In the cytosol there are the ribosomes that create all proteins like structural proteins or enzymes. The endoplasmic reticulum (ER) is the place where everything is created for the outside. All reactions in the ER are accelerated by enzymes that are created in the cytosol. This separation of cytosol and ER makes the processes in both compartments so effective.[125]

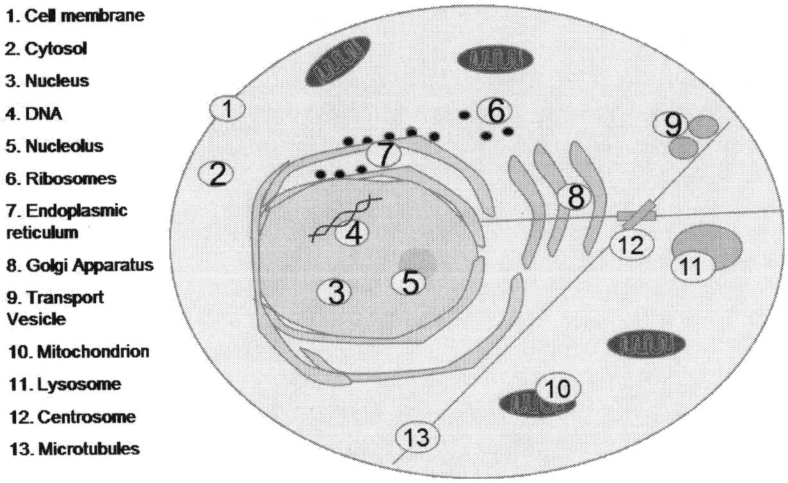

Figure 4-3: Eukaryotic cells. They have specialized organelles like the nucleus or endoplasmic reticulum to provide the best environment for each reaction.

All multi-cellular organisms that consist of specialized tissues (especially all animals, plants and fungi) are based on eukaryotic cells. The high specialization of tissues requires that each organ is extremely effective. A cell is only able to produce something in high quality with low resources if all reactions are highly accelerated by catalysts. By separating the specific functions of a cell type from the self-creation processes, the eukaryotic cell is the perfect component for the multicellular organism.

123 See (Alberts, et al., 2015 pp. 23-30)
124 See (Alberts, et al., 2015 pp. 30-32)
125 See (Alberts, et al., 2015 pp. 28-30)

Business

Teams consist of more than one person. And the persons typically have different strengths and weaknesses. It makes sense that everybody focuses on his strengths. This leads to roles and substructures in the team.[126]

Structured teams organize their processes. Every process has its time, its location and its responsible person. This specialized environment increases speed of every business process.

Structured teams (Figure 4-4) manage their roles, time and their resources more precisely. This structuring helps to provide the best working conditions for every process.

There must be one environment for value creation, one for work preparation and one for learning. The most important point is the ability to explicitly separate work preparation from doing. Value creation processes need to have the best working conditions. These best working conditions are created during a separated work preparation. And the excellent work preparation is based on know-how.

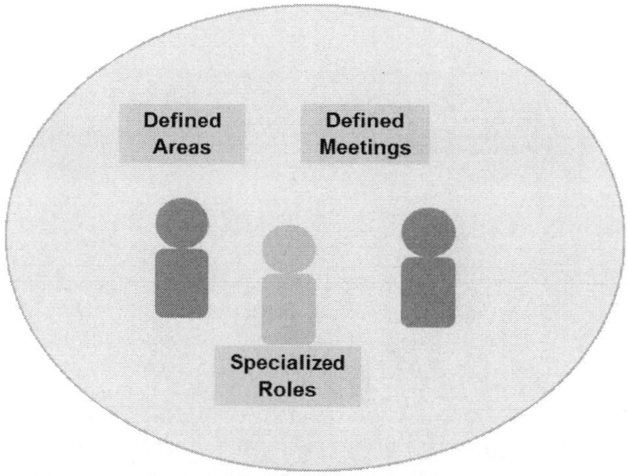

Figure 4-4: Complex teams. Teams can have defined substructures like roles to organize processes

A team is only able to produce something in high quality with low resources if everything is perfectly prepared. Large organizations depend on this ability.

In this work we only focus on the teams that are able to structure their work. In other words: we will translate the eukaryotic cell to the Viable Team.

---

[126] Roles have been introduced in sociology by Weber. A good introduction provides (Vester, 2009)

4.1.3 Multilevel Viable Systems

Biology

Animals and plants consist of many cells. Every cell contributes to the overall success of the multicellular organism. In return it gets a perfect environment. This is like a mutually beneficial symbiosis between the whole and its parts (Figure 4-5).

Multicellular organisms provide the best conditions for their cells.
Every cell contributes in a specialized role.

Well-known examples are plants. Plants consists of billions of cells. Each cell has a defined cell type that is responsible for a specific task. Leafe cells have to import carbon dioxide and produce carbohydrates, root cells import water and nitrates. All the cells are highly specialized for their tasks, which makes them very efficient. This high specialization is only possible for one reason: all the cells can reduce most of their other activities because the environment in the multicellular organism is kept very stable. This kind of stability is called homeostasis.[127]

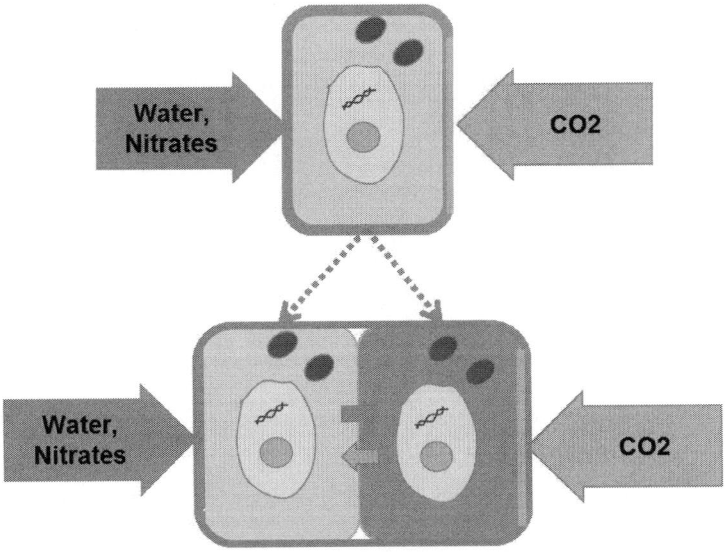

Figure 4-5: Multicellular organisms. Multicellular organisms descend from single cells. They collaborate by mutually exchanging needed resources.

If we compare trees with humans, we can see that they have very different structures. The functions are similar: import food, distribute food and export waste, but a human has completely different tissues and organs than a tree. So, there are many structures / forms possible to fulfill the same functions. That is the reason that it makes no sense to translate structures ("What is the heart of the organization?"), because even a tree has no heart. All large multicellular organisms have a distribution function. Our question will be "How is the distribution of resources organized?" And the answer will describe forms like "Cardiovascular system with a heart" or "Xylem and Phloem" in case of a tree.

[127] See (Alberts, et al., 2015 pp. 32-33) for the arabidopsis thaliana as the model organism for plants. Descriptions of the structure of plants can be found at (Sadava, et al., 2011) or (Weiler, et al., 2008)

Business

Large organizations consist of many teams that are highly specialized (Figure 4-6). This high specialization is only possible because all teams are in a stable environment.

Large organizations provide the best conditions for their teams.
Every team contributes in a specialized role.

What is the benefit of the large organization? The organization allows the teams to be much more successful than if they were independent. If the large organization develops and sells power plants, it is clear that this cannot be done by a single team. There is a market for power plants, and organizations adapt to this opportunity and grow until they can do such projects.

The challenge of the organization is to ensure that every team contributes to the overall success. Thus, all the teams need to be very productive. This is possible if teams are highly specialized, but it requires that the large organization ensures that every team gets all needed resources. [128]

The large organization needs an architecture that ensures that every team has the best working conditions. This allows specialization and high productivity.

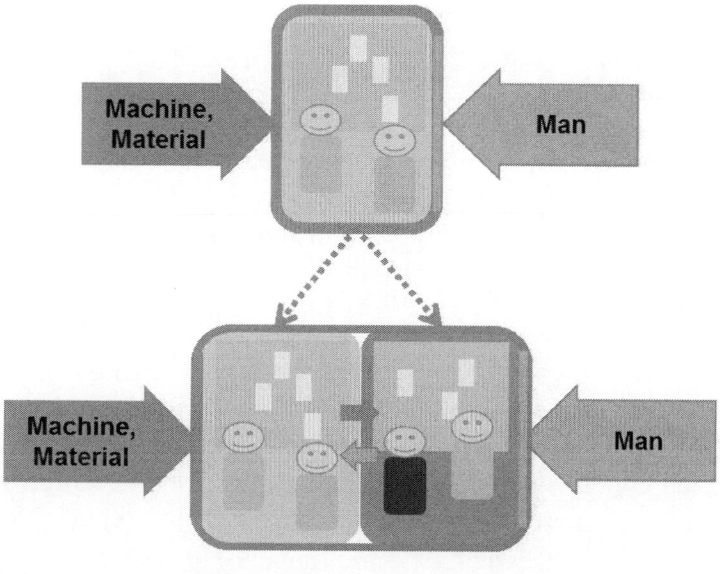

Figure 4-6: Large organization. Large organizations have started as a single team. Their contained teams exchange resources such that each team has best working conditions.

We will analyze the challenges of multi-cellular organisms in Chapter 1 and derive some requirements for the Large Viable Organization.

[128] Organization theory introduces several organizational designs like the functional, project or matrix structure. See (Krajewski, et al., 2016 S. 260)

4.1.4 Summary

4.1.4.1 Insights

This section introduced the protagonist of this story: The **Viable Organization** is an organization that wants to achieve sustained success by implementing the system dynamics of life.

In biology we can see life on the level of cells and on the level of multi-cellular organisms like humans. On the cell-level we can see prokaryotic cells, which have no organelles, and eukaryotic cells, which have them. Multi-cellular organisms are based on these eukaryotic cells, which are connected and interact with each other.

Since life exists on various levels, we should have clear definitions on the business side as well (Table 2Table 2: Definitions related to forms).

4.1.4.2 Definitions

ID	Definitions	Comment
4.1.1	A **Viable Organizations (VO)** is an organization that implements the system dynamics of life to achieve sustained success.	**Viable System** like an organism
4.1.2	A **Large Viable Organizations (LVO)** is a Viable Organization that consists of other Viable Organizations	**Multi-level Viable System** like a multi-cellular organism
4.1.3	A **Viable Team (VT)** is a Viable Organization that has no contained Viable Organizations.	**Simple Viable System** like a single cell.

Table 2: Definitions related to forms

The first definition refers to the implementation of the system dynamics of life, which has not been discussed yet. During this work we will discuss these system dynamics and translate them to business. They will be described as requirements that have to be implemented by Viable Organizations. Some of the requirements only refer to Viable Teams; others refer to Large Viable Organizations that consist of Viable Teams.

4.1.4.3 Existing Solutions

Organization Theory already describes organizations and how they are composed of smaller units Project Management and Agile Methods like SCRUM focus on the team level.[129]

[129] See (Schwaber, 2004)

4.1.5 Annotations

Abstract Terms

In this section we have introduced some abstract terms like Viable Systems, Simple Viable Systems or Multi-level Viable Systems. In general, this is problematic, because it introduces a third glossary beside the glossaries of biology and business. In this case, the terms are needed very often. They allow referring to Viable Systems, when both organisms and Viable Organizations are meant.

Simple Viable Systems

This section showed the difference between prokaryotic and eukaryotic cells. It would have been possible to translate them separately.

E. coli has no organelles. It can be translated to the Viable Team without specialized substructures; Brewer's yeast, which has many organelles, can be translated to the Viable Team with substructures. Such a translation would be much more precise, even if it would require writing some sections (like replication) twice.

For a practical purpose this is not needed. Large organizations are based on structured teams. Thus, the Viable Teams are a translation of eukaryotic cells like brewer's yeast.

Multi-level Viable Systems

Project Business is a good example for a Large Viable Organization. The organization consists of many teams, for example the project and subproject teams, which have to be Viable Teams.

One of the best understood multicellular organisms is the worm *C. elegans*. Theodore has published an early description about the neural network of *C. elegans*. It contains a description of all neurons and their connections via synapses or gap junctions.[130] The Open Worm Community uses the latest information about the neuroanatomy to produce a precise virtual copy. [131]

---

[130] See (Achacoso B. Theodore, 1991)

[131] See (Open Worm, 2018).

4.2 *Functions of Life*

Business

Now we have seen the range of special organizations. Organizations have different sizes and every organization is specialized to markets and products, so the structures are very specific.

However, if we look at the functional side, they have similarities. The overall goal of many organizations is to have sustained success. The goal of having sustained success requires some main functions. What are these main functions? Classically we would assume that it is creation of customer value, making profit and maintaining a stable situation for keeping success for a long period. Both functions are very important, and we will discuss them, but in our analysis of the needed functions we want to strictly follow nature. We will see that the example of nature gives us a new perspective on organizations.

Biology

In nature we also have very different life forms. What is their common denominator? They want to live and to grow. In other words, they want to "live long and prosper".

In this work we breakdown the root function "live long and prosper" via four main functions to the needed sub-functions. The four main functions are:

1. Create Self
2. Protect Self
3. Adapt Self
4. Replicate Self

These main functions allow us to categorize the observations we can find in modern biology books. They also create the bridge to the forms that are necessary to implement the functions. By getting more and more understanding about functions and the needed forms, we can create a picture to which "self" the functions like "create self" or "protect self" refer.

The "self" is the sum of the forms that are needed to realize the four main functions

We will develop this picture step by step. This allows us to translate them step by step and to get a strong logical explanation of the organism and finally the Viable Organization.

Translations

First of all, we discuss the root function: the organization has the purpose to exist over a long time and grow. Then we will discuss why the main functions are necessary and how they contribute to the root function (Table 3).

	Translation	Biology	Business
1	**Root Function** Live long and prosper	Live long and prosper	Live long and prosper Have sustained growth and profit
2	Create self	Create Organism	Create Organization
3	Protect Self	Protect Organism	Protect Organization
4	Adapt Self	Adapt Organism	Adapt Organization
5	Replicate Self	Replicate Organism	Replicate Organization

Table 3: Translation of viable functions

The four main functions will be detailed then in Chapter 5 for the Simple Viable System and in Chapter 1 for the Multi-level Viable System.

4.2.1 *Live Long and Prosper*

Biology

What is the success of life? What can organisms do well? They are able to survive and even grow in a permanently changing environment. And they are able to replicate themselves. The surface of the entire planet earth is full of them.

Now it is difficult to say what the meaning of life is. But obviously all organisms have functions and mechanisms that let them live long and prosper. It seems to be the goal of all organisms.

Organisms have the goal to live long and prosper

That is the characteristic of life. It's what all life forms do, otherwise they wouldn't exist. For our purposes we focus on four main functions that are implemented by every cell. They can be found in every bacterium and in every human cell.

1. **Create self**

Every cell exists in a physical world and thus consists of molecules. If we look inside, we see that beside water the majority of molecules are proteins. The important point is that all proteins have been created by the cell itself. If a bacterium can replicate itself every 30 minutes, it has to create all these proteins in this short time. So, the first function is the self-creation. The cell creates everything inside with high speed. This leads to a positive gross growth rate.[132]

2. **Protect self**

Unfortunately, sometimes the self-created molecules get damaged. So, a cell must be able to protect all self-created objects.[133] Good protection minimizes the losses and thus contributes to a positive net growth rate. Each cell has the plasma membrane, which acts as a selective barrier and keeps the self-created proteins together.[134]

3. **Adapt self**

Since the environment is always changing, the cell has to adapt its self-creation and self-protection mechanisms to the changing situations. This keeps the effectiveness of these mechanisms high and this keeps the net growth rate high. Every cell monitors itself and its environment, processes the information in order to respond adequately.[135]

4. **Replicate self**

If the cell has reached its maximum size, it splits itself. Even if one descendant dies, the genetic information is not lost, because many siblings survive and continue growing.[136]

---

[132] See (Alberts, et al., 2015 pp. 1-40)
[133] Some objects like DNA can also be repaired. See (Alberts, et al., 2015 pp. 266-276)
[134] See (Alberts, et al., 2015 pp. 8-9)
[135] See (Alberts, et al., 2015 p. 813)
[136] See (Alberts, et al., 2015 p. 963)

Business

What is now the goal of a business? When can we say that a business is very successful? It is the case if the business exists for a long time and has a sound organic growth.

Most organizations have the goal of sustained success.

The most important aspect of sustained success is to continue to exist. Every organization that cannot reach this goal disappears. The goal of sustained success is based on the same main functions.

1. **Create self**

Organizations try to earn money by creating customer value. This requires a lot of know-how, trained people, the right tools, and good relations to stakeholders. This work environment does not exist by pure chance. Everything has been created by processes of the organization itself. In other words: the organization creates itself. All the businesses processes that contribute to the creation of this work environment can be seen as self-creation processes. If the organization wants to duplicate itself in short time, these processes have to run fast to duplicate this production capacity.

2. **Protect self**

Unfortunately, employees can get ill or demotivated or even leave the organization, and tools or material can get damaged. So, a good organization has to protect all its self-created assets. Protection reduces losses and contributes to the net growth rate.

3. **Adapt self**

In a changing world the organization has to constantly adapt its self-creation and self-protection mechanisms. Expectations of stakeholders are changing more often. Existing products have to be changed; new products have to be developed. The organization permanently has to observe the inside and outside world to find the current and future bottlenecks of the organization and to react accordingly. This adaptation helps to keep the effectiveness of self-creation and self-protection high and to produce a high net growth rate.

4. **Replicate self**

Finally, the organization will pass the point where it has reached the maximum size. Every team can split when it has reached the maximum size and create two new teams that collaborate. The same thing is possible for large organizations that split large business units into smaller ones.

On the next pages we will discuss these functions in a little more detail to understand their relations to each other and their contribution to the root function.

4.2.2 Summary

4.2.2.1 Insights

Business organizations are similar to biological organisms. It they want to survive or even grow under changing conditions; their goal is to "Live long and prosper". In business terms this can be described as "sustained success."

Sustained success requires that an organization is able to build its own production capabilities. This includes the organizational structures, workplaces, products, tools and processes, the know-how of employees and the relations to all stakeholders. All these factors are needed to provide the best working conditions and to earn money. The organization can reinvest and grow.

A good organization also has to protect everything against threats. Many activities like cyber-security, health and safety, insurances and the classical security guard are necessary to protect the assets of an organization. Good protection mechanisms increase the growth rate, because damaged or lost assets do not have to be created or imported again.

Since everything changes, the organization has to perceive changes in the inside and outside world, analyze the data and decide to adapt its self-creation and self-protection activities. The last step of adaptation is to influence processes with decisions; thus, it requires some discipline to adhere to decisions.

All this contributes to continuous growth. If the organization wants to pass the maximum size of the organizational structures, it has to organize replication and to split itself.

4.2.2.2 Requirements

A Viable Organization (VO) shall implement the following requirements (Table 4):

ID	Requirements	Comment
4.2.1	The VO shall have the goal to have sustained success	Live long and prosper
4.2.2	The VO shall create itself	To be detailed in sections 5.1, 5.2, 5.3 and 5.7
4.2.3	The VO shall protect itself	To be detailed in sections 5.4 and 5.8
4.2.4	The VO shall adapt itself	To be detailed in section 5.6
4.2.5	The VO shall replicate itself	To be detailed in section 5.5

Table 4: List of requirements related to objective and viable functions

4.2.2.3 Existing Solutions

On this high level the functions do not map to lean and agile methods. All the requirements have to be broken down to be more specific.

4.2.3 *Annotations*

Autopoiesis: The Component That Creates Itself (and Others)

Especially the sections about self-creation showed that everything is very abstract. From a purely functional perspective we could not say anything about this component. Obviously, we need to understand how this function is realized by a form. From our literature review in section 2.5 we know that autopoiesis is based on auto-catalysts.[137]

In section 5.1 we will discuss this auto-catalyst in detail. But before we can do that, we have to understand the foundations of space and time and translate them. These translations will give us enough information to understand the nature of objects and processes that are needed to understand catalysis and auto-catalysis.

Creation and Protection

In the later sections we see how all these functions can be realized. Very interesting is the exact difference between creation and protection on the molecular level. Both functions are contributing to the same: they provide the highest concentration of production factors.

Self-creation contributes to that by creating catalysts and by bringing them to the right location. Self-protection contributes by removing all unnecessary objects to maintain concentration and by removing all threats that could destroy the catalysts.

On a molecular level it is difficult to draw a clear line between both mechanisms. Moving something to the best location implies removing the unneeded things first. Lean production has the method 5S, which brings everything to the right position after it has removed everything else. Is 5S part of self-creation or part of self-protection?[138]

Sequence of Main Functions

Is the sequence of the four main functions logical? Of course, something must be created before it can be protected. There must be something that creates, before one of many creation mechanisms can be selected, but there is no necessary sequence that says that protection has to come before adaption or vice versa. And replication can be seen as a special form of creation.

For now, the main functions fulfill their purpose: we will use them to sort all the knowledge in biology. We can now always ask to which main function a biological concept might contribute.

---

[137] See (Maturana, et al., 1980) and (Luhmann, et al., 2017) for autopoiesis. Luhmann pointed out the relation to auo-catalysis.

[138] See (Gorecki, et al., 2013 S. 110) for a description of 5S

4.3 Foundation of Life

The functional decomposition of the root function "Live long and prosper" led to four main functions that have to be implemented by Viable Organizations. Unfortunately, already the first of them "create self" leads to problems. From a functional point it seems to be impossible to detail it, because we do not know what the "self" really is.

There are two problems to understand autopoiesis.

1. We cannot understand form without understanding function.
2. We cannot understand function without understanding form.

This chicken-egg problem can only be solved if we step backwards and try to understand the relation between form and function, long before it comes to autopoiesis.

In this section we focus on the elements the autopoietic systems consist of and the environment in which they exist. This gives a good foundation to understand the relation between form and function. In section 5.1 we will use this knowledge to understand how a self-creating system can be realized.

Biology

Organisms consist of components that realize all the viable functions. Before we try to understand these components and how they work, let us focus on components in general.

Organisms exist in a world with three space dimensions and a single time dimension. At a low level they consist of atoms and molecules. At the end our self-creation component consists of a set of molecules which has been created by chemical reactions.

If the organism is able to grow fast, it means that these chemical reactions run with a high velocity. In this section we want to discuss the factors that influence the speed.

Business

For Viable Business we have to identify the basic components of organizations and to discuss them under the same perspective as we do with biology.

The main challenge is that some relevant objects like rights, information or feelings have no location in the three-dimensional space. Without having the space as the background, it is impossible to speak about distances or concentration of objects; both are very important concepts in biology. Thus, the target is to understand the basic rules and concepts which are underlying the Viable Business.

This perspective can give first insights what the "self" of the organization consists of, and thus allow the function "create self" and their realizing components to be described in more detail.

Translations

The following translations are maybe the most important ones because all other translations are based on them. They explain the nature of processes as a change of relations between objects and highlight the relation between flow, concentration and process speed.

On the next pages you can find the following translations (Table 5):

	Translation	Chemistry/Biology	Business
1	**Objects and object types** Objects are the units that are changed by processes	Matter (atoms, molecules), energy, information	Matter, energy, information, emotion, rights
2	**Time and Space** All objects have a position and form in some dimensions. Events happen at a specific time. Form requires a kind of space as background. Function requires time as background.	Time + Space (Physical world)	Time Conceptual Space Physical world Psychological world Social world
3	**Relationships** Objects can form stable relationships	Chemical bond	Relationships between business objects
4	**Processes** Relationships can be created or removed by processes	Chemical reactions change bonds	Business processes alter the relationships between objects
5	**Speed, concentration and flow**	Reaction speed depends on concentration	Productivity depends on concentration

Table 5: Translation of foundations

4.3.1 Objects and Object Types

Chemistry/Biology

Let us make a short inventory of the objects a life form consists of (Table 6).

	Object Type	Examples	Comment
	Physical World		
1	Matter	Atoms and molecules (water, proteins, DNA, carbohydrates)	Atoms cannot be created or lost. However, structures like molecules can be created or lost.
2	Energy	Solar energy, energy stored in molecules...	Energy cannot be created or lost. Matter can store energy
3	Information	Genetic information in DNA molecule	Information can be copied! Matter can store information.

Table 6: List of object types in chemistry

Organisms consist of atoms and molecules.[139] Albert et al describe that a bacterial cell is approximately composed of 70% water, 15% proteins, 6% RNA, 1% DNA, 2% polysaccharides, 2% phospholipids, 3% small molecules and 1% inorganic ions.[140] This shows that, beside water, the cell mainly consists of four types of macromolecules: proteins, (deoxy-)ribonucleic acids, polysaccharides and phospholipids.

The cell also contains energy. Energy can be found in the chemical bonds, like the ionic bonds, covalent bonds, hydrogen bonds and van der Waals attractions. This energy keeps the component of the cells together. The molecules also have kinetic energy.

Obviously, the constituents and the energy alone are not sufficient to explain the organism. There is more: the organism contains information. DNA is a long polymer of nucleotides. The genetic information is stored in the molecule. It is the exact sequence of the 4 nucleobases adenine, guanine, cytosine and thymine. Matter is the carrier of information. The information is stored in the sequence of chemical bonds. [141]

How can an organism grow? Matter and energy cannot be created, because they follow the conservation of mass and energy.[142] Thus an organism must be about information. Organisms import the necessary matter and energy from outside and use them as carriers for a copy of their blueprint. They copy their genetic information and their structure.

[139] See (Alberts, et al., 2015 pp. 43-51)
[140] See (Alberts, et al., 2015 p. 48)
[141] See (Alberts, et al., 2015 p. 2)
[142] The Emmy Noether Theorem shows that conservation laws are necessary consequences of the structure of space and time.

Business

What are the objects and object types in social systems like organizations? (See Table 7)

	Object Type	Examples	Comment
	Physical World		
1	Matter (Physical objects)	Notebook, mobile phone, hammer, nail, pencil, paper...	
2	Energy	Energy for the notebook ... Energy for humans: Physical Energy	
3	Information	Descriptions, software, ...	Information can be copied
	Psychological World		
4	Feelings	Fear, stress, happiness, Mental energy / Motivation	Feelings are the subjective perception of an emotion.
5	Personal Capabilities, Character and Habits	Ability to keep the mind focused, ability to reflect, discipline, creativity...	
6	Competences	Awareness, knowledge and skills	Information that is stored somewhere in the brain
	Social World		
7	Shared Values and beliefs	Culture, principles, purpose	Values ensure that people can live and work together
8	Rights and duties	Rights, property, roles, organizations	Social structures allow people to specialize

Table 7: List of object types used in organizations

Human beings live in the three worlds. As biological organisms they consist of matter and have physical energy, but they also have a mind with feelings, they have knowledge, and they are part of the social world. Information connect the worlds. Humans can perceive the physical world and can express their feelings. The communication between people creates the social world. [143]

Beside this property to connect the worlds, information has a second property: It can be copied. This allows the organization to use information non-exclusively, whereas other resources are scarce and have to be allocated.[144]

What connects these objects in an organization? The organization is a legal entity and carrier of authorities and responsibilities. [145] It is able to own objects or information in the physical world. It is able to close contracts with people, which have feelings, competences and shared beliefs. All the physical objects, the company's site, the information, and the people are connected by the legal entity.

The psychological world and social world contain object types that cannot be located in three-dimensional physical space. All three worlds together are a conceptual space. Before we are able to

---

[143] See (Luhmann, et al., 2017)

[144] This is a fundamental difference that has influence to the design of Viable Business in section 7.

[145] The definition of an organization in ISO 9000 requires responsibilities and authorities (ISO 9000, 2015 p. 27)

understand relations between objects, we need to understand some properties of the conceptual space that contains all these objects.

4.3.2 *Time and Relative Dimensions in Space*

Physics

Space and time have topological and geometric properties like dimensions and distances. These properties allow us to speak about positions of atoms and the distances between them (Figure 4-7). Stable distances and angles are needed to speak about form and size, and they are needed to understand the terms concentration and flow. All these terms are important in biology. So, it makes sense to study the space itself and its relation to the contained objects. Space and time are the background in which all matter, energy and information are located.

Let us focus on the position of objects first. Every atom has a position in space relative to an inertia system. Two atoms have a distance, which is the difference of their coordinates.[146]

With the fixed speed of light, the distance in space is equal to a distance in time. This distance can be understood as the time the light needs to travel from A to B.[147] In astronomy it is usual to express space distances in a category of time, the lightyear.

Any movement of atoms has a lower speed than light. If one atom is moving towards another atom, its kinetic energy determines its speed and the time it takes until they collide. High kinetic energy means high speed and short duration.[148]

In normal life, it is possible to express a distance in such a time category. We say the distance to a location is ten minutes by car or one hour by foot. Both values are calculated with an expected speed. In the case of the car it depends both on the car itself and the used roads. It is the same principle as with the speed of light, except the facts that speed of light is a natural constant and the upper limit of speed.

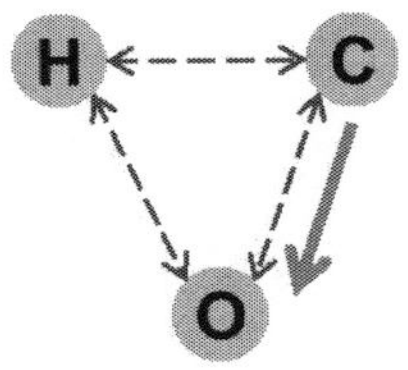

Figure 4-7: Space and time in the physical world. Atoms have a distance in space (purple) that is equivalent to the time light needs to travel. Moving objects (gray arrow) have kinetic energy and a related speed. Using this speed, it is possible to convert distance to travelling time.

The translation of space to time can help us to express distances in the conceptual space.

---

[146] See (Bartelmann, et al., 2015 S. 1-13)
[147] See (Bartelmann, et al., 2015)
[148] See (Bartelmann, et al., 2015 S. 1-13)

Business

We can connect the physical, psychological and social world by calculating every distance in time instead of space (Figure 4-8). The main problem here is the missing natural constant for each world. For our purposes it is sufficient to use the speed of movement.

What is the distance of an information? Here we do not speak about the information carrier, which might be a server or a piece of paper. The server in China may be far more distant than the paper in the next office, but it may be faster to get the information from the server.

Let us see this from the time perspective: How long does it take to get information? Like in the physical world, this depends on the travelling speed and the physical distance. The travelling speed of the internet connection is much higher than the speed walking to the next office. For organizations the duration is important, not the physical distance.[149]

How long does it take to get a permission? When you are empowered to decide, the duration is zero. If you have to ask someone for permission it depends. If the other person sits in the neighbor office but does not want to be disturbed, this may take longer than if he has a cellular phone on the other side of the planet. At the end, it is about time.

Physical objects have a travelling time. The human mind has a travelling time as well. How long does it take to focus on a new task? How long does it take to learn new competences?

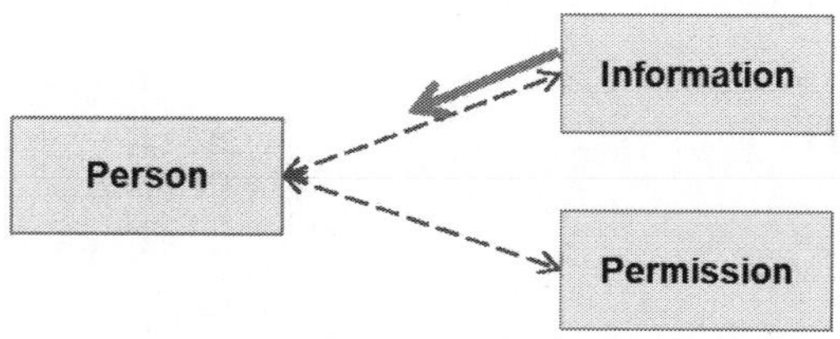

Figure 4-8: Space and time in organizations. The distance of information and permission can be understood as the amount of time it takes to get them. This duration depends on speed.

The distance in time integrates the different worlds: If a process needs physical objects, information, permission and competences, then it needs time to get each object. This time can be understood as their distance. Thus, the different dimensions can be compared.

Many methods in business try to reduce these distances. The lean method 5S wants to place all physical tools at their best place.[150] In an office environment, all the other factors are important as well. Information should always be available at the fingertip. Empowerment reduces the duration of getting permissions to zero. Employees are trained in advance, in order to already have the competences, when they are needed.

Now we have seen that, for our purpose, time and distance can be used equivalently. This equivalence helps us to translate biological concepts that are expressed in space to business concepts that need to be expressed in time.

---

[149] Digitization creates its benefits by reducing the time until the right information is at the right place.

[150] See (The Productivity Press Development Team, 2002)

4.3.3 Relationships

Chemistry

Chemical bonds connect atoms and hold them together with a fixed distance (Figure 4-9).

Chemical bonds connect atoms to larger compounds. The bond requires energy.[151]

Examples: Covalent bond (connecting H, H and O in Water), ionic bond, hydrogen bond

Chemical bonds keep stable distances between atoms, and they influence the geometry. This is the reason for the typical form of molecules and their specific properties. The chemical bond is stable because it represents an energy minimum. It costs energy to break a bond (which is not always freely available). The form remains until enough energy is available to remove the bonds.

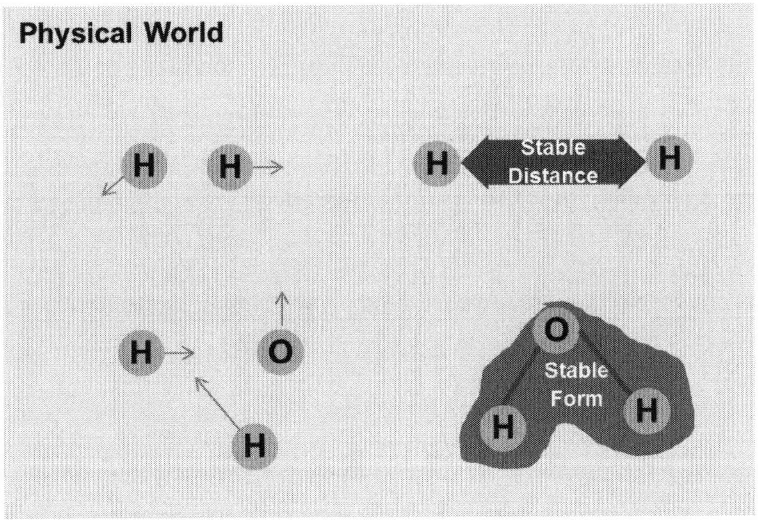

Figure 4-9: Chemical bonds. Chemical bonds lead to stable distances and a form. The form is stable since it costs energy to change distances or angles.

The relationships between the atoms are the carriers of energy: energy is stored in the chemical bond. ATP carries around 50kJ/mole in the bond between ADP and phosphate.[152]

The relationships can also carry information. The connected atoms can have different relative positions in a molecule, which is the base of storing information. DNA, for example, carries genetic information by connecting the nucleotides in a specific sequence.[153]

---

[151] See (Alberts, et al., 2015 pp. 43-50)
[152] See (Alberts, et al., 2015 pp. 65-66)
[153] See (Alberts, et al., 2015 pp. 1-6)

Business

In business we also have compound objects that consist of some basic objects.

Relationships connect simple business objects to more complex ones. This requires energy.

Relationships between business objects bind them together (Figure 4-10). They are more or less stable, so there is some energy or motivation needed to release the relationships again.

In the physical world this is similar to chemistry. Tools and material are located in three-dimensional space. This allows sorting everything to the right place to minimize the distances. If every tool has its fixed place, then this creates a stable form.

In the psychological world the objects are bound by associations. For an organization it is good if Would, Could and Should are related to the same process. The worker would like to do the job, has enough competences to do it and is also empowered to do it. If one factor is missing, then the situation is not ideal. Thus, it makes sense to keep them together. The three factors are not in the physical world, they are subjective.

The empowerment also relates to roles and the design of the entire organization.

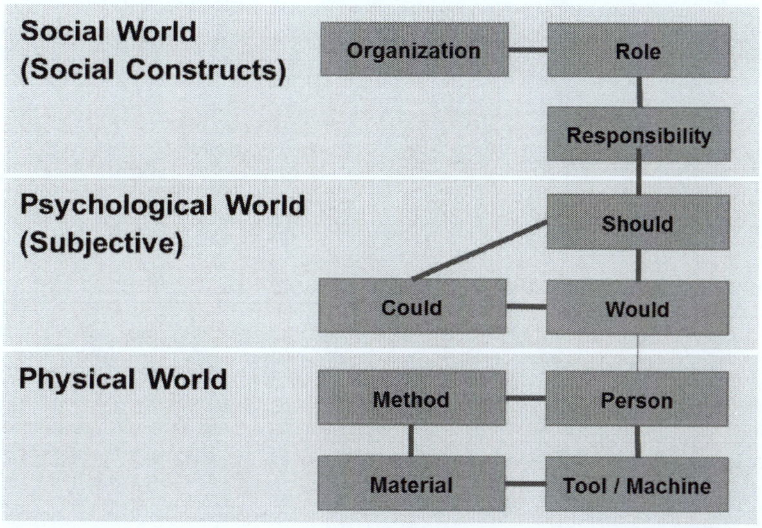

Figure 4-10: Relations between business objects. Business objects have stable relations, if it costs energy to change them. Relationships are not only in the physical but also in the psychological and social world.

It costs energy to bring all the factors stably together. Motivation of employees can get lost, management can strengthen or weaken empowerment, competences are changing. The organization can build up complex behavior only if there is some stability in the relationships.

4.3.4 Processes

Chemistry

Chemistry is not only about the different characteristics of substances. It is also about the chemical reactions that change substances by forming or removing chemical bonds (Figure 4-11).[154]

Chemical reactions form or remove chemical bonds.

Everything that is created in biological organisms is created by chemical reactions. Reactants are processed to products. A chemical reaction creates new chemical bonds, which keep the atoms together. These new bonds contain some energy. This makes the product more or less stable because it costs energy to remove these bonds.[155]

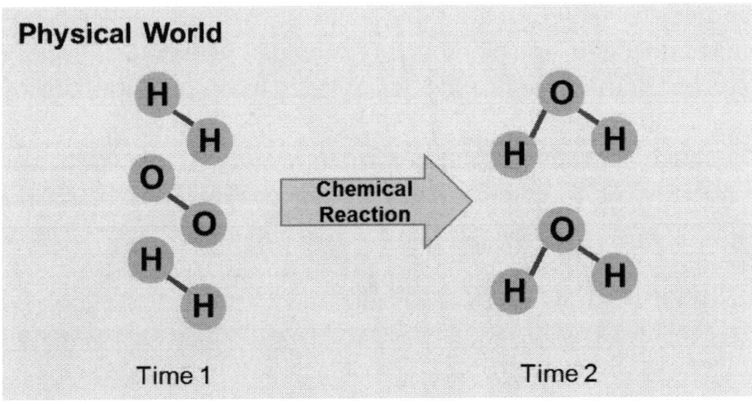

Figure 4-11: Chemical reactions. Chemical reactions change chemical bonds and thus create new forms

The bonds determine how stable the product is. The strength of the bond is the amount of energy that is necessary to release it.

[154] See (Sadava, et al., 2011 S. 38)
[155] See (Sadava, et al., 2011 S. 32-40)

Business

Analogously to chemistry we are interested in changing things. The most obvious example is the creation of products based on material, but business contains many other processes that change something. We create contracts, motivate people and assign budgets. All the complex objects are created by creating or removing relations between more simple ones.

Business processes change relationships between business objects.

The process as the transformation is one of the most important concepts (Figure 4-12). We are especially interested in knowing which processes are necessary, and how does the process speed depend on factors that we can influence.

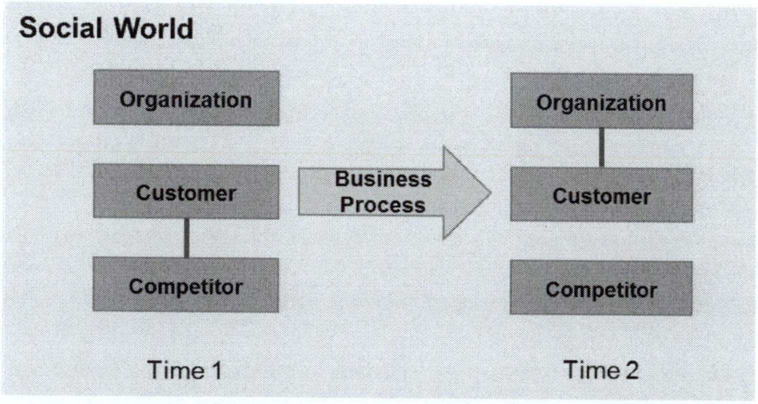

Figure 4-12: Business processes. Business processes change relationships between objects

The strength of the relationships determines how stable the output is. This is not only true for physical products but also for social constructs. If we use business processes to work on the organizational culture, we want to ensure that the new values are strongly anchored and that it costs a lot of energy to destroy them again.

Change management methods refer to that as unfreeze and freeze. In the unfreeze phase we need energy to remove a relationship. In the freeze phase we have to ensure that a new relationship is strong enough to remain stable.

4.3.5 *Speed, Concentration and Flow*

Chemistry

Now we can build the first relationship between function and form: It is the relationship between speed and concentration (Figure 4-13). Speed relates to processes, concentration relates to the objects.

The speed of chemical reactions depends on the concentration of reactants.

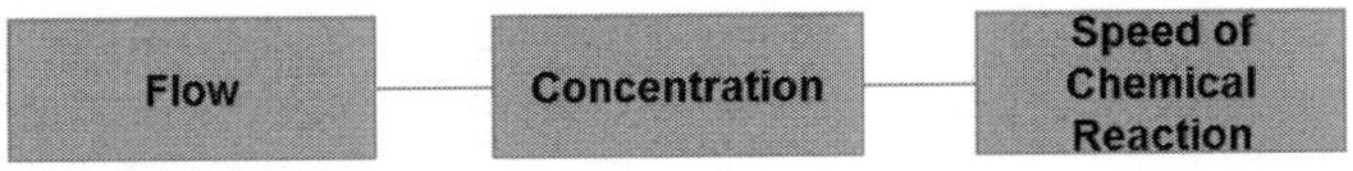

Figure 4-13: Speed, concentration and flow in nature

Reaction speed is the number of created objects per time (e.g. mole/s)

For a reaction the speed is interpreted as the number of newly created objects per time. For a movement the speed is the distance per time.

Concentration is the number of objects per volume
Example: Molar concentration of ions in water (mole / m^3)

Concentration is also about the number of objects but in relation to the space they need. This number of objects links it to the reaction speed. Chemical kinetics analyzes the order of reactions.

In many chemical reactions the reaction speed is proportional to the concentration of each of its reactants. High concentration leads to short distance between reactants, and a short duration of movement, which results in high reaction speed. [156]

In nature there is a natural tendency to reach a thermodynamic equilibrium. High concentration always gets lost by diffusion if there is no boundary that limits movement. [157]

Flow is the number of objects that enters or leaves an area in a given time.
Examples: Volumetric flow (m^3/s) or mass flow rate (kg/s), energy flow, information flow

In biology many mechanisms are about the limitation, generation or restriction of flow. Their purpose is always to maintain the concentration of molecules in a volume. Reactants are needed inside, thus they should have high concentration. Other molecules that are not needed are pumped outside. The purpose of that is – of course – to increase reaction speed and with that to increase the growth rate, which is an evolutionary advantage.[158]

---

[156] See (Berg, et al., 2015 p. 225)
[157] See (Berg, et al., 2015 p. 225)
[158] See (Berg, et al., 2015)

Business

The terms for concentration and flow can be used immediately in the business context (Figure 4-14). The main problem is to get a precise idea of volume

Speed of business processes (productivity) depends on the concentration of production factors

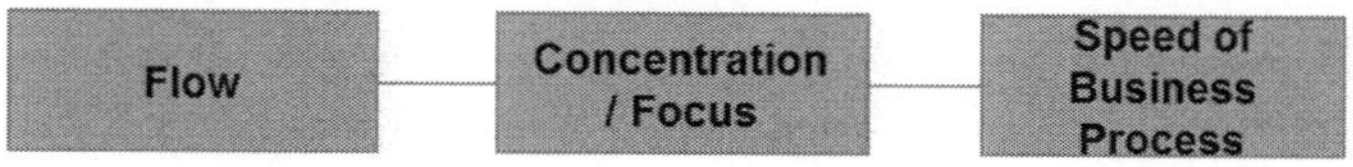

Figure 4-14: Speed, concentration and flow in business

Productivity can be measured as the number of created products per time (products/day)

In an organization it is good that every process has a high productivity. This can be measured by the created output per time.

Concentration measures the number of objects per volume

The whole idea of the lean production methods 5S and Best Point are to create high concentration. Everything that is needed should be at the right position. This implies removing other things. If all material is at the right position, then processes are faster. If they are at the wrong position, they have to be transported (which is one of the seven wastes lean production wants to eliminate).[159]

So, if everything is at the right location, then processes are faster. And this not only includes physical objects, it also includes psychological and social objects: The mind of the operator has to be in the here and now. The empowerment for the process has to be assigned to this operator not to others; otherwise the operator always has to ask for permission.

Keeping everything at the perfect location is difficult. There is a natural tendency of chaos that takes all the objects and distributes them to other locations (especially in children's rooms).

Flow measures the number of objects that enters or leaves an area per time.

Flow expresses the speed that brings something to the right location or removes it from there. Throughput of processes can be compared with flow. Its unit is number of created objects per time.

Many business processes focus on managing flow. Bring something to the right location, remove something, ensure that nothing leaves the location. All this has the purpose to keep the concentration of needed factors high and to increase productivity.

---

[159] See (Gorecki, et al., 2013 S. 110-119)

4.3.6 Summary

4.3.6.1 Insights

Let us repeat the main messages about the foundation of life.

1. The growth of a self-creating system depends on the speed of its processes
2. Process speed depends on the concentration of necessary production factors. The closer the inputs are to the point of processing, the shorter is the time of movement (transport) and the higher is the probability to have all inputs at the same time.
3. Systems have a natural tendency to balance different concentrations. Clear boundaries are needed to separate spaces with different concentrations.
4. Systems generate, restrict or limit flow in order to create a high concentration of production factors. This leads to high process speed and faster growth.

Many mechanisms in organisms and also many business methods are based on these concepts. This already starts with elementary self-organization.

Every object should have its defined place

That is the reason that the first Japanese Lean production experts always started with 5S. Every tool, every material must have the best position. This is only possible if everything not needed is removed. This makes processes already faster.[160]

In an office environment this first step is about necessary information. Where is all the information that is needed for a process? This information has to be available in high concentration. And high concentration means having only the needed information and nothing else. Information in the internet like Wikipedia has exactly that problem. Information is available, but not in high concentration; it is embedded in the knowledge of the entire world.

Every process should have its defined time

This second basic rule is the basis for demand-based production. Produce something only, if it is needed, because this avoids storing products, which costs space and reduces the concentration of those things that are really needed.

4.3.6.2 Requirements

A Viable Organization (VO) shall implement the following requirements (Table 8):

ID	Requirements	Comment
4.3.1	The VO shall be able to manage concentration of production factors	Concentration increases speed
4.3.2	The VO shall run processes at the right time	
4.3.3	The VO shall have all production factors at the right location	

Table 8: List of requirements related to the foundations of viability

[160] See (Gorecki, et al., 2013) for 5S. Gorecki underscore that 5S does not only improve speed. It is also a prerequisite for continuous improvement with PDCA.

4.3.6.3 Existing Solutions

The focus on concentration can be seen in many methods of Lean production or Agile Management: Lean production uses the 5S-method to have workplaces where everything is at the right place.

Agile Methods speak about collocated teams and trying to avoid split heads. This also results in a high concentration of production factors. The ranked backlog creates a strong focus on the next features the team has to implement.

In both cases there is a natural tendency to lose this strong focus. Thus, the methods invest energy to keep that stable. Lean production makes 5S-audits to check whether the workplaces are still well-organized.[161] In SCRUM it is the SCRUM master who checks that the focus is still there.[162]

4.3.7 Annotations

About the Foundation

The translation of the foundation is maybe the most important one. The foundation is the background that contains all the rules of nature that can lead to Viable Systems.

In biology it is the velocity that connects the dimensions space and time. Rules of kinetics that relate concentration of objects in space with the velocity of reactions have the potential to create auto-catalytic systems.

If we understand the nature of the background, we understand the deeper reasons of Viability.

A System Is More than the Sum of Its Parts

Since Einstein we know that $E = m*c^2$, which means that the mass and energy are equivalent. Why is the mass of the proton much higher than the contained quarks? The difference between the energy of the proton and the energy of the three quarks must be in the relations between them.[163]

This could be an interesting aspect when we speak about value. The value of a workplace is higher than the value of its parts. The additional value is in the relationships between the parts. Since it is already a problem for accounting to see know-how as an asset, it is even more difficult to see the relationships between objects as an asset. This is one of the reasons that the best working conditions are very difficult to find.

---

[161] See (Gorecki, et al., 2013 S. 110-119) for the fifth S: Shitsuke

[162] See (Schwaber, 2004)

[163] See (Bartelmann, et al., 2015)

5 Simple Viable Systems

Business

After the detailed discussion about form and functions we are prepared to translate the system dynamics of eukaryotic cells to requirements for Viable Teams. Viable Teams are the building blocks for Large Viable Organizations, like cells are the building blocks of multicellular organisms.

In the previous section we have seen that Viable Teams have to be able to self-create, self-protect, self-adapt and self-replicate if they want to have sustained growth. The self-creation is the core for exponential growth. It contains the reinforcing loop. All other functions support that by protecting, adapting or replicating the system, if it becomes too big. Without self-creation nothing would exist.

In business it is not necessary that the Viable Team duplicates itself every 30 minutes like *E. coli*, but an annual net growth rate of 10 – 20% would be great. Maybe a startup would like to duplicate its size every year. Independent from the growth rate, we have to realize exponential growth. It is good to understand the problem of sustained growth in detail. Let us learn from our best-of-best benchmarking partner:

How is it possible that cells have such a high growth rate? How do they manage to accelerate chemical reactions by a factor of billions? Can it be translated to business? Is it possible to realize it with methods and tools that are already available?

Biology

The literature review already showed that self-creation is well understood in biology. It is realized by an autocatalytic system, where chemical reactions produce the catalysts which in return accelerate exactly these chemical reactions. This will be the starting point for our journey. We will translate the auto-catalyst to business.

Then we will add biological concepts step by step. For that we use a sequence where every new function or form has a clear logical explanation. By translating biology to business in exactly this sequence we can copy not only the concepts but also their logical relationships. This leads to requirements for a Viable Team with strong logical explanations for every concept.

Let us start with the puzzle piece that already contains the secret of growth: autocatalysis.

C. Dachs, *Viable Project Business*, Contributions to Management Science, https://doi.org/10.1007/978-3-030-62904-5_5

5.1 Autocatalysis

Business

In the previous section a business process was described as a transformation where inputs are transformed to outputs. Now we want to do this process as effectively and efficiently as possible. An effective process is a process that produces exactly what the customer wants to have. An efficient process produces that result with a minimum of resources.

In reality we know that different people have a significant difference in effectivity and efficiency, even if the inputs are identical.

Here is an example:

Two software developers get the same task with exactly the same requirements as inputs. One developer is much more experienced, he uses other tools, or he has other techniques to ensure quality. At the end his delivered quality is higher than the results of his colleague. Since the requirements were exactly the same, there are additional factors like know-how, experience, used tools or communication that led to the difference in quality, speed and cost. This section wants to study these factors in detail.

We want to classify these factors and ask where they come from. We want to know why they affect the effectivity and efficiency. We want to know how they interdepend.

If all these factors come together, we can speak about the **best working condition**. The best working condition for a process can be defined as the sum of all factors that make the process highly effective and efficient.

We will see that these needed factors do not exist by accident. Instead they are the results of work preparation and support processes. How could it happen, that the work preparation process was not perfect? Here we can use the lean method "5*Why" to find the root cause.[164]

Let us take an example:

1. Process X is not efficient. Why?
2. X is not efficient, because the tool is not sufficient. Why?
3. The tool is not sufficient, because the tool selection process was not effective? Why?
4. The tool selection process was not effective, because the person has not been trained to look at usability. Why?
5. The person has been trained, but the training was ineffective. Why?
6. The trainer did not check whether the trainee understood the importance of usability? Why?
7. Because it was not part of the best practice for trainings, to check that. Why?

In this example it is obvious that work preparation processes require other work preparation processes. This is not a sequence that ends at a point. It is a circular system of processes, which depend on each other.

---

[164] See (Gorecki, et al., 2013 S. 94)

The deeper cause could be:

8. There is no budget available to write and train best practices for trainers. Why?
9. It is because the organization does not earn enough money. Why?
10. Because all the processes (like process X) are inefficient...

Here we can see how the loop closes itself. In a loop there is neither an ultimate singular root cause nor a final effect. In a loop the effects of problems create the root causes of the problem.

The good news is that these loops are not necessarily negative, they can also be positive. Excellent work preparation creates the best working conditions for value-creating processes. This allows earning money which can be reinvested to improve work preparation. The question now is: how can a set of processes be created that provides the best working conditions for everybody and leads to a positive reinforcing loop?

Biology

Let us take a look at the benchmark. In a cell every chemical reaction is accelerated by catalysts. All catalysts are created by the cell itself. A cell is able to provide the best working conditions for all of its chemical reactions.[165]

Translations

On the following pages we translate the concepts of catalysts and auto-catalysts (Table 9):

	Translation	Chemistry	Business
1	Processes	Speed of chemical reactions, reaction, order, kinetics	productivity, efficiency of business processes
2	Catalysts	Biocatalysts accelerate reactions	Best working condition accelerates business processes
3	Auto-Catalysts	Cells accelerate catalyst-creation	Organizations provide best working conditions for work preparation
4	Modular Auto-Catalysts	Cells can create many catalysts based on blueprints and building blocks	Organizations can create best working conditions for many processes

Table 9: Translations of auto-catalysis

[165] See (Alberts, et al., 2015 p. 6)

5.1.1 Processes

Biology

Every organism can grow. What is growth? The organism itself consists of molecules. Besides water it mainly consists of complex polymers like proteins, RNA and DNA and their monomers, the amino acids ribonucleotides and deoxy ribonucleotides. These complex polymers are not freely available in the environment of the cell. Instead, they have been created by chemical reactions which run inside the cell. [166]

Let us take these chemical reactions as the starting point for our translation.

Chemical reactions transform reactants to products

In a chemical reaction the reactant molecules collide. If they have enough kinetic energy, they can be transformed to a product molecule (Figure 5-1).

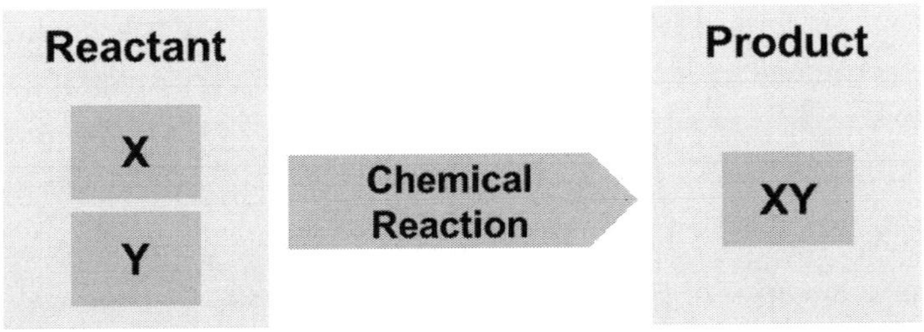

Figure 5-1: Chemical reactions. A chemical reaction converts inputs X and Y to output XY by creating a chemical bond.

Bacteria can duplicate themselves every twenty minutes.[167] This means that the chemical reactions run fast enough, to duplicate the number of polymers, which are in the cell, in twenty minutes. This would not be possible in the inorganic world. Inside the cell the reactions run faster than other chemical reactions that could run alternatively.

The speed of chemical reactions is the subject of the theory of chemical kinetics. Reaction speed depends on the concentration of reactants and temperature. The higher the concentration is, the higher the probability is that two reactants collide. A high temperature means that there is high kinetic energy available which is needed for the reaction.[168]

Both factors, energy and concentration, are independent from cells. They influence reaction speed also in the inorganic world. What leads to the high reaction speed inside cells? In the next section we will see that there is one major difference that accelerates reactions dramatically.

---

[166] See (Alberts, et al., 2015)
[167] See (Alberts, et al., 2015)
[168] See (Alberts, et al., 2015)

Business

What makes an organization successful? Organizations, which grow over a long time, need to hire new employees, train them, buy new tools and material and build trustful relationships to their stakeholders. These things are important production factors for the organization. They are not there by pure chance. They are all the results of business processes.

Business processes transform inputs to outputs[169]

The process is a transformation of the inputs (Figure 5-2). For physical objects this is similar to biology. The inputs are consumed when the outputs are created. This makes it easy to differentiate between material and tools. Materials are inputs that will become part of the output, whereas tools remain unchanged.

In the case of information, it is more difficult. Information is not necessarily lost, when it is used, because it can be copied. For our purposes we can see all information that is not reused anymore as input.

Figure 5-2: Business processes. A business process converts inputs X and Y to output XY by creating a stable connection.

An organization growing faster than others has processes that are more efficient and effective. Let us focus on efficiency first. If we only focus on the inputs (the material) of a process, we see that the speed of processes depends on the availability of the inputs. If no material is available, then also no output can be generated. If the material has to be searched or moved, then the business process also takes longer, because the search and movement time extends the process duration.

Material / input is only one factor. In the next section we see what makes the real difference.

---

[169] See the definition in ISO 9000 (ISO 9000, 2015)

5.1.2 Catalysts

Biology

Chemical substances do not know whether they are inside a cell or outside. Chemical reactions also run independently from life. Why do the reactions inside a cell run much faster than in the inorganic world? The different speed has one reason.

Biocatalysts can accelerate chemical reactions by a factor of 100 billion.[170, 171]

One example for biocatalysts are ribozymes. Ribozymes are RNA molecules with catalytic behavior.[172] Another example are enzymes, which are proteins with catalytic behavior.[173] Both are able to accelerate reactions. In section 5.2 we will discuss their relation in detail. For the beginning it is sufficient to refer to them as biocatalysts.

Biocatalysts provide the best environment for a reaction (Figure 5-3). They have a physical form that acts as a negative template for the reaction. This reduces the needed activation energy for the reaction and thus accelerates it.

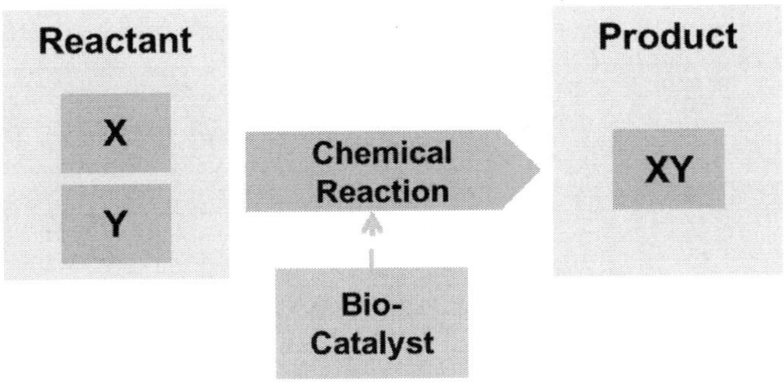

Figure 5-3: Catalyzed chemical reaction. A biocatalyst accelerates the chemical reaction without being consumed.

The high acceleration is only possible if enough biocatalysts exist. In other words: reaction speed depends on the concentration of reactants, the temperature and the concentration of biocatalysts.[174]

Biocatalysts like ribozymes are normally much bigger than the reactants. They are big macro-molecules.[175] It is extremely unlikely that they exist in high concentrations by pure chance. Why is their concentration so high?

---

170 An extreme example is the OMP decarboxylase. Radzicka and Wolfensen showed that the enzyme accelerates the reaction by a factor of 10^{17} from 78 million years down to 18 milliseconds (Radzicka, et al., 1995)
171 See (Alberts, et al., 2015 pp. 140-146) for the description of enzymes
172 Ribozymes are RNA with catalytic behavior. (Alberts, et al., 2015 p. 51) A famous example is the ribosomal RNA.
173 All cells use proteins as catalysts. See (Alberts, et al., 2015 p. 5)
174 The Michaelis-Menten-equation shows the reaction speed as a function of the catalyst and substrate concentration. The higher the concentration, the higher the speed. See (Alberts, et al., 2015 pp. 142-143)
175 RNA consists of nucleotides, enzymes consist of amino acids. See (Alberts, et al., 2015 p. 48)

Business

What can dramatically accelerate processes? If we have all the inputs, and all the know-how and tools we need for a process, then the process is much faster compared with the case that something is missing. This situation can be called the **best working condition** for a process.

This best working condition consists of the inputs, which will be processed to outputs, and some other factors. These other factors are not consumed. Their only purpose is to make the process happen. In the ideal case, they accelerate the process. Thus, they can be called business catalysts.[176] The **business catalyst** for a process is the part of the best working condition that is not consumed (Figure 5-4).

The business catalyst can accelerate business processes.[177]

Business processes need several factors that have to be simultaneously available at the best location. This does not only include all tools and information, but also the operator has to be focused on the here and now. If everything is available at the best place, it accelerates the process.

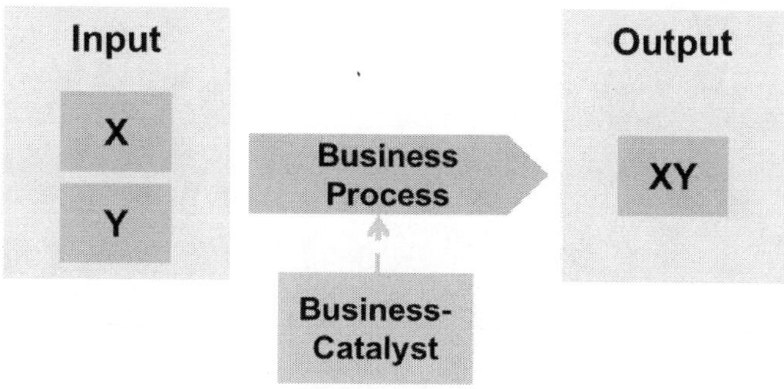

Figure 5-4: Catalyzed business process. A Business Catalyst accelerate business processes without being consumed.

Every business process has a different set of factors. An innovation process is different from a production process, but there is also a best working condition for innovation processes. Every innovator needs information, tools, know-how that helps to accelerate the idea generation.

Even if the factors are different, it is possible to use a fishbone diagram to classify them. Typically, we use the classification Man, Machine, Material and Method, Milieu to identify all factors that help to provide best working conditions.[178] Materials are the inputs that are to be transformed by the business process. Man, and machine together form the business catalyst. The method is the information that describes how the other factors have to be organized in the best way.

In many organizations, people would say that they do not have best working conditions for every task. How is it possible, that all business processes have best working conditions and not only some of them? Why is every single chemical reaction in a cell accelerated by catalysts?

---

[176] See the glossary in the appendix for definitions of best working conditions and business catalyst. Section 5.1.5 shows how the terms are related.

[177] We use the term working condition instead of work environment, which is used in ISO 9000. (ISO 9000, 2015) The best working condition can be understood as the working condition that creates the best process results. Thus, it is tautologic that it accelerates the business process.

[178] Section 5.1.5 shows the mapping of best working condition to man, machine material, method.

5.1.3 *Auto-catalysts*

Biology

Biocatalysts accelerate processes, because they are formed like a negative template for the reaction and thus reduce the needed reaction energy. Where do they come from? It is very unlikely that they exist by pure chance.

Biocatalysts are only available because they are created by the cell itself.[179, 180]

The cell contains biocatalysts that accelerate the chemical reaction which produces these biocatalysts (Figure 5-5). This makes the cell to an auto-catalyst.

Many scientists believe that RNA (ribonucleic acid) was the first biocatalyst.[181] RNA has two important features:

- RNA is a polymer (long sequence) of nucleotides. The exact sequence of the different nucleotides is genetic information. So, we can say RNA stores information.
- The RNA molecule can work as a catalyst. Ribozyme is just another word for RNA that works as a catalyst. Some Ribozymes catalyze their own creation. So not only the molecule is duplicated but also the information is copied.

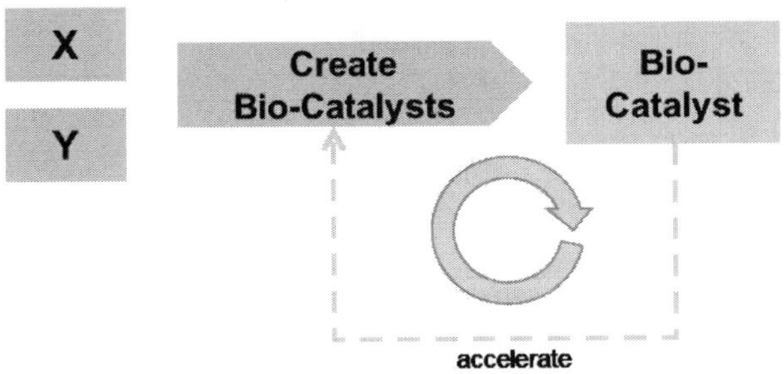

Figure 5-5: Autocatalysis in biology. Some biocatalysts can accelerate their own creation. The more biocatalysts are available, the faster is the chemical reaction, the more biocatalysts are produced.

This is one of the core concepts of life: Living organisms run chemical reactions that create catalysts. The catalysts accelerate exactly these chemical reactions. This forms a reinforcing loop. The more biocatalysts are available, the higher the reaction speed, the more biocatalysts will be produced. This reinforcing loop is only limited by the available reactants. Such an auto-catalytic system converts all existing reactants to catalysts, because the reaction speed is much higher than that of alternative reactions. [182]

---

[179] In this section we are very generic to show the basic principle of autocatalysis. In cells we do not have a single auto-catalyst. Instead, we have two catalysts that depend on each other. RNA catalyzes the creation of Enzymes. Enzymes catalyze the creation of RNA. This will be explained later.

[180] Enzymes are created by the translation process. The translation process is catalyzed by the ribosome (in particular the ribosomal RNA).

[181] The RNA world hypothesis proposes that everything started with RNA (Alberts, et al., 2015 p. 362)

[182] The reinforcing loop is a consequence of the Michaelis-Menten-Equation that shows that reaction speed depends on catalyst concentration. If a process creates catalysts, it increases its concentration and thus accelerates the process. See (Alberts, et al., 2015 pp. 142-143)

Business

How do we get the best working conditions? Are they available by chance? No! They are the results of a business process "Create Best Working Condition". This process needs its own product (Figure 5-6).

Best working conditions are only given if there is a process that creates them

What is the scope of the last step of work preparation? It is the step that ensures that all preconditions for a process are given. All the preconditions must be identified and must be available when a process or task is executed.

The three simple steps of work preparation are:

1. Systematically identify all resources that are needed
2. Get all these resources
3. Place all resources at the best location to minimize search and movement time. (This applies also to information and rights: Everything is about concentration)

With these three steps the best working condition for a process can be produced. The process is faster, because all resources are available immediately.

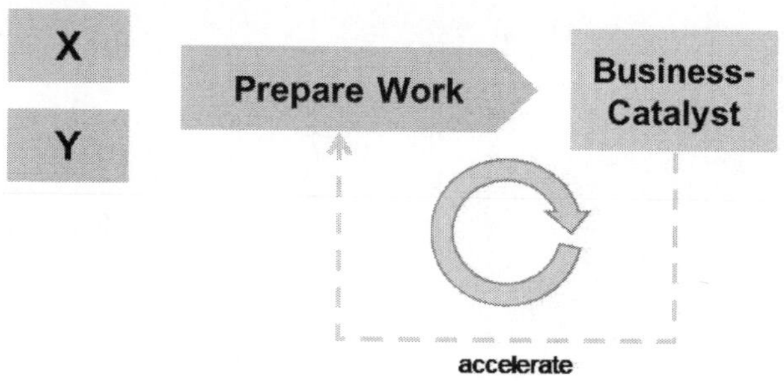

Figure 5-6: Auto-catalysis in business. Work preparation creates the business catalyst as part of the best working conditions. Work preparation itself needs a catalyst as well.

How is it possible that a process depends on its output? It is quite simple: it only has to be a cyclical process, which has to be repeated. Every day you can make work preparation for the coming tasks. During work preparation you ensure that each of the next tasks have the best working conditions.

And one of the next tasks will be work preparation again. In simple words: during work preparation on Monday morning, you have to ensure that the Tuesday work preparation will have the best working conditions.

In other words: the auto-catalyst in business is to ensure that the next daily work preparation has the best working conditions. This leads to excellent work preparation and, as a side effect, to best working conditions in other processes. Many self-organization methods are based on such weekly and daily work preparation meetings.[183] However, the improvement of work preparation is not the target. How can the system help to create best working conditions for all processes?

---

[183] One example is Getting Things Done by David Allen (Allen, 2013)

5.1.4 Modular Auto-catalyst

Biology

The auto-catalyst would be very limited, if it could only catalyze its own creation. In every cell we can need hundreds or thousands of different enzymes and ribozymes[184]. This is only possible if they are based on a modular system (Figure 5-7).

Cells use a modular approach to create a high number of catalysts based on a small set of monomers.[185]

The auto-catalytic system contains genetic information. Genes are the blueprint to construct catalysts based on monomers, which are used as building blocks. The entire system has the monomers as inputs and creates catalysts for other processes as outputs.

Since the autocatalytic system also creates its own catalysts, the system contains a reinforcing loop. The more monomers that are available, the more catalysts will be produced and the faster the system itself works.

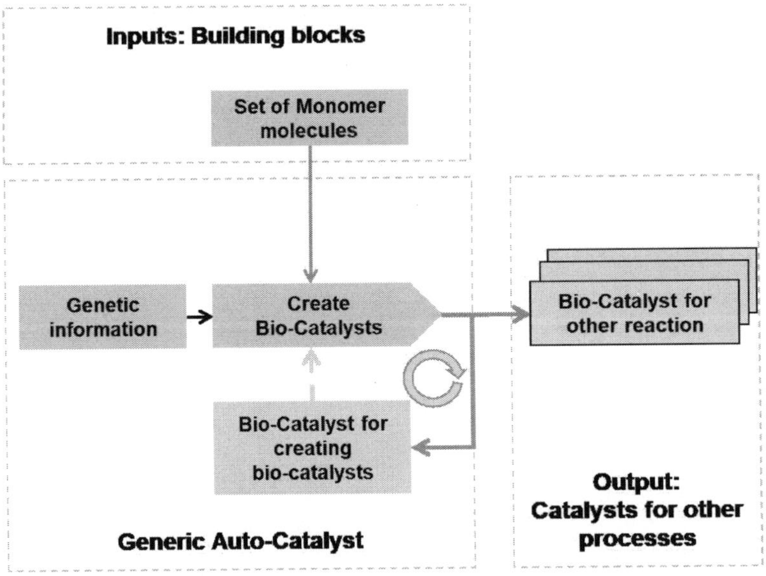

Figure 5-7: Modular approach for creation of Bio-Catalysts. The modular auto-catalyst can create many different catalysts based on a small set of building blocks. The genetic information contains the information how to combine the building blocks for each catalyst.

In section 5.2 we will detail the auto-catalytic system and explain where the genetic information comes from. The detailing of the auto-catalyst also gives a clear picture about the required monomers. In section 5.3 we discuss the creation of the monomers.

---

[184] Some bacteria have less than 500 genes (Alberts, et al., 2015 p. 9) , but most bacteria have 1000-6000 genes (Alberts, et al., 2015 p. 16).

[185] This is true both for ribozymes (RNA) and enzymes (proteins). Both consist of monomer molecules and are created based on genetic information. For creation of RNA see (Alberts, et al., 2015 pp. 301-333), for creation of proteins see (Alberts, et al., 2015 pp. 333-362).

Business

An organization also has to execute many processes. If all of them need to have the best working conditions, then we need an auto-catalyst that can create many business catalysts for different purposes (Figure 5-8).

Work preparation needs to use a modular approach to create a high number of catalysts (BWC) based on a small set of building blocks.

Work instructions or SOPs (standard operating procedures) contain the know-how. They explain step by step which tools and skills are needed, and which step has to be performed. This allows a wide set of processes to be explained even if the toolbox and the basic processes are limited.

The important point here is the reference to building blocks. Know-how that refers to unknown processes or tools is useless. Instead it has to clearly reference things that are already known.[186]

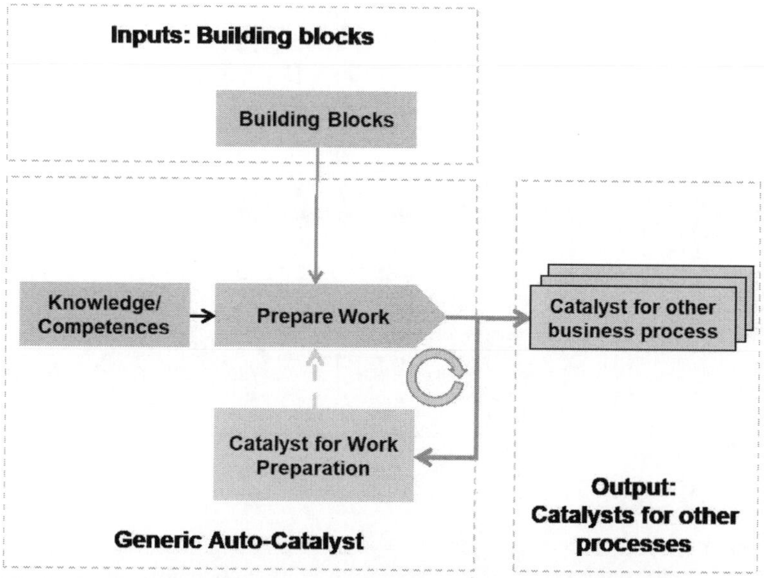

Figure 5-8: Modular approach for creation of a business catalyst. The work preparation process needs to be based on a small set of building blocks that can be combined and the know-how how to combine them.

This modular approach allows the creation of many catalysts (e.g. best working conditions for getting new contracts or creating customer value). It leads to a system which is not only auto-catalytic, but also able to produce catalysts for many other purposes.

The next sections are about the two necessary inputs of our auto-catalyst: Section 5.2 will explain where the knowhow comes from; Section 5.3 will explain where the building blocks come from.

Before we discuss that, let us summarize the insights first.

[186] The naming of elementary processes, skills or tools allows to describe know-how. The mapping is analog to the codon / anticodon used in the translation of RNA to protein.

5.1.5 Summary

5.1.5.1 Insights

Biology teaches us, that all processes of cells can be accelerated if they are all catalyzed. This is only possible, because the cell creates these catalysts all by itself. It is an auto-catalyst. Even more, it is a modular auto-catalyst, which can create a huge number of catalysts based on a small set of components.

The first insight is that the best working conditions accelerate processes dramatically (Figure 5-9). It consists of the material that is consumed during the process, and other factors: the business catalyst.

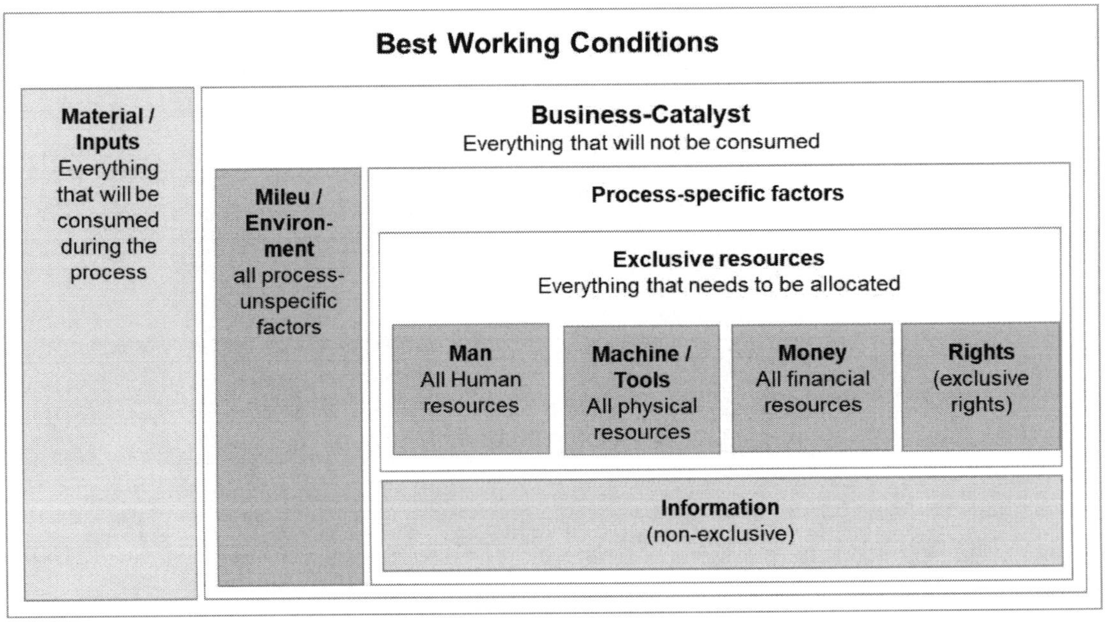

Figure 5-9: Best Working Conditions. Best working conditions includes having all material and the business catalyst, which contains all factors that contributes to the fastest possible process execution.

The best working conditions require that all the factors (man, machine, material and method) are not only available, but also located at the best point they can be, so that only minimum relocations are necessary. Thus, the recommendation is to understand the best working conditions for the different processes and for the people. Non-physical factors also have to be considered.

The second insight is that the best working conditions never come by accident. There are work preparation processes needed to procure all the factors and to bring them to the best position.

The most important insight: the entire system dynamic of Viable Teams starts with the reinforcing loop: The best working conditions have to be given for the work preparation processes that create them. This loop is the fundamental core of acceleration.

The work preparation processes are well-understood in lean production. If you visit a factory, ask for the best working conditions for the work preparation team, and then you see the core of autocatalysis. The work preparation team creates the best working conditions for itself.

5.1.5.2 Requirements

A Viable Team (VT) shall implement the following requirements (Table 10):

ID	Requirements	Comment
5.1.1	The VT shall separate work preparation from execution	Work preparation creates the best working conditions and thus accelerates the execution of a process.
5.1.2	The VT shall have work preparation processes for every process	The work preparation process has to be modular
5.1.3	The VT shall have a modular approach for work preparation	This allows the creation of best working condition for many processes

Table 10: List of requirements related to the creation of best working conditions

5.1.5.3 Existing Solutions

Especially lean production knows many principles and methods which are related to best working conditions. This section shall not describe every detail. Instead, the focus is on the question, how they work together, and how they realize a modular auto-catalyst.

Process Approach

One of the basic principles of quality management (as described in the ISO 9000 series) is the process approach. The deep understanding of a process allows to create good process results with minimum resources. [187] It is important to train people as well as to improve processes.[188]

SIPOC

A simple black-box description of processes can be given with SIPOC, which stands for Supplier – Input – Process -Output – Customer. Suppliers deliver inputs to a process which transforms them to outputs for customers. The process takes some time. Lean production methods like time studies help to understand this duration in more detail, in order to shorten it.

Muda/Waste

If we study the duration of a process, we can see that not every step of the process contributes to the transformation of material. In other words: Not every step creates value. By following the material on the time-axis, it is possible to see, that the duration consists of storage times, transportation time, processing times incl. over-processing times. If we observe the worker, we can see waiting times, movement and processing times.

Toyota uses the concept of the 7 mudas to categorize the waste of time in processes (Table 11).[189]

---

[187] See ISO 9000 (ISO 9000, 2015)

[188] See (The Productivity Press Development Team, 2002)

[189] See (The Productivity Press Development Team, 2002) and (Gorecki, et al., 2013 S. 16-18) and (Bicheno, et al., 2009 S. 20-28)

Muda	Extends Time for Material	Extends Time for Resources	Creates unneeded Repetitions	Creates additional Processes	
Overproduction	x			X	Creates a stock of unused products that have to wait
Defects			x	X	
Over-processing	x	x			
Transport	x				
Storage	x				
Waiting		x			Waiting time of employee
Movement	x	x			Movement of employee

Table 11: Effect of 7 mudas on duration

Toyota sees the root cause of many problems in the overproduction of processes, which create stocks, that requires space, additional resources like warehouses and additional processes. Overproduction directly lead to storage (and maybe to transport) of material and extends process duration. It also creates demand for additional processes and resources.

A second aspect are the defects. If defect products need to be repaired, it costs additional time for workers. The overall duration of production is even longer. It contains the duration of the original (defective) production, the analysis and the repair processes including all the waiting time. If the defect product has to be thrown away, the entire production has to be repeated.

Over-processing includes every activity that does not create value for customers and thus will not be honored. The root-causes can be a poor understanding of customer value, too complex design in engineering or in production itself.

As a summary: Mudas are wastes, because they increase duration of processes. Thinking positive: Elimination of Muda leads to an acceleration of processes.

The search for Waste / Muda in Processes includes the question whether everything is at the right location.

Fishbone Diagram

This fishbone diagram is one method to identiy and classify all production factors needed for the best working condition of a process.[190] By classifying all production factors according to the Ms (man, machine, material, method, management, milieu, money, measurement,...), the fishbone diagram shows the components of the catalysts.

With that knowledge it can be checked, whether everything is available and at the right place.

5S

[190] See (Gorecki, et al., 2013 S. 83)

Set in Order means to bring all necessary objects as close as possible to the point of the business process. This increases productivity because the waste of tool and material transport can be avoided.

5S is a typical method to support standard processes. It stands for:

1. Sort
2. Set in Order
3. Shine
4. Standardize
5. Sustain

In the first step "sort" necessary tools are separated from the un-necessary tools. This allows a clear arrangement of the necessary ones. In step two the tools are set-in-order. The best location is the one, where the movement is at a minimum. This is also called best-point. Typical tools are the shadow-boards, which shows the operator, where every tool has to be. The shadow boards also help to see, when a tool is missing. For non-physical processes the checklists with hyperlinks have the same purpose. They give access to all the factors that are available and highlight the missing resources. In the third step everything in cleaned, to provide a work environment where employees feel comfortable and deviations can be seen easily. Standardization includes the description of the detailed process and the usage of tools. It includes 5S-audits to ensure that the standard is really kept. In the last step "sustain" the organization has to develop a habit[191].

PDCA

All the preparation work has to be done before the work is done. A well-known method PDCA contains the 4 phases plan, do, check and act. The plan-phase contains the best-possible preparation of the do-phase. This includes to provide the best working conditions.[192]

In other words: This plan-phase contains the work preparation process that leads to accelerated processes in the do-phase. It creates the catalyst.

5.1.6 Annotations

Other Aspects That Have Not Been Translated

This introduction of catalysts explains that they provide the best location for chemical reactions. What has not been discussed is their ability to reduce the necessary energy needed to make the reaction run.[193] The typical energy diagram that is used to show how catalysts work can also be translated to the business context. It needs mental energy to execute some processes. The best working conditions reduce the necessary mental energy. This is obvious if people are not trained in a process. In such a case the stress level is high, and it costs much more energy than if they are trained. Know-how lowers the energy barrier.

---

[191] See (The Productivity Press Development Team, 2002) and (Gorecki, et al., 2013 S. 110)
[192] See (Sobek, et al., 2008), (Rother, 2010) and (Gorecki, et al., 2013 S. 49) for PDCA.
[193] For the ability of catalysts to reduce the activation energy see (Alberts, et al., 2015 p. 57)

5.2 Self-Reference

Business

In the previous section we saw that every process needs the best working conditions, including the right people, to run fast. They are the result of work preparation, which requires best working conditions of its own. The focus on work preparation creates an auto-catalyst. If it is based on a modular platform, then it can provide the best working conditions for many processes.

Unfortunately, this description is still quite abstract. It shows the system dynamics clearly, but it is difficult to see how to implement that in detail.

In this section we want to go more into the details of implementation. The fishbone diagram already showed that the best working condition have several aspects: Man, Machine, Material and Method.[194] All these factors have to be integrated precisely.

The material is the input of the business process which is converted to the output. Man and machine form the business catalyst. Their purpose is to provide the excellent environment which accelerates the process. The method describes how all the other factors work together.

Obviously, there is a dependency between them: Man, and Machine both influence the process directly. People use tools and have certain competences. But the Method is something abstract, it is more the way of doing it, rather than a part of the process environment. During the execution of the process either the man or the machine contains the knowledge, how to do it.

How do all these factors work together? How are the best working conditions constructed based on them? How can they be improved?

Biology

Let us have a look at our benchmark:

In recent cells there is not only one type of catalyst. Instead there are two of them: ribozymes (RNA) and enzymes (proteins).[195] Today they both depend on each other, but many researchers believe that RNA existed first before proteins or DNA came up.[196]

We will follow the history and start 4 billion years ago in the RNA world, where RNA was an auto-catalyst, which catalyzed its own creation. Then this system extended itself to a system where the auto-catalysts of RNA and proteins were connected. This big picture shows the entire path from DNA via RNA to proteins.

This chain from DNA via RNA to proteins is the core process in every cell. It can be found in every life form from bacteria to humans. The basic mechanisms have not been changed much in billions

[194] See (Gorecki, et al., 2013 S. 84)
[195] See (Alberts, et al., 2015)
[196] See (Alberts, et al., 2015)

of years. This long-term stability of a concept that has survived billions of years of evolution is obviously very successful.[197]

We will see that these biological concepts correlate to some methods in the business world which are also extremely powerful. They helped to increase productivity of military production in USA during World War II. [198] After that they were used and extended by a Japanese automotive company: Toyota. Now they are important concepts of the Toyota Production System. Is this pure coincidence?

Translations

The following translations can be found on the next pages (Table 12). Starting with the description how RNA and proteins work together as a two-component auto-catalyst, we go backwards and ask where the genetic information came from.

This sequence is quite unusual in biology books; they would explain transcription before translation. We use this inverse sequence to highlight cause-effect relationships. The starting point of every growth is the self-creation, which is generated by the auto-catalyst. All other mechanisms support this by delivering material, increasing concentration or contributing other benefits.

	Translation	Biology	Business
1	Two-component autocatalysis	RNA + protein	Competences + tools
2	Translation	Creation of proteins at the ribosome	Creation of tool environment
3	Transcription	Creation of RNA at the DNA	Creation of competences
4	Evolution	Changes of genetic information	Continuous improvement, Kaizen, A3

Table 12: Translation of self-reference

There are many other interesting aspects that cannot be discussed in the same detail. The most interesting ones are explained in the annotations.

---

[197] See (Alberts, et al., 2015)

[198] A description of Training within industry can be found at (Liker, et al., 2008) and (Training within Industry Program, 2009a), (Training within Industry Program, 2009b), (Training within Industry Program, 2009c), (Training within Industry Program, 2009), (Training within Industry Program, 2009)

5.2.1 *Two-Component Auto-catalyst*

Biology

How did life start on earth? Most scientists follow the RNA world hypothesis, where RNA was the first biocatalyst.[199]

RNA was the first auto-catalyst

RNA is a polymer of ribonucleotides that has two features at the same time: it can store information and it can catalyze chemical reactions – especially its own synthesis. This makes RNA an auto-catalyst. It accelerates its own creation and the creation of other molecules. [200] The ribosomal RNA catalyzes the creation of proteins. Enzymes are proteins with catalytic properties. They extended the range of reactions that could be catalyzed.

- Ribozyme (RNA) = "Primary catalyst"
- Enzyme (protein) = "Secondary catalyst"

Today RNA molecules do not catalyze their replication anymore. Instead their creation is catalyzed by enzymes called RNA-polymerase based on genetic information stored in DNA.[201]

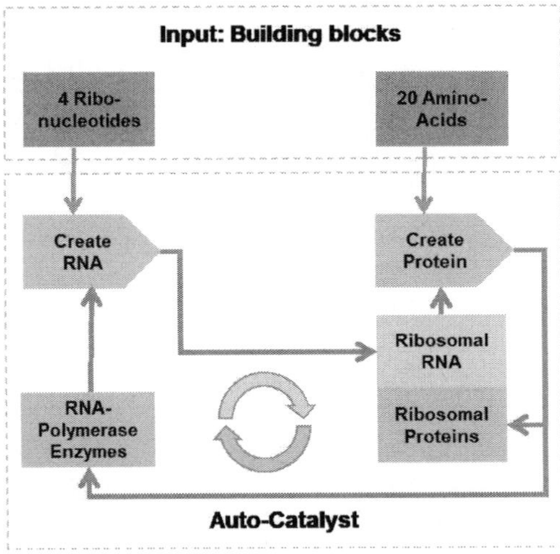

Figure 5-10: Ribozymes and Enzymes. Both molecules mutually catalyze their own creation. Ribozymes catalyze the creation of enzymes, enzymes catalyze the creation of ribozymes.

Ribozymes catalyze the creation of enzymes.[202] Enzymes (RNA-polymerase) catalyze the creation of ribozymes. The auto-catalyst consists now of two components which mutually create each other (Figure 5-10). This shows how an auto-catalytic system can extend itself. RNA had two purposes at the beginning. Later it created two tools that are specialized for only one task. Enzymes are only focused on catalysis, whereas DNA is specialized for the storage of information.

---

[199] For the RNA world hypothesis see Alberts et al (Alberts, et al., 2015 pp. 362-366)

[200] For the catalytic behavior of enzymes and ribozymes see Alberts et al (Alberts, et al., 2015 p. 51) and (Alberts, et al., 2015 p. 69)

[201] For the transcription mechanism from DNA to RNA see Chapter 6 of Alberts et al (Alberts, et al., 2015 pp. 299-317)

[202] The translation mechanism from RNA to proteins is used in all biological organisms (Alberts, et al., 2015 p. 6)

Business

In business we also have a two-component auto-catalyst. Let us start with the first component:

All social systems start with humans

Humans have two capabilities: they can store information (know-how) and they can accelerate changes in the outside world (influence their environment with their hands to make things happen). Humans had this ability already before the invention of tools.

Humans were able to create tools to extend their capabilities (e.g. hand axe, spear, wheel and fire). Some of the tools (e.g. hand axe) accelerate the creation of other tools.

- Human =" primary catalyst"
- Tool = "secondary catalyst"

One of the tools was the invention of writing, to conserve and transfer know-how. This helped humans to avoid the loss of information. Know-how can now be retrieved by reading books.

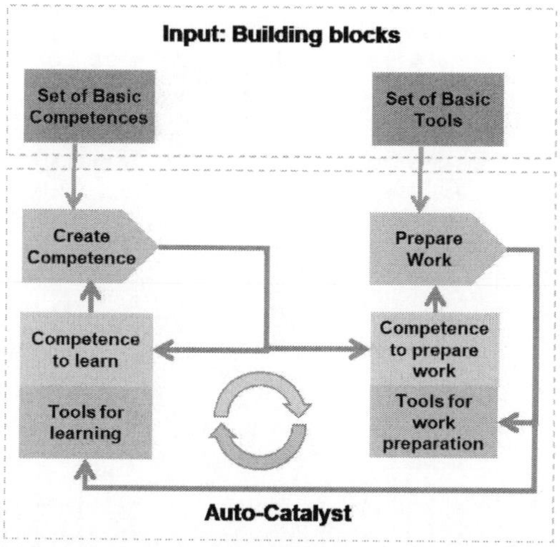

Figure 5-11: Competences and Tool Environment. Know-how is a catalyst to create tools and tools catalyze the creation of know-how (e.g. books).

The auto-catalyst has two important parts that depend on each other: competences and the tool environment. First of all, a competent person is needed to prepare a perfect tool environment. Secondly, it requires a good tool environment to create competences (Figure 5-11).

In the last decade the tools and the information store have become more and more sophisticated. Robots have taken over complex tasks and information is stored in the cloud.

Let us go through the different creation processes step by step.

5.2.2 Translation

Biology

We have already seen that a modular system separates building blocks from the blueprint. Proteins are large polymers consisting of a small set of amino-acid monomers that are connected in a defined sequence given by the (protein-coding) genes.[203]

Translation creates proteins by connecting amino acids in a sequence that is defined by the genetic information stored in the mRNA molecule.

Every protein of the cell is created by the translation process that happens at the ribosomes (Figure 5-12). The materials for the translation process are 20 types of amino acids which have to be connected in a specific sequence. The sequence is stored in the RNA molecule, called messenger RNA (mRNA).

The entire translation process is catalyzed by the ribosome which itself consists of ribozymes (rRNA) and additional proteins.[204]

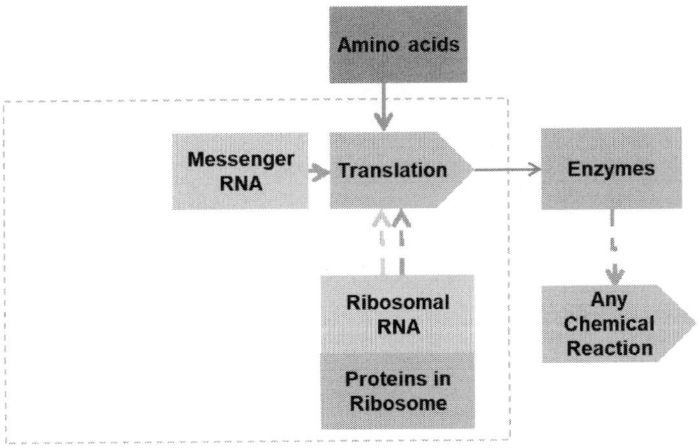

Figure 5-12: Translation process. Translation process creates proteins (orange) by connecting amino acids. RNA (yellow) is used as the catalyst (rRNA) and as the carrier of genetic information (mRNA)

RNA is used for two purposes. Messenger RNA (mRNA) contains the genetic information that describes the sequence in which the amino acids have to be connected. Ribosomal RNA (rRNA) is the catalytic part of the ribosome. A third purpose is not shown in the picture: Transfer RNA (tRNA) contains the mapping between the genetic code on RNA and the 20 types of amino acids.

This explains how the secondary catalyst is created. In the business environment the secondary catalysts are the tools.

---

203 For translation see Alberts et al (Alberts, et al., 2015 pp. 333-362)

204 For the structure of ribosomes see Alberts et al (Alberts, et al., 2015 pp. 346-347)

Business

How is it possible that an organization can do many different processes efficiently? It is possible if all processes depend only on a small number of tools and there is know-how available how to use them.

Work preparation creates the best tool environment by arranging tools in a defined way.

Experts need competences to execute processes. They are able to select and arrange the right tools. Competences and the best tool environment contribute to the best working condition.

If experts have the competences to execute a process, they also know which tools are needed and have to be used in which situation. Thus, they can prepare the best tool environment based on a set of available tools. After preparation it needs both factors: the competence of the expert and the right tool environment. If both factors are available, then the business process has the best working conditions and is much more effective and efficient.

The work preparation itself is also a business process (Figure 5-13). It needs a catalyst as well: the competence to do work preparation and some supporting tools. Both together deliver the best working conditions for work preparation itself.

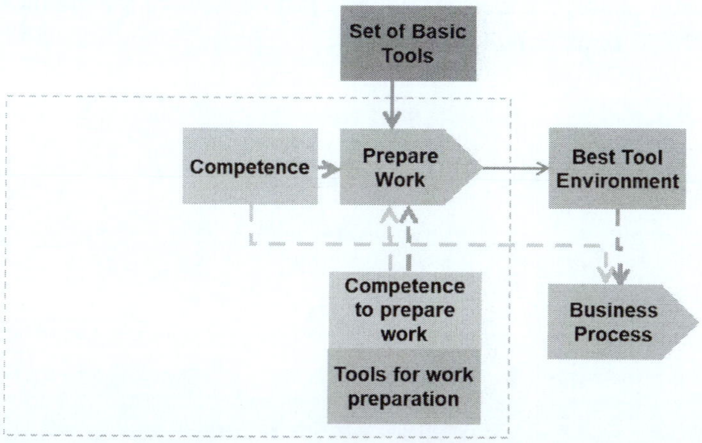

Figure 5-13: Work preparation process. Work preparation provides the best tool environment by combining available elementary tools. This preparation is based on the competence to execute the specific process.

The competence always has to refer to tools that really exist. There must be a clear reference between knowledge and skills on the one hand and the toolkit on the other hand.[205] In other words: the expert must be able to identify the tools and know how to use them.[206]

If competence of people is the translation of RNA: What is translation of DNA?

---

[205] The naming of tools is analog to the mapping of RNA to protein via codons. The triplet of ribonucleotides that makes the codon identifies the correct amino acid.

[206] Liker and Meier emphasize, that the training process of Toyota is not only focused on execution of processes. Trainees have to learn the correct names of tools as well. (Liker, et al., 2008)

5.2.3 Transcription

Biology

RNA molecules contain the know-how that is needed to create proteins. Unfortunately, they are very unreliable data storages. DNA is much more stable due to its different chemical characteristics.

DNA is a reliable carrier of genetic information. Temporary RNA copies can be made in the transcription process.[207]

DNA is the unique source of genetic information. The transcription process creates a temporary copy by connecting ribonucleotides to an RNA molecule using the same sequence as in DNA. This sequence is the genetic information that is copied from DNA to RNA (Figure 5-14).

The transcription process is catalyzed by an enzyme, the RNA polymerase. Cells have different RNA polymerases to produce messenger RNA, which is used to create proteins, or ribosomal RNA, which is part of the Ribosome.[208]

This RNA-polymerase closes the loop of the two-component auto-catalyst. RNA-polymerases are enzymes which catalyze the creation of RNA. Ribosomal RNA catalyzes the creation of enzymes. This two-component auto-catalyst requires DNA as the data storage for the blueprint and ribonucleotides and amino acids as building blocks.

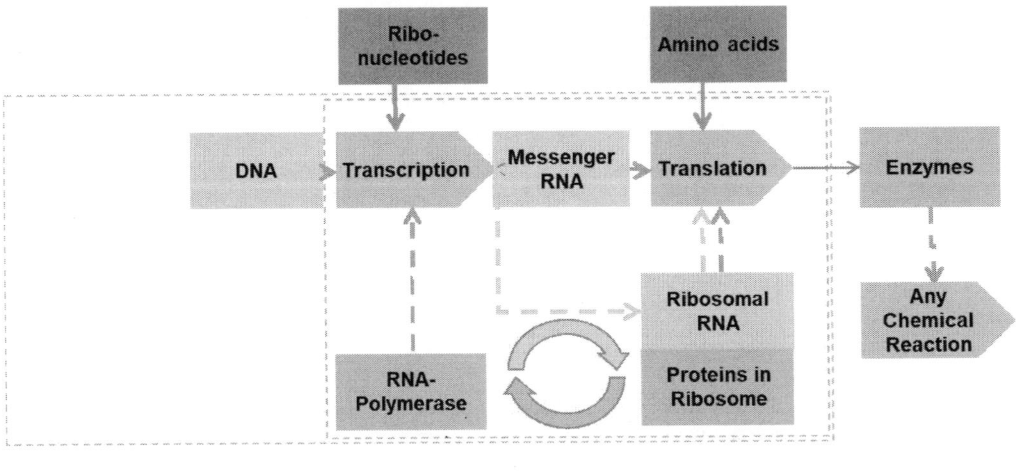

Figure 5-14: Transcription process. DNA as the stable storage of genetic information

This makes DNA the information master. It works only if the RNA molecules are always destroyed after a period of time.[209]

Now the RNA has two specialized partners. Proteins are specialized on catalysis and DNA is specialized on data storage. RNA has still both properties.

---

[207] For the transcription process see (Alberts, et al., 2015 pp. 301-332)

[208] See (Alberts, et al., 2015 pp. 301-332) for differences between mRNA, rRNA and tRNA

[209] This is fundamental for gene expression, where cell signaling decides, which gene of the DNA has to be expressed or not.

Business

Experts know what to do. But two experts often have three opinions. Thus, it is better that they share their best practices and write down all the knowledge.

Best practice descriptions are reliable carriers of know-how. They can be used to train many employees.

A simple example is a cooking recipe. It describes all the needed ingredients and explains step by step how to create a fantastic meal. The recipe is a best practice description that conserves know-how and makes it available to many people.

Work instructions describe the best practice of how to prepare and execute processes[210], but they are only written words on paper or on the computer. They only help if people apply the documented knowledge in a process; and the first step to use them is to read and learn them (Figure 5-15). The ideal result of learning is the competence. It is more than knowledge. It is the capability to prepare and execute a process according to the best practice.[211]

This learning of a new competence happens in the brain of the trainee. Experience in learning and a good environment helps. A best practice for learning has to describe that.

But it is not the job of the employee to become a good learner on his own. Trainers can help here to create competences, especially the competence of self-learning.

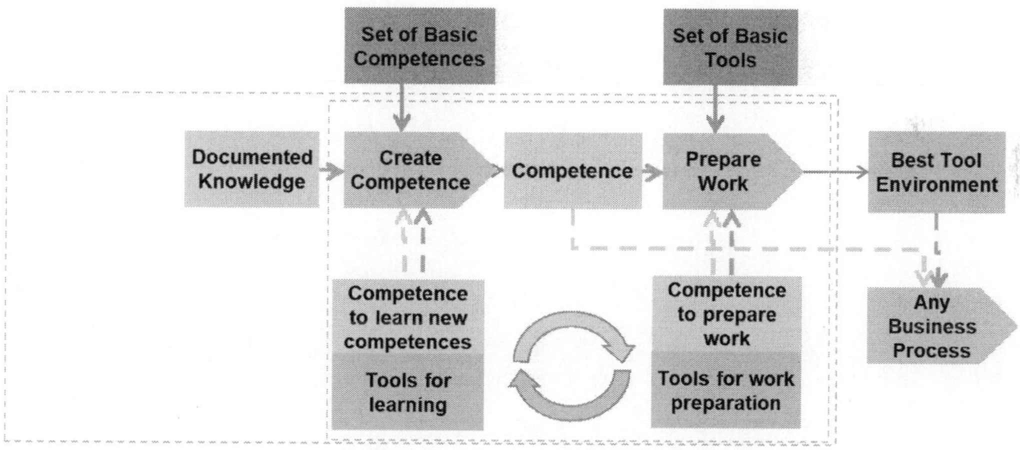

Figure 5-15: Learning process. Work instructions are the stable storage for know how

One of the problems in large organizations is the inability to forget competences. Every employee has his own way of doing a process. This makes it difficult to improve processes. Some professions like pilots are trained to use checklists as the written standards. This allows the continuous improvement of knowledge, like evolution improves the DNA. How does this work in detail?

---

[210] See (The Productivity Press Development Team, 2002 S. 28) for an example of a standard operating sheet, (Training within Industry Program, 2009a) for job instructions and (Liker, et al., 2008)

[211] See (Liker, et al., 2008) for a description of training processes and (Training within Industry Program, 2009a) for job instructions

5.2.4 Evolution

Biology

Where do all the blueprints come from? At the beginning there was only some RNA molecule that contained some genetic information and catalyzed reactions. [212] Later, the RNA created proteins, which had catalytic properties as well, and they created DNA molecules, which could store the genetic information more reliable. Even this information is not stable. Experiments with *E. coli* show a mutation rate of 3 nucleotide changes per 10^{10} nucleotides per cell generation.[213]

Mutations change genetic information. This changes catalysts and the metabolism.[214]

The genetic information can be modified during the evolution of organisms (Figure 5-16). The natural selection process prefers changes that have a benefit for survival in a given situation.[215]

DNA molecules can mutate. This leads to changed genetic information in the DNA that is copied to RNA. This changed genetic information leads to changed proteins. Maybe the changed protein has changed its catalytic properties. This can be an advantage or disadvantage for the organism. Natural selection decides whether the mutation remains in the genetic pool or not.[216]

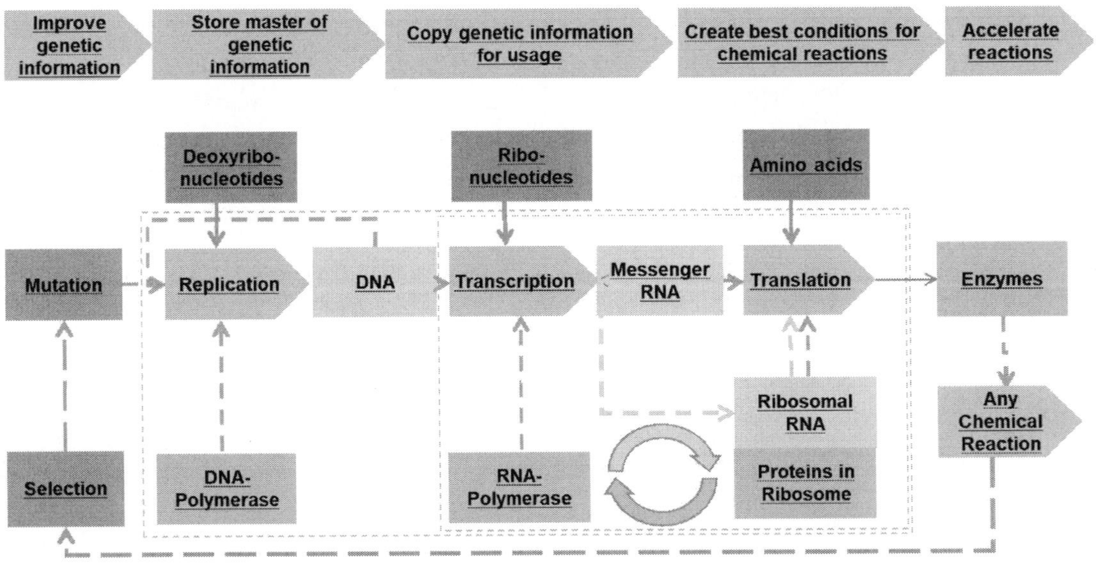

Figure 5-16: Evolution

Evolution is like the continuous improvement of DNA, but there is no plan behind it. It can also produce some proteins that have no meaning at all.[217] But the selected ones can produce something new.

---

[212] See (Alberts, et al., 2015 p. 364) for the RNA world hypothesis: RNA can store information and catalyze reactions
[213] See (Alberts, et al., 2015 pp. 237-238) for mutation rates of DNA
[214] See (Alberts, et al., 2015 pp. 15-16) for the description of natural selection
[215] See (Alberts, et al., 2015 pp. 216-222) for the description of evolution
[216] See (Alberts, et al., 2015 pp. 216-233) for the description of evolution
[217] See (Alberts, et al., 2015 pp. 216-233) for the description of evolution

Business

Where do the written best practices come from? The first version must have been written by a experienced employee, who documented his way of working as a best practice. Such a description is the starting point for improvement.

Process improvement changes work instructions. If they are used, this influences the best working conditions for processes.

Organizations can change and optimize work instructions based on necessities following the PDCA (Plan, Do, Check, Act) process (Figure 5-17).[218] The existing best practice can be the starting point for the improvement. Additionally, it needs know-how about the business process that has to be improved. The improver tries to understand the problem in the process and the necessary changes of the working conditions. The changes can be documented as a proposal for a new best practice. All this happens in the plan phase.

Now the changes have to be tested in an experiment. The new best practice is used to prepare and execute the process, and the results are checked. If the proposed new best practice is better, it can become the new standard.[219] By doing the PDCA process regularly, all processes can be systematically improved.

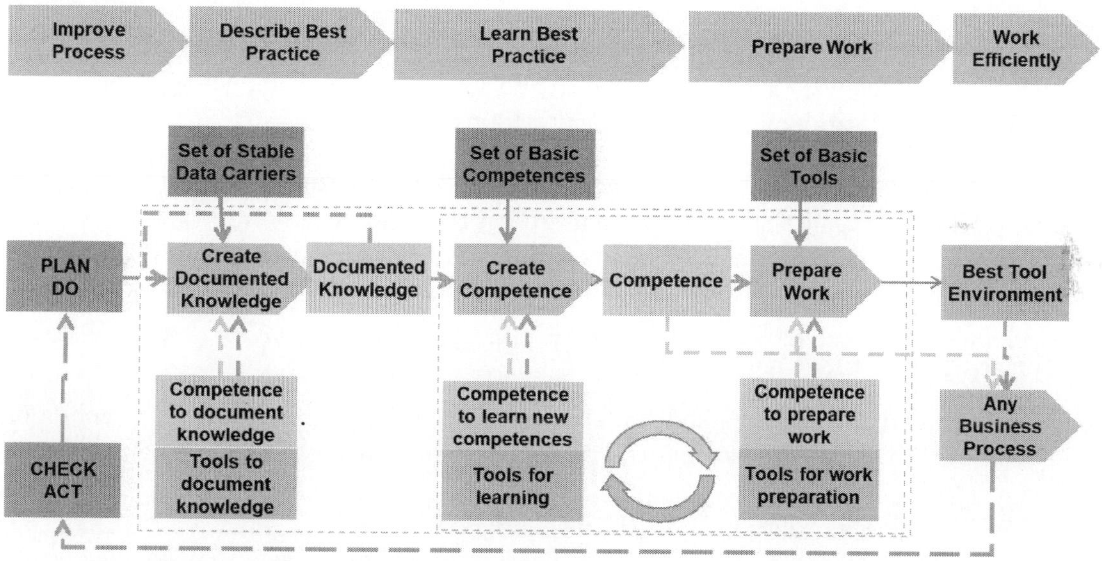

Figure 5-17: Improvement of know-how with PDCA

The main difference to biology is the planning. Process improvements are not pure mutations that might be better or not. Instead the plan phase of the PDCA process tries to select those changes that have a high probability to be a real improvement. As in nature, the real selection process is at the end, after the best working conditions have been changed and the results of the affected business process can be observed. Let us now summarize the findings about the blueprints of catalysts.

---

[218] See (Shook, 2008), (Rother, 2010) and (Sobek, et al., 2008) for PDCA and continuous improvement with A3

[219] See (Sobek, et al., 2008)

5.2.5 Summary

5.2.5.1 Insights

Biology teaches us how to create a modular auto-catalyst that is based on a small set of components, the amino-acids and nucleotides, and a stable genetic blueprint, the DNA, that describes how to combine these components to create catalysts. The DNA can change during evolution and thus create new blueprints for new catalysts with new properties, which might be beneficial.

In business our driving force is the creation of the best working conditions for every process. They include the business catalysts that accelerate everything. If we provide them for the creation of best working conditions, then we have a positive reinforcing loop.

Such a system is only possible if we have a modular approach, where we can create the best working conditions for a huge number of processes based on a small set of building blocks.

The combination of the basic building blocks is the know-how that explains how to combine these factors in details. A Viable Team must have such know-how.

If possible, it should be written down as a work instruction that contains the best practice for a process. Everybody can then read and follow the work instruction. Training within Industry (TWI) shows how to write good work instructions that can be used to conserve knowledge, and it shows how to make the knowledge available by conducting trainings.

It is important that everybody uses the current work instruction to prepare the best working conditions. Consistently using a standard is the prerequisite for improvements. If the work instruction is a unique source, then it can be improved with A3 or Kaizen processes. They change the work instruction following a PDCA cycle which allows the evolution of the organization.

5.2.5.2 Requirements

This section showed the main processes that are necessary to create the best working conditions. A Viable Team (VT) shall implement the following requirements (Table 13):

ID	Requirements	Comment
5.2.1	The VT shall have a process to provide the best tool environment	The process creates the catalyst that accelerates the processes
5.2.2	The VT shall have a process to create competences	The process creates the competence to prepare and execute a process. The competence is result of a learning process that can be supported by a trainer.
5.2.3	The VT shall have a process to create documented process knowledge	
5.2.4	The VT shall have a process to improve processes	Process improvements lead to improved work instructions that are used to prepare and execute processes.

Table 13: List of requirements about work preparation, learning and improvement

5.2.5.3 Existing Solutions

Lean management already provides many methods that realize these requirements.

Standard Work

Lean management is based on the principle to always use the best practice as a standard. This standard can then be continuously improved. This standard implies that for a process always the same material, tools, qualification and methods have to be used. Especially for tools and material there is a best position where they should be placed. Workplace design creates such an environment, where everything is at the right pace.[220]

Work Instructions

Work instructions document knowledge about processes to make it permanently available. They explain the necessary preconditions of processes, their execution step-by-step and the expected post-conditions (definition of done)[221].

Tailoring

Work instructions can be very complex if they can be used in various situations. Tailoring for a specific purpose helps here. An example is a large checklist for projects that can be simplified before usage in a small project.
(This is the translation of post-transcriptional modification, which has not been discussed in detail)

Training & Coaching

Documented knowledge needs to be understood by people. Training within Industry shows how to transfer the written information to the user. A very good description of training processes at Toyota is given by Liker and Meier.[222]

Process Improvement (PDCA, Kaizen, A3)

Process improvement results in the change of a work instruction. Sophisticated methods like A3 are tools that lead the process improver through a process that results in improvements of high value[223].

---

[220] See (The Productivity Press Development Team, 2002)
[221] See (Training within Industry Program, 2009), (Training within Industry Program, 2009a) (Training within Industry Program, 2009b) and (Liker, et al., 2008)
[222] See (Liker, et al., 2008)
[223] See (Shook, 2008) (Sobek, et al., 2008)

5.2.6 Annotations

Other Aspects That Have Not Been Translated

There are some biological concepts that have not been translated in detail:

Post-transcriptional modification / alternative splicing eliminates sections of the RNA before its usage in translation. This leads to different proteins.[224] This is the equivalent of tailoring of work instructions for a specific situation. A translation may be helpful.

tRNA contains the mapping between RNA-codons and amino-acids.[225] This is equivalent to the identification of tools by name. Every object needs to be referenced by an identifier. This is prerequisite for a communication about tools. Here a precise translation may be helpful.

The genetic information on the DNA molecules is organized in chromosomes[226]. This could be translated to packages of work instructions. The packaging of work instruction makes sense, but the genes on the chromosomes are not sorted by their purpose or other categories. Thus, a detail translation of this concept does not help in the business context.

Training Within Industry (TWI)

Beside the perfect workplace, the operator needs a high qualification. Lean production is based on the works of the War Manpower Commission, which developed the concept "Training within industry" during World War Two.[227]

Dinero describes TWI as the foundation of Lean. A very interesting section is the detailed explanation how the principles of TWI has been invented already in World War One, when the United States had to increase their production capacity for ships by a factor of 10 in a couple of months. This required training of 500.000 people in a short time to produce high quality ships. The War Manpower Commission has then developed TWI for the usage during World War Two, where it allowed that unexperienced, but systematically trained workers had higher productivity than the experienced ones. After WW2 these concepts have been transferred to Japan and especially to Toyota. Dinero describes "What is clear is that since 1945 the Japanese have incorporated TWI in its companies, while the use or even awareness of TWI and its contributions have nearly disappeared in the Unites States."[228]

Graupp and Wrona point out in Implementing TWI, that the need for training is of high relevance. Especially in an environment where everything is changing faster, organizations need a good environment for learning.[229]

---

[224] See (Alberts, et al., 2015 p. 415) for alternative RNA splicing
[225] See (Alberts, et al., 2015 p. 334) for a detail description of tRNA, codon and anti-codon
[226] See (Alberts, et al., 2015 pp. 179-216) for the description of chromosomes
[227] See (Training within Industry Program, 2009a) (Training within Industry Program, 2009b) (Training within Industry Program, 2009c) (Training within Industry Program, 2009) (Training within Industry Program, 2009)
[228] See (Dinero, 2005 S. 21)
[229] See (Graupp, et al., 2011)

5.3 Metabolic Network

Business

The previous sections explained the importance of work instructions, the development of complex competences and the preparation of tools. All of these processes can be summarized as work preparation. They have the goal of providing the best working conditions for other processes. These work preparation processes run even better, if they have best working conditions themselves. This system is the auto-catalytic system.

Work preparation requires some basic competences and some basic tools that are used as building blocks. The work instructions explain how to combine these basic competences and tools to get the best working conditions for specific processes. Where do the basic competences and basic tools come from?

If the basic competences and tools would be available on the labor and supplier market, then nothing would have to be done. An organization only has to hire the people, who already have all the competences and buy the needed standard tools from suppliers. Finished!

Unfortunately, the basic competences are not directly available. Thus, the organization needs to run processes to develop the basic competences based on the previous knowledge that is available on the labor market. If the organization is good in these processes, it can select between many potential employees that have learned some competences at school and the university, and systematically train them until they all have the set of basic competences which are needed as input for work preparation. The same thing applies to tools. The organization can buy standard tools from a supplier, but it needs processes to configure and integrate everything until it has the needed basic toolbox it can use.

All these processes have one purpose: they connect the ability to create best working condition with the real world. The best working conditions can now be created based on competences and tools that are readily available outside. The organization can select the best of many potential employees and the best of many potential tools because it can do all processes to create basic competences and basic tools.

Biology

Nature has had exactly the same challenge. Cells can create DNA, RNA and proteins as long as enough (deoxy)ribonucleotides and amino acids are available. Nucleotides and amino-acids are already very complex molecules. They are not available in high concentrations. This limits the speed of self-creation dramatically.

The best feature of the generic auto-catalytic system of DNA, RNA and proteins is that they can produce enzymes for other purposes. Evolution can try out various enzymes. If some of them catalyze a reaction which creates nucleotides or amino-acids based on food that is available in higher concentrations, then this would be an evolutionary advantage.

The concentration of nucleotides and amino-acids is higher, and the auto-catalyst has more input and can grow faster.[230,231]

This mechanism leads to the growth of metabolic pathways from the auto-catalyst backwards to a source which has practically unlimited concentration. In nature this is reached if the inputs are something like water, carbon dioxide, nitrates or phosphates and the energy source is sunlight[232]

Now the auto-catalyst is connected to the real world and can grow exponentially without any resource limitation.

The network of processes which connects the available food to the auto-catalyst is called the metabolic network.[233] Let us study and translate its topology in detail.

Translations

The following subsections describe some core concepts of the metabolic network (Table 14). The focus is on the question how the auto-catalyst manages to get unlimited food for unlimited growth. Other aspects that are not mentioned here are described in the annotations.

	Translation	Biology	Business
1	Feed the auto-catalyst	Produce components of DNA, RNA and proteins	Provide basic competences and basic tools
2	Pathways	Metabolic pathways, intermediates	Process chains, work in progress
3	Direct connection	Multi-enzyme complexes	One-piece flow, U-cell
4	Anabolic pathways	Synthesize amino-acids or nucleotides	Synthesize the building blocks for work preparation
5	Catabolic pathways	Breakdown molecules to get energy and input for the anabolic pathways	Decompose objects for reuse
6	Energy	ATP as renewable energy carrier	Motivation, energy

Table 14: Translation of process network

Let us start with the big picture first: The metabolic network feeds the auto-catalyst.

---

[230] The cell uses inhibition mechanisms to control the reaction speed.

[231] The relation between resource concentration and reaction speed shows an asymptotic behavior. A reaction speed increases with the resources concentration, but the speed increases get smaller (Saturation).

[232] Databases like BioCyc and EcoCyc have a detailed documentation of these metabolic pathways. (BioCyc, 2018)

[233] See (Berg, et al., 2015 p. 423) for an introduction to metabolic networks

5.3.1 Feed the Auto-catalyst

Biology

The auto-catalyst described in the previous sections would already create itself as long as sufficient (deoxy-)ribonucleotides and amino acids are available. The speed of self-creation is limited by the availability of these molecules. If the organism can synthesize nucleotides and amino acids based on material that is available without limit, then the organism would grow much faster.[234]

The metabolic network creates the building blocks of DNA, RNA and proteins based on available food (Figure 5-18).[235]

Since an auto-catalyst can produce catalysts also for other chemical reactions, this is no problem. If evolution leads to enzymes that catalyze the creation of the building blocks, such a system would grow much faster, because it has unlimited resources.[236]

The metabolic network grows backwards from the auto-catalyst. If evolution produces a new enzyme which connects the network to a source of food of a higher concentration than the food before, then the entire system has fewer limitations and can grow faster.

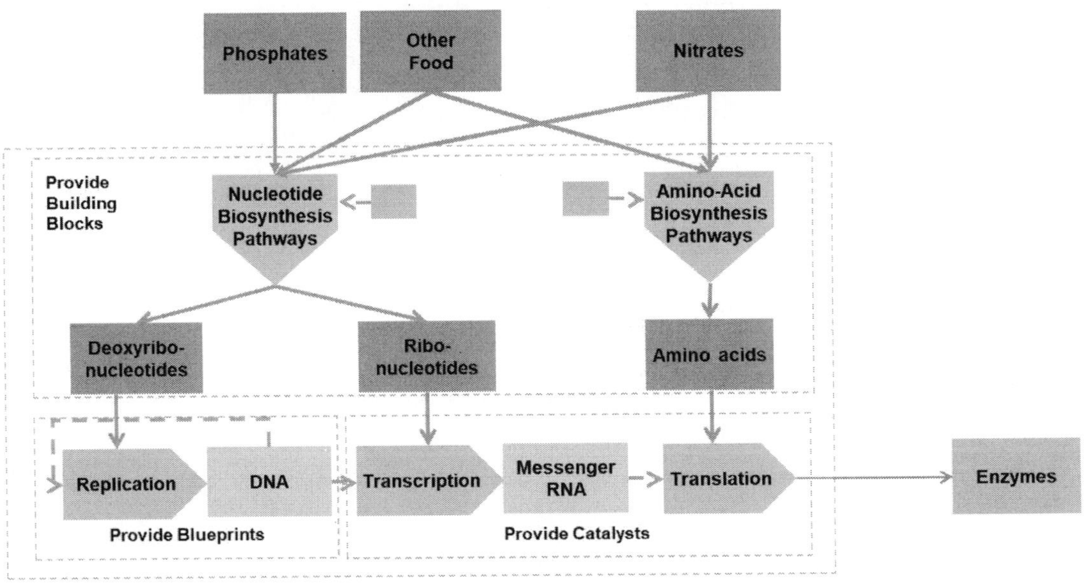

Figure 5-18: Metabolic network. The metabolic network feeds the auto-catalyst by creating all building blocks based on available food.

One aspect should be mentioned here: mutations in DNA result in new enzymes. These trials of mutations are not in a specific sequence because there is no plan to extend the metabolic network in a direction. Changes which connect to a new food of higher availability increase the speed of self-creation and are thus selected by evolution.[237]

---

[234] Availability is the same as concentration.
[235] See (Alberts, et al., 2015)
[236] This is a consequence of concentration
[237] See (Alberts, et al., 2015 pp. 216-232) for an explanation of evolution

Business

For an organization it would be very limiting if it has to wait for employees that have the perfect competences for work preparation and to search for perfect tools. Instead the organization has to accept that there is a certain qualification level available at the labor market and that there is a market for tools. The organization can use the available inputs from outside as a starting point if it has processes and ideas to create the necessary basic competences and tools itself.

The process network creates the building blocks for work instructions, know-how and the best working conditions based on available inputs (Figure 5-19).

All processes of the process network can be very efficient and effective because they have the best working conditions.

Such a process network can grow backwards from the auto-catalyst to the markets. If the organization becomes better at training employees, it has many more degrees of freedom when selecting them.

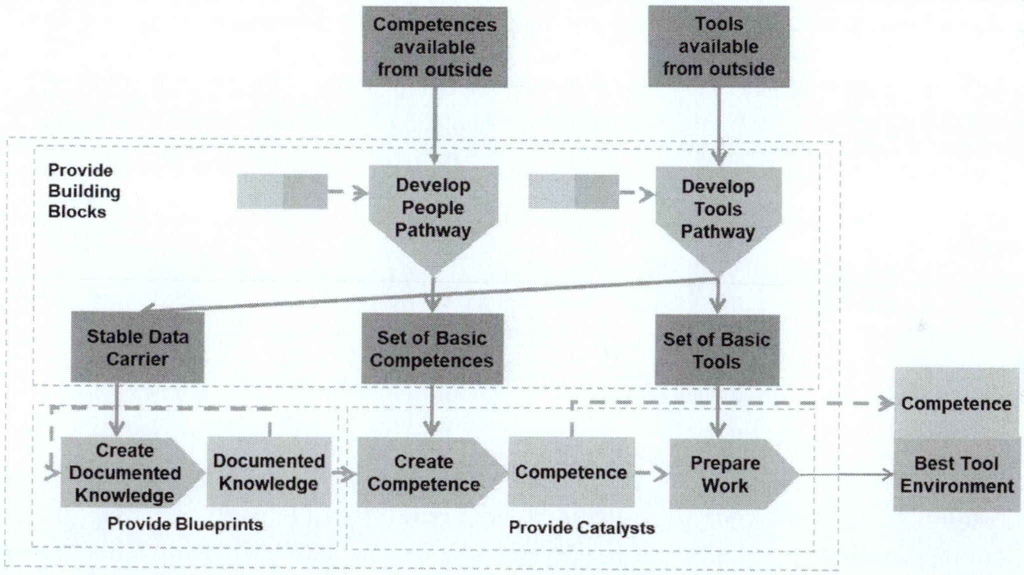

Figure 5-19: Process network. The process network feeds the auto-catalyst by creating all building blocks based on highly available inputs.

This picture is now extremely simplified because there is not only one process "Create building blocks." Instead every building block is created by different business processes. These processes have many dependencies between each other.

This makes the process network very complex. Which concepts can help to structure this network to make it understandable?

5.3.2 Metabolic Pathways

Biology

The metabolic network consists of thousands of chemical reactions. Very often these chemical reactions are connected to pathways. In a pathway the output of one reaction is the input of the next reaction. Thus, the metabolic network can also be described by the pathways rather than by the single reactions (Figure 5-21).

Metabolic pathways are chains of chemical reactions where the output of one chemical reaction is used as the input to the next. [238]

The metabolic pathway has one or more inputs and produces one or more outputs. In the middle it creates so-called intermediates.

Pathways can be categorized by their reaction types:[239]

1. Anabolic pathways build up (synthesize) complex molecules.
2. Catabolic pathways break down (degrade) complex molecules to smaller ones.

Anabolic pathways are typically named by their product (e.g. alanine biosynthesis pathway), whereas catabolic pathways refer to the input they degrade (alanine degradation pathway).

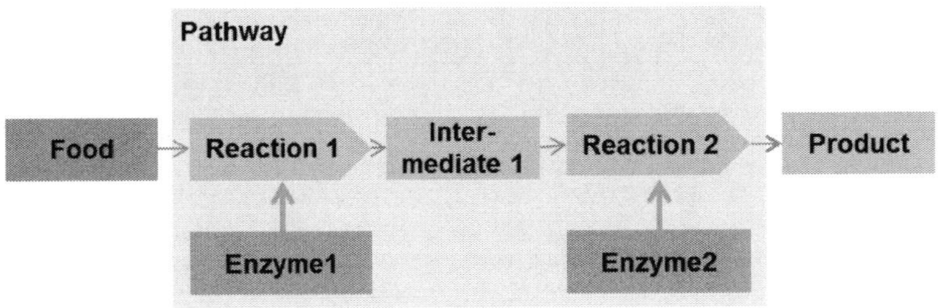

Figure 5-20: Metabolic pathway. The pathway is a chain of reactions.

Since the metabolic network contains a high number of pathways, it makes sense to group them by the types of their outputs (or inputs in case of degradation). Thus, we speak about amino acid biosynthesis pathways, which groups all pathways that produce amino acids. This group contains the alanine biosynthesis pathway, which produces the amino acid alanine, and all the other pathways which create the other amino acids.

This grouping now significantly reduces the perceived complexity in the metabolic network.

---

[238] See (Berg, et al., 2015 p. 424)
[239] See (Berg, et al., 2015 p. 424)

Business

The process network consists of a large number of business processes. These are often connected to process chains, where the output of one process is the input to the next one (Figure 5-21). This allows the simplification of the description of a process network by expressing it by the contained process chains.

Process chains connect business processes where the output of one business process is used as the input to the next.

Process chains have a set of inputs and create a set of outputs. In the middle they produce some intermediates which are only used internally. These intermediates are also known as work in progress.

In most cases the processes use some inputs to build up something more complex that is needed.

Sometimes processes break down something. This can be used for recycling where complex work environments are decomposed, so that their parts can be reused. Search processes also start with a big input (maybe the entire internet) and narrow the search down, until the wanted information is found. The output is part of the input.

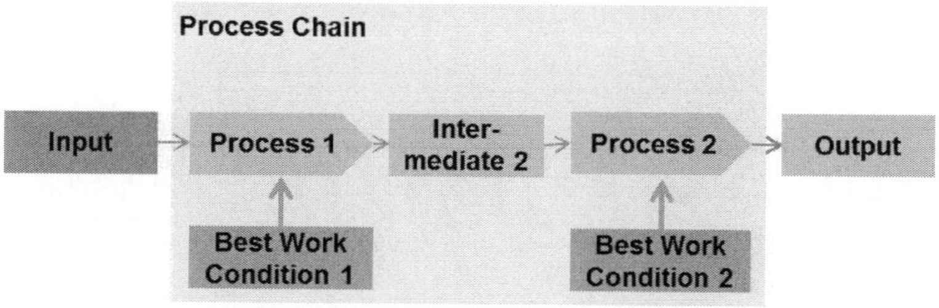

Figure 5-21: Process chain

Like in biology we can build groups of process chains by the type of their outputs (or inputs for degradation). We already know two groups of process chains that are absolutely necessary:

1. Basic competence synthesis process chains (people development pathways)
2. Basic tools synthesis process chains (tool development pathways)

Their process chains start with available competences and tools as an input and produce the set of basic competences and tools that are needed by the organization. This connects the ability of self-creation with the outside world.

At the end of the day, the organization is only interested in the results of the process chains, not in the intermediates. How can it be avoided to produce too many intermediates that are not used at the moment?

5.3.3 *Direct Connection*

Biology

How long does it take to produce the product? The duration of the conversion from food to product is not the sum of the durations of every chemical reaction in the pathway. Every chemical reaction has to bind the inputs to the catalyst, let them react and then release the created product. Then the created intermediate has to move on to the next catalyst.

Especially the last step may take a long time if the concentration of enzymes is low, and the intermediate has to travel around, colliding with many other molecules until it finds the right enzyme. The entire pathway is much faster if the enzymes are directly connected (Figure 5-22).

Connection of enzymes increases overall speed by eliminating the travel time of intermediates and avoiding reducing concentration.[240]

This movement is much faster if there is a short distance between the catalysts because they are connected directly. This leads to an increased speed of the entire pathway and thus to a faster self-creation.

A good example for the direct connection of catalysts is the ribosome, where many ribozymes and proteins are directly connected. This allows performing several steps of the translation process to be performed directly one after the other.

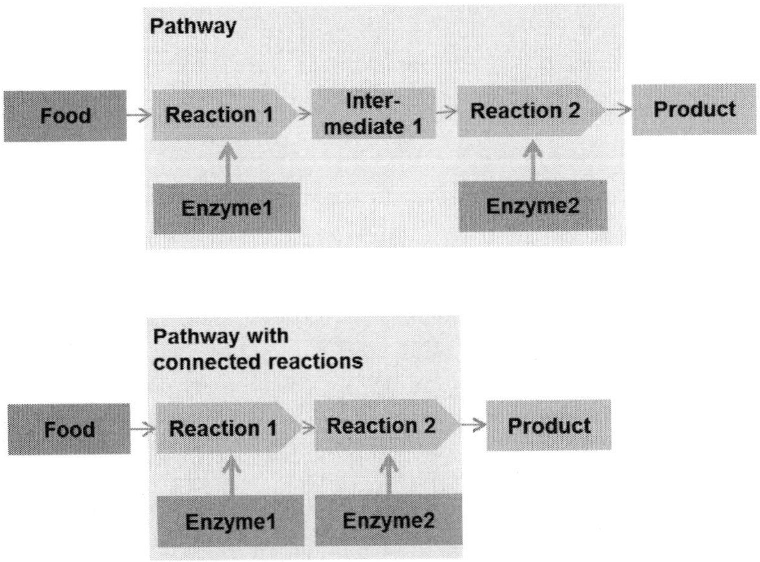

Figure 5-22: Connection of enzymes. Connected enzymes increase speed of metabolic pathways because the outputs of reaction 1 have short distance to the catalyst of reaction 2.

There is a second aspect beside the travelling time of the intermediate. Every produced intermediate needs space and contributes to an additional reduction of concentration. By connecting catalysts, this can be avoided.

---

[240] See (Alberts, et al., 2015 p. 148)

Business

In business we have the same situation: the duration of the process chains depends not only on the duration of processes, but also on the time between the processes.

Direct connection of two processes increases overall speed by eliminating over-production and the resulting waste of time and space (Figure 5-23).

The direct connection does not really eliminate the intermediates.[241] They are only consumed directly by the next process. The intermediates are not waiting for the next process and if they are physical, they do not have to be stored.

The simplest case of such a connection is a set of processes which is done by one role. There is no handover necessary. If one process is finished, then the intermediate is available, and the resources are free at the same time.

If the processes are done by different people, a handover is needed. And this should happen in the shortest possible time. For physical objects this requires that the workplaces are located close together. For information objects this requires that information flows fast and that the next process is prepared to start immediately.

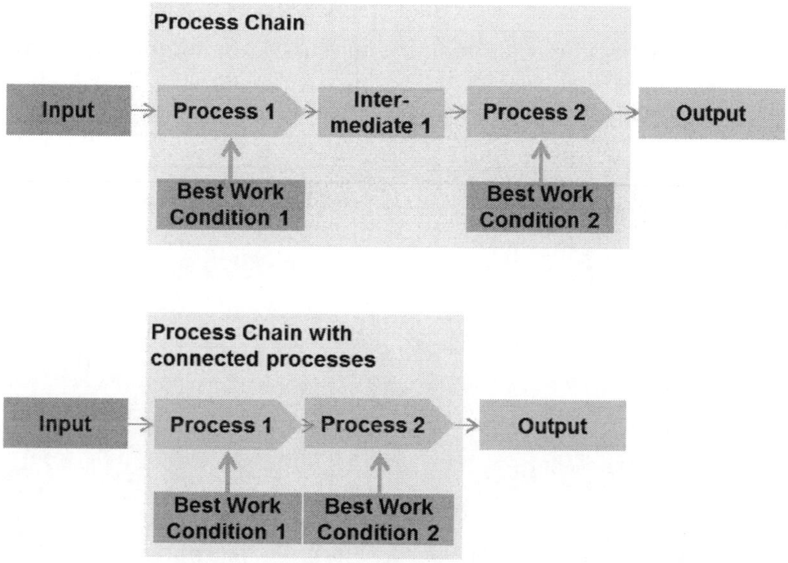

Figure 5-23: One-piece flow. Direct connection increases speed, because the output of process 1 can directly be sued as input of process 2.

Lean management uses Value Stream Analysis to make the waiting time transparent. One-piece flow and U-cells are the realization of such connected processes.[242]

After this focus on one-piece flow let us get back to the process network and look how this network develops all the resources that are needed for best working conditions.

---

[241] Intermediates are outputs of a process that are immediately consumed by the next process. Compare biology.

[242] See (The Productivity Press Development Team, 1999) for a description of one-piece flow

5.3.4 *Anabolic Pathways*

Biology

Now that the general considerations about pathways have been described, we can now focus on anabolic pathways.

Some of the anabolic pathways create the building blocks of DNA, RNA and proteins

Anabolic pathways create these building blocks based on smaller molecules which are available in higher concentrations (Figure 5-24). The anabolic pathways always start with a first molecule, which was part of the food or produced by another pathway. Then several chemical reactions add smaller molecules to it until the product is created. [243]

Every product needs its own anabolic pathway to create it. For our purpose it is sufficient to refer to them not individually but in groups of pathways.

The purine / pyrimidine synthesis pathways create the four nucleobases adenine, guanine, cytosine and uracil which are used to create nucleotides, the monomers of RNA and DNA.[244]

The amino acid synthesis pathways create the 20 amino acids.[245]

In addition to the building blocks, the cell is able to build up other complex molecules with anabolic pathways: carbohydrates, lipids or co-factors. All of the processes are catalyzed by enzymes produced in the auto-catalyst.

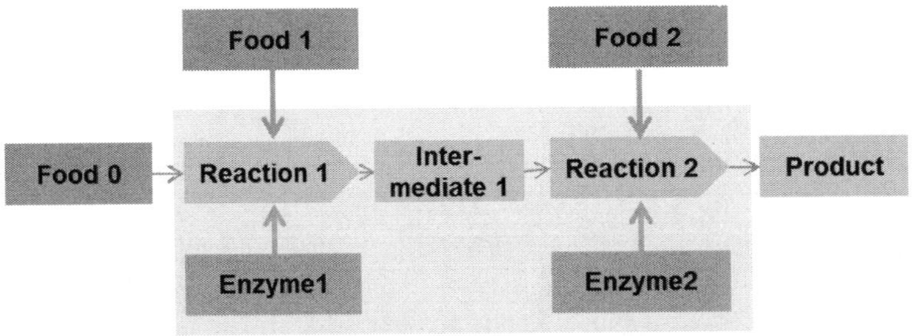

Figure 5-24: Anabolic pathways. The processes build-up the building blocks of the auto-catalyst.

In human cells not every amino acid can be synthesized. Such amino acids are called essential and have to be imported from the food directly.[246]

---

[243] See (Berg, et al., 2015 p. 524)
[244] See (Berg, et al., 2015 pp. 743-766) for a description of the purine / pyrimidine pathways and creation of RNA and DNA
[245] See (Berg, et al., 2015 pp. 719-733) for a description of the amino acid biosynthesis pathways
[246] See (Berg, et al., 2015 p. 719)

Business

In business it is the same. The basic competences and basic tools have to be created based on the competences and tools that are available outside (Figure 5-25). This requires pathways to build them up.

Some of the build-up process chains have to create the basic competences and basic tools that are needed to create the best working conditions.

The build-up process chains can start with competences and tools that are available on the labor and supplier market. On this basis they can build up what is needed.

People development process chains have to create the set of basic competences. Their input is the previous knowledge and skills of the trainee. During training more and more know-how and skills are added until a required basic competence is developed.[247]

Development of tools also starts with tools that come from outside. In the case of software tools something can be added: tools get plugins that connect them with other tools, they have to be configured, data has to be added. Finally, the tool is usable.

Besides the competences and tool for work preparation, there are many other competences and tools needed. They also need build-up process chains to create them. These process chains will be highly effective and efficient because they are catalyzed by the best working conditions.

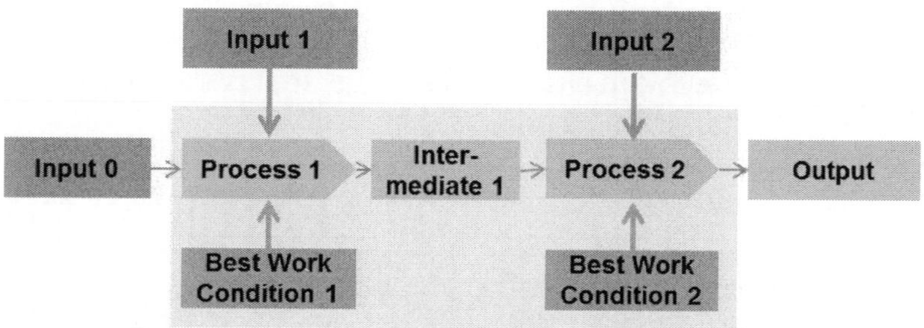

Figure 5-25: Development processes. The processes create the needed building blocks of the auto-catalyst based on inputs that are highly-available outside.

As in biology, it is not always necessary to build up something. If the needed tools or competences are directly available, then they can be used directly. The make-or-buy decision determines the starting point of the network of own processes. A good example in biology are essential amino-acids: Human do not produce them, because it is not needed. They are directly available in the food.

After this discussion of the build-up processes, let us focus on those processes that deconstruct something. What is their purpose?

---

[247] See (Liker, et al., 2008) for the development of people

5.3.5 Catabolic Pathways

Biology

If a cell uses organic material as food, it does not directly get the inputs that are needed for the anabolic pathways. Cells mainly consist of proteins, carbohydrates, DNA, RNA and fat. All these polymers have to be broken down to monomers and to even smaller molecules.

This breakdown has two evolutionary advantages. Firstly, it increases the set of molecules that can be used as food. A cell can use inorganic material and create everything from scratch, or it can use organic material and break it down. Secondly, the organic material contains energy that can be used.

Catabolic pathways decompose complex molecules to get energy and inputs for the anabolic pathways (Figure 5-26).[248]

The first step of catabolism is to decompose the polymers to their monomers, and then the monomers can be broken down to smaller molecules.

One example is the degradation of proteins. Proteins are long chains of amino acids. The proteolysis process breaks the connections to create amino acids. Other pathways can degrade the amino acids to their components.

Another example is the glycolysis pathway which decomposes glucose to pyruvate. This pathway not only produces pyruvate, which is needed by the pyruvate dehydrogenase pathway to produce acetyl-CoA for the citric acid cycle. It can also use the energy that is freed during breakdown and store it in ATP.[249]

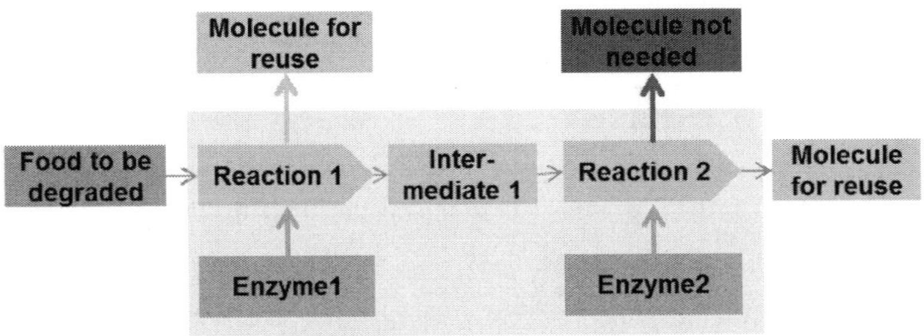

Figure 5-26: Catabolic pathways. They breakdown food to smaller molecules that can be used by anabolic pathways

Some pathways like the citric acid cycle are not purely catabolic. Some of their processes add something, others subtract something. They are called amphibole pathways. As a cycle they can connect many other pathways.

---

[248] See (Berg, et al., 2015 p. 424)
[249] See (Berg, et al., 2015 p. 449)

Business

What is the translation of organic food to the business context? As in biology it is about complex structures that have been built by other organizations.

This typically happens when organizations merge. If a smaller organization is bought by a bigger one, then the smaller one has to be integrated. This is called post-merger integration. The departments of the smaller organization are integrated to the organization chart of the bigger one. Additionally, they have to adapt to the new rules and processes.

All this is a catabolism (Figure 5-27). Existing complex structures are deconstructed, to have the inputs to create something new.

Some process chains decompose complex tools and competences to get inputs for the composition of new ones.

Catabolic pathways decompose everything, to make the parts available for other processes. The outputs are the inputs of the anabolic pathways. In our case, the provide inputs for people development and tool development.

In our example above, it might be necessary to train the new colleagues. Thus, a trainer has a process to find out the existing knowledge of the trainee. This reveals previous knowledge that can be used immediately. It can also reveal previous knowledge that has to be replaced by the training.

The decomposition can also be found internally. If tools are not used any longer, they can be decomposed to reuse parts in other contexts. A simple example is to reuse notebook computers by uninstalling all software to make it usable for another employee. The removal and recycling of tools can also be necessary, if a better tool is available, but the employees hesitate to use it because they feel comfortable with the old one (Skip the old one out).

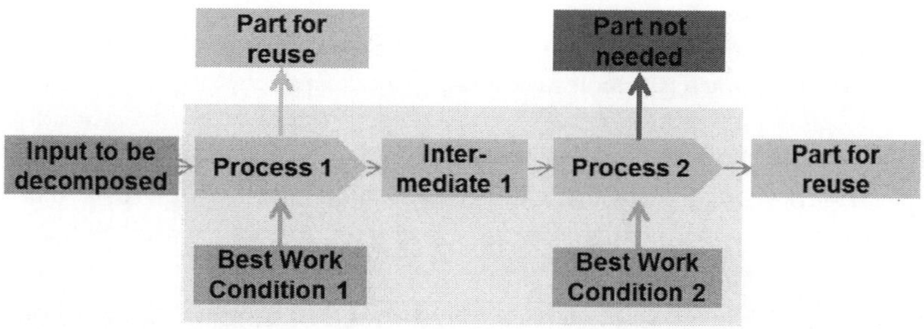

Figure 5-27: Recycling processes. Process chains decompose existing tools and competences into their parts to get inputs for a new synthesis.

Many processes require energy to be driven. How does a cell manage energy?

5.3.6 *Energy*

Biology

Every chemical reaction needs energy to run. In the simplest case this energy comes from the kinetic energy of the colliding molecules. Catalysts can lower the needed activation enthalpy, but they do not allow running endergonic reactions to run. These reactions require an additional energy source to run at high speed. Where does the energy come from (Figure 5-28)?

The three possible alternatives to get energy from outside are:

1. Photosynthesis using the energy in the sunlight
2. Catabolic processes freeing the energy that is bound in food
3. ATP as energy carrier imported from other life forms.

Cells import energy from outside and make them usable in small portions to drive chemical reactions.

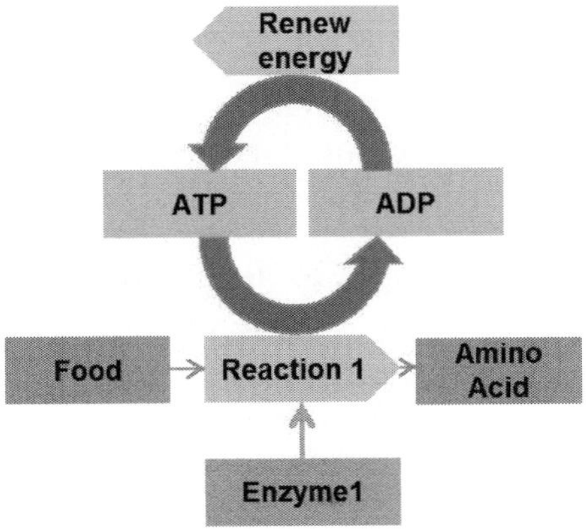

Figure 5-28: ATP as energy carrier. The energy of ATP is used in chemical reactions by splitting ATP to ADP and phosphates. It needs a process to renew ATP molecules as energy carriers.

Cyanobacteria can use photosynthesis to create carbohydrates based on carbon dioxide and water. The carbohydrates can store a high amount of energy, which is too much for most reactions. The glycolysis pathway breaks down the carbohydrates and uses the freed energy to create ATP (adenosine-tri-phosphate) from ADP (Adenosine-di-phosphate). The ATP molecules can then bind to enzymes where they provide the energy for the catalyzed reaction by degrading ATP back to ADP. This is called energetic coupling. The ADP can be used to create ATP again. Energy carriers are recycled by the cell.

An alternative would be to live in symbiosis with another life form that provides ATP. The mitochondrion in a human cell is such an endo-symbiont that provides ATP for the cell. [250]

---

[250] See (Alberts, et al., 2015 p. 25)

Business

Business processes also need energy to drive them. The energy is necessary to make tools or move people (Figure 5-29).

Organizations import energy from outside and make them available to drive business processes.

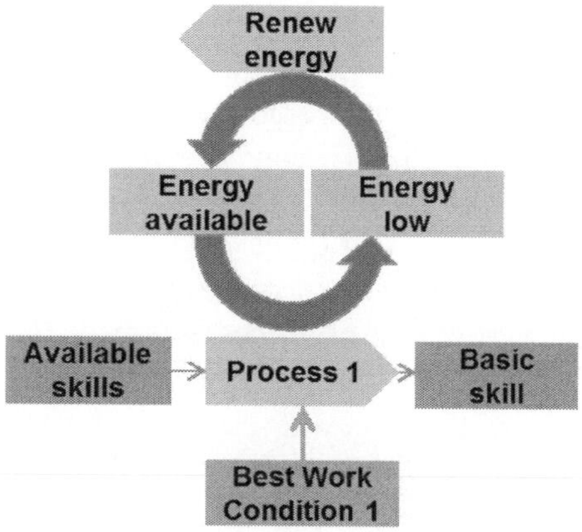

Figure 5-29: Energy for driving business processes. Many processes require energy. It needs other processes to renew this energy

Business processes are driven by different types of energy. Machines may need electrical energy. If a worker does the same process without a machine, then the worker has to provide this as physical energy. Office workers do not need so much physical energy; instead, they need the mental energy of motivation. Motivation drives them to do a job.

In all cases the energy comes from outside the organizations.

1. Photosynthesis / wind power can be used to get energy from nature
2. Energy can be bought at the energy market.
3. Physical energy from workers is provided by the workers.
4. Motivation is provided by employees.

In all cases except the first one, the energy comes from a supplier or employee of the organization. We will see in section 5.7 that these stakeholder relations are similar to a symbiosis.

The organization depends on the ability of the stakeholders to renew this energy. It can help to increase the motivation level of employees by giving a clear shared target, or by celebrating successes. Good coaches can help to increase the energy level. Additionally, the organization has to ensure that the energy is used carefully and not wasted.

Let us now digest the new insights about the conversion or raw material to components for the auto-catalyst.

5.3.7 Summary

5.3.7.1 Insights

Biology teaches use, that the components for the autocatalyst, the amino-acids and nucleotides, are not there by fortune. Instead, they are imported from the environment or self-created in the metabolic pathways. Anabolic pathways convert the material in the environment step-by-step to components of the auto-catalyst, whereas the catabolic pathways destroy large molecules to make their parts available for anabolism and to gather energy. ATP is one of the energy carriers that allows to drive endergonic reactions. This energy carrier is recharged by the mitochondria.

In the business translation we have seen that the basic competences and the basic tools for creating the best working conditions have to be developed by the organization itself. This makes the organization more independent from competences and tools that are available on the market. Additionally, it allows the organization to get a competitive advantage if its processes are better than those of the competitors.

The process network can be understood easier by focusing on process chains (value streams) or groups of process chains. One group of process chains provides basic competences. Another group of process chains provides basic tools.

The duration of process chains not only depends on its contained processes, but also on the time between them. Direct connection of processes removes this delay and thus accelerates the process chain.

Some process chains have to create the building blocks for the best working conditions. Others are necessary to decompose existing structures and provide their constituents.

All of the processes require energy. One of the forms is motivation that drives people to do something; other forms are classical forms of energy. In all cases energy can be a limiting factor. It has to be used carefully and renewed.

5.3.7.2 Requirements

A Viable Team (VT) shall implement the following requirements (Table 15):

ID	Requirements	Comment
5.3.1	The VT shall contain all processes to create the building blocks of the best working conditions based on factors that are available outside.	The Viable Team is always responsible for the last mile of work preparation.
5.3.2	The VT shall connect processes directly to reduce intermediates	This leads to one-piece flow, where outputs of processes are immediately used by the next process.
5.3.3	The VT shall have processes to identify usable parts in complex capabilities.	This is necessary to integrate experienced people from outside who are not trained from scratch.

5.3.4	The VT shall take care about the usage of energy (including motivation as mental energy)	
5.3.5	The VT shall contain processes to renew energy	Most processes require physical or mental energy, which have to be renewed by the employee.

Table 15: List of requirements related to the process network

5.3.7.3 Existing Solutions

The process network connects the outside world with the auto-catalyst. Since human resource management hires people and procurement buys tools, there are various management disciplines involved in these processes. Lean also contributes many methods that can help to understand this network.

Value Stream Mapping (VSM)

Value streams describe chains of business processes with a defined output. Typically, they are used for customer value creation processes, but they can also be used for all other process chains. Value Stream Mapping extends the perspective. Whereas work instructions focus on a single process, the value stream is focused on the process chain.[251]

It is important to mention that a cell has no idea of metabolic pathways. Every protein-coding gene is a work instruction to create a protein (e.g. an enzyme), but there is no description for an entire pathway or a group of pathways. These concepts have been introduced by biologists.

Muda, Muri, Mura

The duration of process chains is mainly determined by the time between the processes. In lean production the transport and storage which result from over-production (intermediate products) are described as some of the seven mudas (waste). Muri refers to the bottleneck of an entire process chain, mura on the unevenness. These concepts are used to identify problems in single processes and in an entire process chain.[252]

One Piece Flow, U-Cells

In one-piece flow the output of one process is immediately used as the input of a second process. In U-cells all the steps are directly connected[253]. This is similar to multi-protein-complexes like the ribosome, where several reactions are directly connected.

Kanban systems[254] work slightly different. This will be explained in section 5.6.

Make-or-Buy Decisions

The make-or-buy decision is well-known in strategic procurement. Should material or tools be produced by the organization or should they be bought from suppliers? Human resource

---

[251] See (Lindner, et al., 2010)
[252] See (Gorecki, et al., 2016) (Bicheno, et al., 2009) (The Productivity Press Development Team, 2002)
[253] See (Womack, et al., 1996)
[254] See (Gorecki, et al., 2013)

management has a similar question. Which qualification of candidates should be a precondition for employment? What is the starting point of people development?

Starting with the pre-qualification of the hired employees and the standard features of tools, the organization needs to have all processes to develop people and tools until they are able to run excellent processes. The better the organization can train employees the less it depends on high pre-qualifications. The better the organization can develop or configure tools the less it depends on standard tools.

Merger & Acquisition and Post-Merger Integration

The post-merger integration adds complete organizations including their structures and process houses. In many cases the structures and processes of one or both partners have to be changed. This creates a high demand for change for many employees. The catabolic process chains try to identify elements that can be reused to build up something new.

Let us focus on energy now, especially on motivation as the driver for processes.

Coffee Breaks

What is the purpose of a coffee break? Beside the welcome and wrap-up, it is the most common agenda point of all meetings. Obviously, the break is necessary to renew the energy after long discussions. Without the break all processes would be slower. Thus, the coffee break is a process that has the target to renew energy.

Motivation

Many other business processes, like the presentation of success stories, are about creating motivation. This motivation is the mental energy that drives other processes.

Many internal communication processes also go in that direction. Why are they needed? They seem to produce nothing, and the information is very often irrelevant for the daily work of employees. The communication can influence people and motivate them.

5.3.8 Annotations

Other Aspects That Have Not Been Translated

In this section we focused on the main concepts only. There are some others that should be mentioned as well.

Amphibole pathways, like the citric acid cycle, have no clear direction. They are not chains, but circles. The citric acid cycle connects some other pathways. This allows an internal balancing of resources. Carbohydrates can be used as sources for the creation of amino acids. If there is more than needed, than it can be stored or retrieved via other pathways. Amphibole pathways allow such a balancing.[255]

---

[255] See (Berg, et al., 2015 p. 425) for amphibolic pathways

About Building Blocks and Process Chains

Since it was not possible to define a set of basic competences and basic tools in the previous section, it is not possible to define a set of process chains that produce them. Our translation is still on the group level:

1. Basic competence synthesis process chains produce all the needed basic competences.
2. Basic tool synthesis process chains produce all the needed basic tools.

For the next steps of translation this is sufficient. A practical implementation is possible, even if the processes are not defined. We can use practical experiences to design such a set of basic competences and tools, even if there is no clear biological correlate.

Platform of Building Blocks

It is not possible to define a set of basic competences and tools in business that can be combined to business catalysts for all processes. Such a list of tools would be very long and individual and include everything from calendar to forklift.

Nature had the same problem 4 billion years ago: It was not possible to create catalysts for all processes based on four nucleotides. RNA could catalyze only some types of reactions, and this limited the extension of the metabolic network. The solution was to invent proteins as an additional type of catalyst produced by ribozymes.

We have to think in the same categories. We need to design the auto-catalyst and its feeding process network, even if it might have limitations. In such a case we need additional competences and tools that are on a separate extended list. Different organizational functions might have a different extended list, but they all need to make work preparation which can be done by combining the basic competences and tools. Work preparation has to be very flexible to apply it to each process individually.

Systems Biology

Klipp et al describe the modeling of biochemical systems. This work shows the bridge between biology and informatics and mathematics. Very important in this description is the detailed explanation of reaction speed, especially the Michaelis-Menten Kinetics. The book also explains how to model metabolic networks mathematically in one compartment or in many compartments (eukaryotic cell). Advanced concepts like the modelling of gene expression are included. All these models are usable to simulate biological systems.[256]

Databases of Metabolic Pathways

Currently all detail researches about metabolic pathways, reactions, enzymes, their gens and all their regulation are integrated to complex models. The Biocyc-Database and its related EcoCyc-database documents the current state of knowledge.[257]

---

[256] See (Klipp, et al., 2012)
[257] See (BioCyc, 2018) and (EcoCyc, 2018)

Today large organizations have so-called process houses that describe all business processes. However, they do not reflect the overall system dynamics; instead, they focus on the most important process chains, like the customer value creation process, and some support processes that are not fully integrated. Maybe this work can help to develop a top-down picture in business as well, which shows a clear system dynamic.

5.4 System Boundary

Business

In the previous sections we have discussed self-organization and the self-creation processes. Self-creating systems can grow exponentially with a certain gross growth rate. This growth rate depends on the speed of processes, which depends on best working conditions.

But this is not the reason for the real growth rate. In reality, many self-created resources get damaged or lost. They have to be created a second time. These losses reduce the gross growth rate to a net growth rate. Self-protection mechanisms contribute to growth by minimizing the losses.

Here is an example: Know-how is a critical production factor for organizations. It might take years to train individuals to an expert level, but if the individuals leave the organization, this knowledge is lost and it can take years to replace it. Without this expert knowledge some processes are no longer catalyzed and slow down.

All the resources of an organization are assets. It takes time and money to produce them; and it takes time to earn that money. Thus, resources have to be protected and they have to be kept together. As a benefit, the organization is able to maintain the concentration of critical resources and avoids that self-creation slows down.

However, how are the protection mechanisms realized in nature? Let us have a look at our benchmark.

Biology

How does a cell ensure that the enzymes are protected and do not leave? The metabolic network cannot be the solution for that. It is only a graph of connected chemical reactions. It has nothing to do with space.

The cell protects all self-created objects by surrounding them with a cell membrane. The cell membrane encloses the interior of the cell and separates the inside from the outside. Enzymes cannot pass the cell membrane.[258]

A strict separation of inside from outside would protect everything. On the other hand: The cell cannot create itself anymore, because no material and energy can enter the cell. Thus, the cell needs some additional mechanisms to import and export molecules. It has membrane proteins which are able to selectively move object from inside to outside and vice versa.[259]

This section shows the details and translates them to business.

---

[258] See (Alberts, et al., 2015 pp. 597-640)

[259] Other mechanisms are exocytosis and endocytosis.

Translations

On the next pages are the following translations (Table 16):

	Translation	Biology	Business
1	Inside and Outside	Cell membrane	Legal entity, property
2	Protection	Cell membrane as passive boundary	Protection of property
3	Import	Import of food for self-creation	Import of resources (workforce, …)
4	Export	Export of unneeded Byproducts	Export of waste
5	Storage	Storage of food for later consumption	Storage of resources for later usage

Table 16: Translation of the system boundary

5.4.1 *Inside and Outside*

Biology

Up to now our cell only consisted of many chemical reactions that are connected and that convert available material to catalysts. Up to now our cell has no form, no size or defined position. It needs a system boundary that keeps all the created objects together (Figure 5-30).

The cell membrane separates the interior from the environment of the cell.[260]

The cell membrane consists of a double membrane of phospholipids which are synthesized by the cell itself. The cell membrane completely encloses a space. This leads to a defined interior of the cell which is separated from the outside.[261]

With the cell membrane as a system boundary everything inside is part of the cell, everything outside is not.

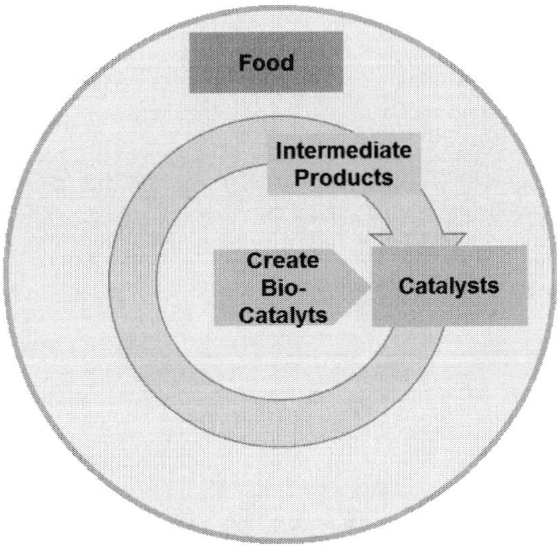

Figure 5-30: Cell membrane separating inside from outside. The membrane encapsulates all molecules.

The concept of inside and outside is different from the metabolic network. The metabolic network is a set of processes that consumes food and converts it step by step to catalysts. This set of processes has no defined size or position. The separation by the cell membrane is only about enclosing a space. It does not contribute to the creation of objects.

This defined space and the control of flow across the membrane allows the concentration of food and catalysts inside to increase, which results in accelerated processes and fast growth. In the next sections we will see how this works in detail.

---

[260] See (Alberts, et al., 2015 p. 565) for the description of the cell membrane
[261] See (Alberts, et al., 2015 pp. 568-569) for phospholipids

Business

In business it must also be decided what part of the organization is and what is not (Figure 5-31). We take the legal entity with its properties, rights and obligations as a starting point.

A legal entity can be defined by the properties, rights and obligations it has.

The legal entity allows an inside from an outside to be separated from a legal perspective. Every property or right that belongs to the legal entity can be considered as inside; everything else is outside.

Two particular types of rights connect this legal definition to the physical and psychological worlds we have discussed in section 4.3:

The first one is the exclusive right to use an area like a room or a building. This right makes the area, a space in the physical world, to an area of the organization. The organization can protect it with a doorman.

The second one is the right of an employment contract. This connects the employee to the organization. Employees are the carrier of skills, know-how and a personality, even if they do not work. On the other hand, they are more or less concentrated if they are working. The organization is interested in protecting the employee and increasing his concentration at work.

This shows that rights allow us to speak about a space and concentration. The rights allow the organization influencing this concentration to increase the speed of processes.

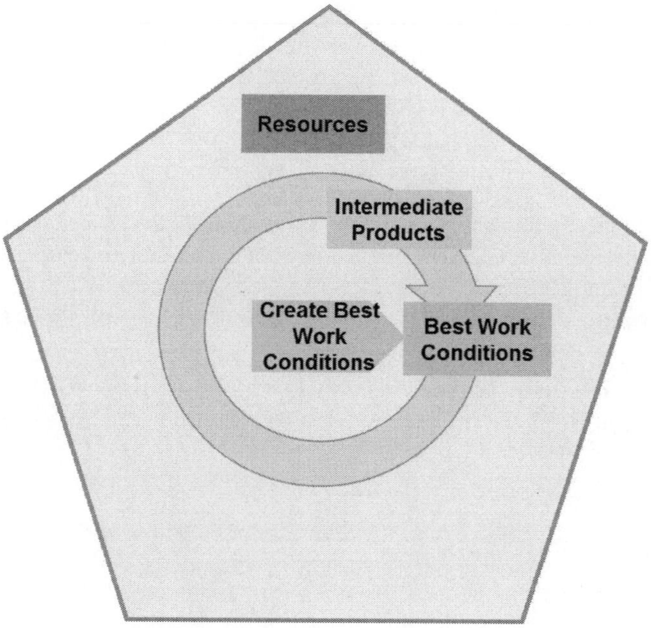

Figure 5-31: Boundary of a team. Everything that is owned by the team is inside, everything else is outside the system. The authorities of the team define the boundary.

Let us discuss the main purpose of the system boundary:

5.4.2 Protection

Biology

The system boundary protects the interior of the cell (Figure 5-32).

Cells are able to restrict the flow between inside and outside. This allows a high concentration of enzymes and needed raw material to be maintained inside.[262]

Enzymes cannot pass the cell membrane. This restriction of flow allows the concentration of enzymes and food to be increased inside. This increases reaction speed. Additionally, no other enzymes can enter the cell.[263]

With this mechanism the cell is able to control exactly, which catalysts are inside. Thus, it is able to control which reactions are accelerated.

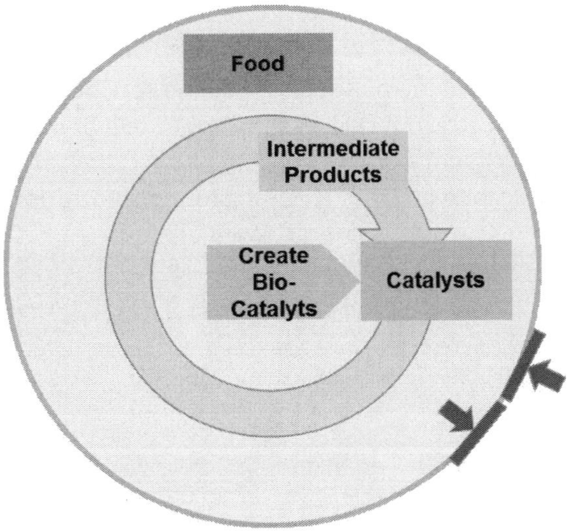

Figure 5-32: Protection. The cell membrane helps to maintain concentration by blocking the transfer of molecules. The red arrows show that objects cannot leave the cell, whereas others cannot enter it.

This limitation of movement is the basis of a central concept of the cell: homeostasis. Homeostasis is the way that the cell is able to maintain a stable milieu inside. It is not any milieu, but the best environment for the processes inside the cell. The homeostasis is realized by the adaptation mechanisms and is based on the ability that the cell can keep concentrations stable.

---

[262] See (Alberts, et al., 2015 p. 565) for a description of the cell membrane
[263] See (Alberts, et al., 2015 pp. 597-640)

Business

If the organization has property and rights on production factors, it makes sense to protect them to ensure that they do not get lost (Figure 5-33).

Legal entities manage their property, rights and obligations.

The most important properties are those that contribute to best working conditions. This enables the organization to run processes at high speed. If the team had no tools or at least the rights to use them, then someone else could take them away, and the best working conditions are destroyed.

With the right to use areas, it is possible to speak about physical concentration as well.

A team does not want other teams to take its laser printer away because that destroys the best working condition. The team also does not want other teams to place their defective equipment in the team's area, because this would reduce the concentration of the needed things. It would also increase searching times and creates bad mood for the team members.

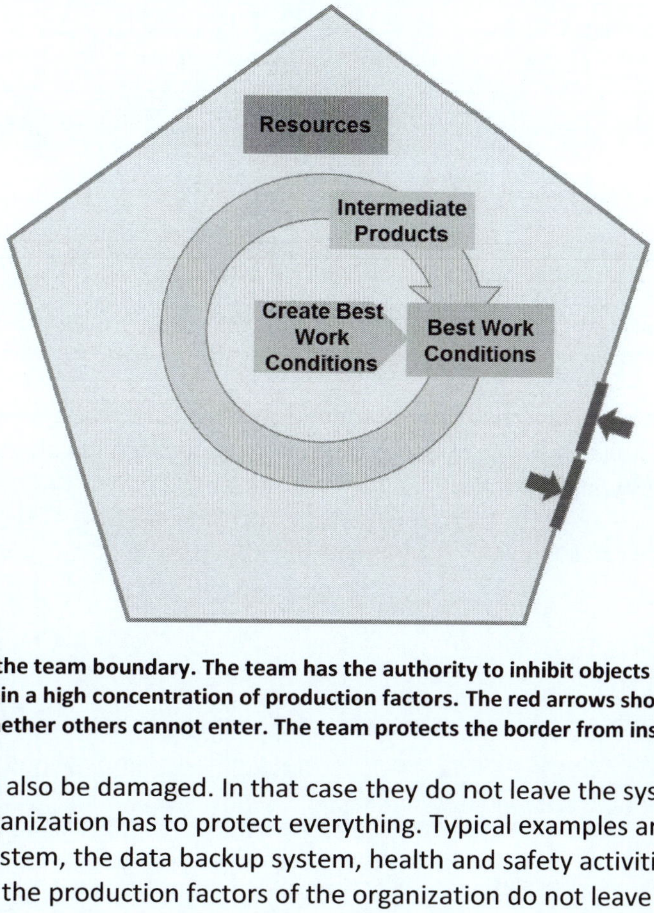

Figure 5-33: Protection at the team boundary. The team has the authority to inhibit objects entering or leaving the team. This helps to maintain a high concentration of production factors. The red arrows show that objects (e.g. tools) cannot leave the team, whether others cannot enter. The team protects the border from inside.

Production factors can also be damaged. In that case they do not leave the system, but the effect is the same. Thus, an organization has to protect everything. Typical examples are the security guard, the firewall in the IT system, the data backup system, health and safety activities. They are all needed to ensure that the production factors of the organization do not leave the system or get destroyed.

The protection with a system boundary has one big problem: It isolates the inside from the outside. Obviously, the organization has to implement some additional mechanisms to be able to grow.

5.4.3 *Import*

Biology

Protecting the system boundary creates a stable environment inside, but a closed system can never grow. A cell has to import food (Figure 5-34).

Cells actively import food from the outside to use it as material or a carrier of energy.[264]

If a cell can duplicate in short time, then it also has to import at least that much food into the interior. The cell membrane must allow this throughput.[265]

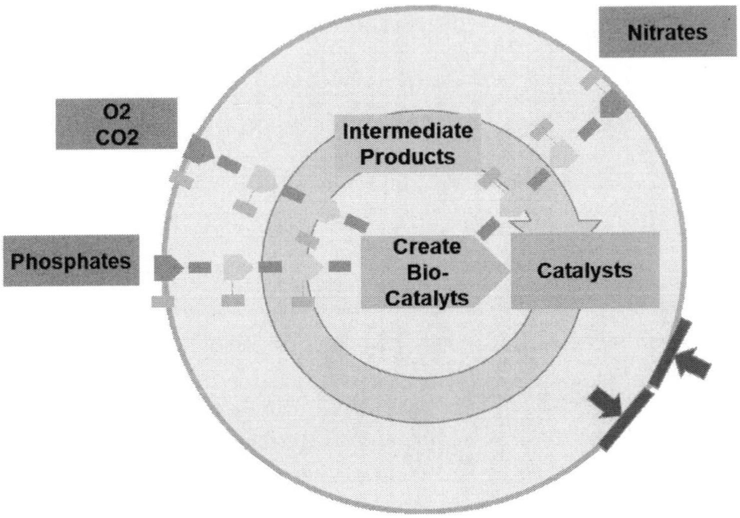

Figure 5-34: Active import of food. Carbo-dioxide, nitrates and phosphates have to be imported from outside in order to synthesize amino acids and nucleotides, which are used to create the catalysts for all these reactions.

The problem of growth is the fact that the cell membrane as a surface grows in two dimensions, the volume in three dimensions. If the cell becomes too big, its membrane cannot import food fast enough for the faster growing interior.[266]

Replication of the entire cell solves this problem of the limited size. [267]

---

[264] See (Alberts, et al., 2015 pp. 597-640)
[265] See (Alberts, et al., 2015 pp. 597-640)
[266] See (Sadava, et al., 2011 S. 102)
[267] See section 5.5 for the cell cycle

Business

All physical resources of an organization have their origins outside. They have to enter the organization. If the organization's boundary has to do with property and rights, then the active import is about adding new property and rights by making contracts (Figure 5-35).

Organizations have to import production factors from factor markets.

The production factors include the right to the work of employees and the property on tools and material from suppliers. Employees or suppliers do not enter the system because they are stakeholders[268] and not property of the organization.

If the organization hires people with the right competences and buys the right tools, it has the necessary inputs for its process network. The better the hiring and purchasing process is connected with the internal process chains, the higher will be the concentration of needed production factors inside.

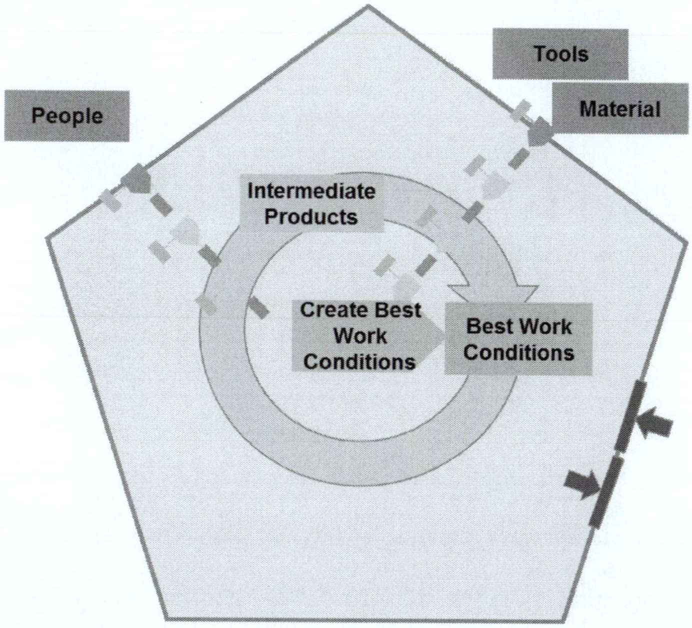

Figure 5-35: Active import of resources

Adding more and more people let the organization grow. It also increases complexity in the communication between people. The split of the organization will be necessary if the amount of people becomes too high.[269]

Beside the import mechanism the organization needs clear mechanisms to remove everything from the inside that is not needed.

---

[268] Stakeholders are other systems that have a mutually beneficial symbiosis with the organization. See section 5.7
[269] See section 5.5 for the splitting of teams

5.4.4 *Export*

Biology

Import of needed objects increases the concentration inside, but there is also a second way to reach the same target.

Cells export everything that is not needed or a threat.

Sometimes molecules pass the boundary unintentionally. Other unneeded molecules are created by the cell itself. The metabolic pathways create some by-products that are not needed anymore. The cell needs to export these molecules to the outside again (Figure 5-36). Otherwise these additional objects would continuously decrease the concentration of catalysts and food.[270]

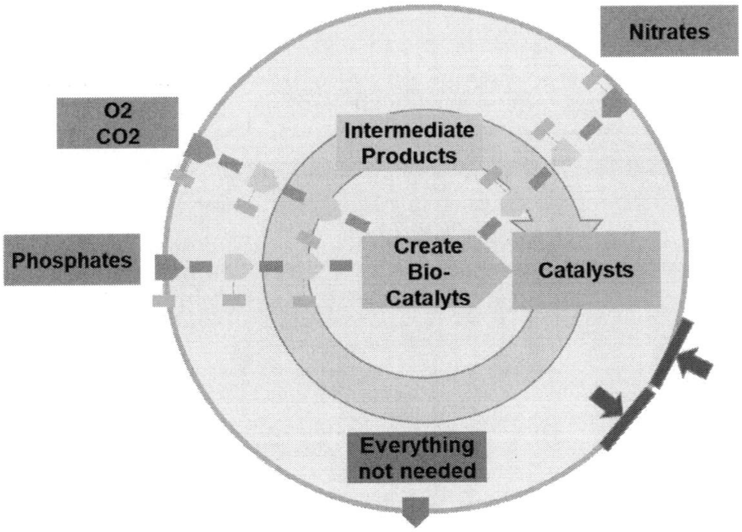

Figure 5-36: Active export of waste. Everything not needed is waste and has to be exported.

This export-mechanism contributes to the self-protection function that was discussed in section 5.4.2. Self-protection is not about creation. It focuses on the unneeded or dangerous objects that are in the same space as the metabolic network. In the physical world of the cell everything is about molecules. This makes it sufficient to protect the cell with the outer boundary in three-dimensional space. We will see that this is more complex in business.

Together with the passive protection and the import mechanism, the cell is able to create a comfortable environment, the homeostasis, which supports the activities of the metabolic network. [271]

---

[270] See (Alberts, et al., 2015 pp. 597-640)

[271] See (Sadava, et al., 2011 S. 106-107)

Business

As in biology the protection mechanism cannot be understood by looking at value streams that create something. Protection mechanisms are more indirect. They focus on the unneeded things or threats.

Organizations have to remove everything that is not needed or a threat.

Threats are everything that could damage the self-created assets. By removing threats, the organization can reduce losses. This contributes to a high net growth rate.

Additionally, the organization needs to increase concentration by removing everything that is not needed (Figure 5-37). The identification of muda (waste) is an important method of lean management. The reduction of muda does not help to create something. Instead it is about removal. It helps to increase the concentration of production factors.

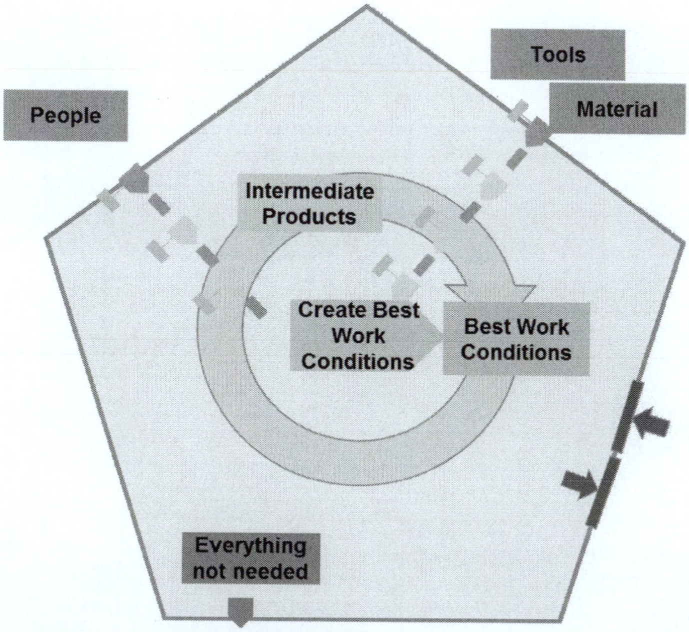

Figure 5-37: Active export of everything that is not needed.

What happens if people are no longer needed in a team? In lean production they have to leave the team, but they get a new job in a different team. It is the responsibility of the organization that employees can rely on a safe workplace and well feeling.

Obviously, it makes no sense to export something, if it is a valuable resource that can be used later. There is a much better solution available.

5.4.5 Storage

Biology

The cell membrane not only keeps all self-created objects like the enzymes together. It also allows food to be stored inside (Figure 5-38).[272]

The cell membrane allows material to be stored for later usage.

This storage decouples the self-creation process (creating enzymes) from the availability of food. If there is more food available outside than needed, then it can be stored for later use.

The storage of food also has a disadvantage. The food is not needed at the moment, but it requires storage space. This storage space reduces the concentration of the currently needed factors. One solution is to use chemical reactions that convert the food to a very compact form.

One example is the glycogen pathway in animal cells that stores carbohydrates as glycogen in very compact form. If there is a demand for carbohydrates in the future, chemical reactions can make them available again.[273] The signaling pathways of the cell determine which process chains have to be used at which time.

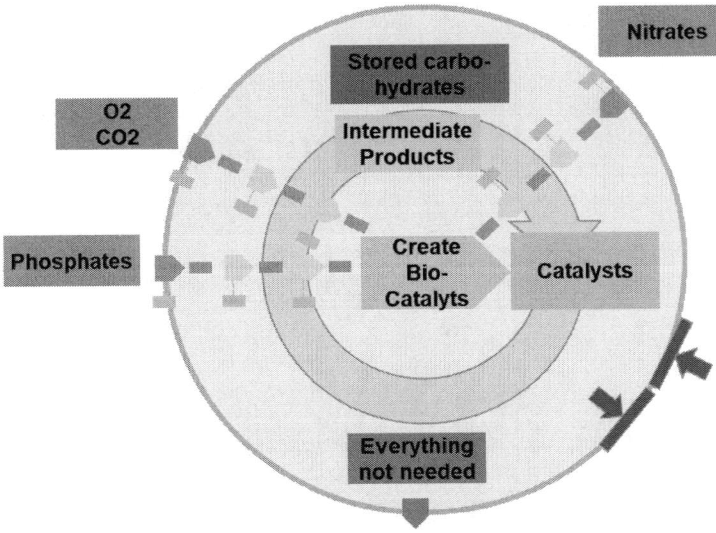

Figure 5-38: Storage of food for later usage

With the ability to store food, the cell is more independent from the changes in the environment and can keep the best environment.

[272] See (Alberts, et al., 2015 pp. 78-81)
[273] See (Alberts, et al., 2015 pp. 78-81)

Business

Organizations can also store resources for later usage (Figure 5-39).

The ability to have property allows resources to be stored for later usage

The best example is the budget: money that can be used later and can be converted in rights or property. Money / budget is very good at storing value. They have no physical extension, they do not expire, and they can be easily exchanged. They give safety for the future.

The organization also stores physical objects: office supplies. This makes the organization more independent from the delivery of suppliers. Normally organizations try to minimize the storage.

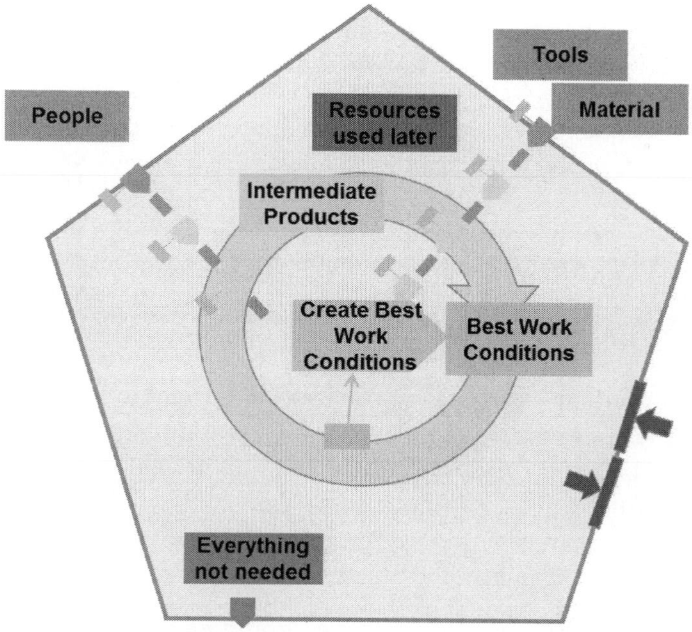

Figure 5-39: Storage of resources for later usage

5.4.6 Summary

5.4.6.1 Insights

Biology teaches us that cells not only convert material via the metabolic network to the auto-catalyst. Instead, the cell membrane provides an environment with the highest concentration of necessary molecules. Membrane proteins import the food and export the waste. The cell has also mechanisms to store molecules and energy for later usage.

The discussion of the system boundary showed the importance of property and rights. They separate the inside from the outside. Property on physical objects and the right for the usage of company site allow us to transfer this legal definition of inside/outside to a physical inside/outside: Owned physical objects can be located at the company's site or not. Employees can be at the company site or outside.

The site can be ringfenced and protected by a security guard. This protects the physical objects in the physical world. Lawyers protect rights in the legal world. Cyber-security protects information of the organization.

If the organization buys physical objects, it adds new properties. On the other hand, rights and properties can also get lost. Selling products, selling unused tools, throwing things away or destroying them results in losing property.

An organization has to ensure having exactly those resources that are necessary. Unused properties reduce concentration and are a disadvantage for the organization. Thus, the organization has to import everything that is necessary, export everything that is not necessary, and store something that is used in the near future.

5.4.6.2 Requirements

A Viable Team (VT) shall implement the following requirements (Table 17):

ID	Requirements	Comment
5.4.1	The VT shall have assigned resources	
5.4.2	The VT shall be empowered to decide which resources enter or leave	Team is managed from inside
5.4.3	The VT shall be able to import needed resources	Hire people and buy tools
5.4.4	The VT shall be able to export all unneeded objects	
5.4.5	The VT shall be able to store resources for later usage	Preferred money

Table 17: List of requirements related to the system boundary

5.4.6.3 *Existing Solutions*

The following methods might help to implement the Viable Team:

Empowerment of Teams

Many modern management approaches like Agile, Holacracy emphasize the importance of empowered teams.[274]

The need for empowerment can be explained as in biology: the high productivity of a team depends on its ability to create the best working conditions for every process. This requires some resources. Only the team itself has the detailed understanding of how everything works together, thus only it can say which resources are needed. It needs to decide what is necessary and what is not. As a consequence, it should not be possible to take its resources away, because this would reduce productivity.

On the other hand, this works only if the team has the responsibility to release all unneeded factors.

Collocated Teams

Agile Project Management recommends having collocated teams. Teams perform better if all persons are at the same location. This improves communication and increases concentration.[275]

No Split Heads

Agile Project Management also recommends avoiding split-heads. Split-heads are people who work for several teams. Employees should work 100% for one team. Inside this team they should work on a minimum number of tasks in parallel. This increases concentration.[276]

5.4.7 *Annotations*

About Space

In this section we saw that the translation of space and time is important. They are the prerequisites to discussing concentration and flow.

The legal entity allows the different spaces of the physical and psychological world to be connected. It creates the form of the organization.

---

[274] See (Holacracy , 2018) and (Robertson, 2014)
[275] See (Schwaber, 2004)
[276] See (Schwaber, 2004)

5.5 *Lifecycle*

Business

In the previous sections we have seen how a team can organize itself to have the best working conditions. It is very likely that such a team is successful and begins to grow.

There is a limitation of growth for every team. If a team has too many team members, then the communication gets more and more difficult. The team has the tendency to found sub-teams.

At the end of the process it might be necessary to create two teams. This phase can be very critical. If a team has a strong team leader and just one good technical expert, then it can be risky to make both to leaders of the two descendent teams. The first team has a good leader without technical experience, and the second team has a good technical expert who has no leadership qualities. Both teams will have severe problems to remain successful.[277]

This example shows that splitting of teams can destroy the viability.

Obviously, there is a demand to train new team members before a team can be created or split. This requires a budget and the culture of knowledge-sharing.

Every small team has the problem when it begins to grow. If you have a successful coffee shop with five employees and you want to start a second one at an additional location, then you can start the second one with new employees from scratch or you can split your existing team before you add new employees. Both alternatives have risks. Creating from scratch has a high risk of failure at the new location. Splitting the team is a risk for both locations.

What are the important steps of splitting a team? Let us have a look at the benchmark.

Biology

Nature shows us how to do it the right way. Bacteria can replicate every 20 minutes and produce viable descendants. This mechanism is the basis for the success of life in general. All existing cells on earth have been created by this mechanism.[278]

The replication of cells follows some defined steps. Each of the steps has to be done in high quality before the next one starts. These leads to a life cycle of cells that consist of defined phases.[279]

---

[277] This problem could be observed in case study one, when the teams for project management and engineering have been separated.

[278] See (Alberts, et al., 2015 p. 2)

[279] See (Alberts, et al., 2015 pp. 963-1020) for a description of the cell cycle

Translations

Let us see why this mechanism is necessary and how it works in detail (Table 18).

	Translation	Biology	Business
1	Replication	Cell Cycle	Team Cycle

Table 18: Translations of replication and lifecycle

5.5.1 Replication

Biology

Obviously, there is not only one cell in the world. A large part of our planet is covered with organisms. This is the consequence of replication.

Cells are able to continuously grow in number by splitting themselves.

Why do cells need to replicate? Why can they not grow continuously as a single cell? The problem of growth is that the surface of a cell has two dimensions whereas the volume has three. For that reason, the ratio between surface and volume becomes more and more unbalanced and the membrane proteins on the surface are not sufficient to organize import and export fast enough. The solution is to split the cell of a certain size into two smaller cells, which again can grow[280].

The mechanism of replication always follows the cell cycle (Figure 5-40). If a cell grows, it creates all of its constituents anyway. Thus, most of the constituents have already been duplicated. However, some parts of the cell are unique and have to be duplicated explicitly. Examples are the DNA with the genetic information and the cell membrane. They have to be duplicated as well; at exactly the same speed. During cytokinesis the cell separates all its constituents, especially the critical parts like DNA. [281]

If the cells still stay connected then they form a multicellular organism; otherwise, we have two independent cells.

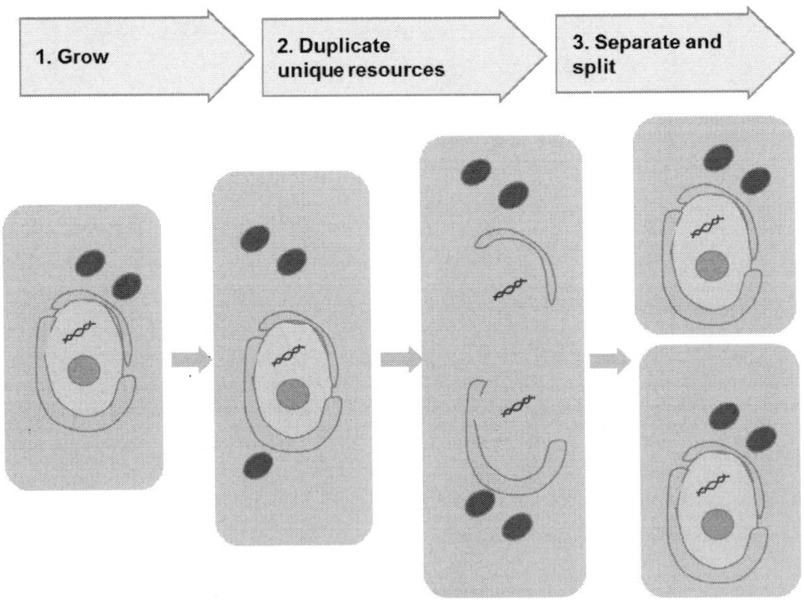

Figure 5-40: Simplified cell cycle. The cell has to duplicate all resources before it can split itself into two descendant cells. Both descendant cells have to be viable.

[280] See (Alberts, et al., 2015 pp. 963-2010)
[281] See (Alberts, et al., 2015 p. 996)

Business

Imagine you have a small startup with an excellent product idea. Your organization grows until you have more than ten persons. Now you face the problem that it is no longer possible to work together in one team. It is better to split your startup into two teams.[282]

Viable Organizations[283] have to split themselves if they want to grow continuously.

Why is it necessary to split teams? It is the complexity of communication which grows in squares with the number of people. Typical teams have around 7 team members. If groups of people grow larger, there is a tendency to create subgroups anyway.

For a growing organization it is crucial to understand the growth and its limitation by the ideal team size. The team has to duplicate the unique resources (like the role of the team leaders, space for visual boards, file-shares for teams, timeslots for regular meetings...) before the maximum size is reached. This requires a lot of training activities to duplicate the available know-how. At the end the resources and responsibilities can be assigned to the descendent teams, and the teams can be split. The best teams should grow and split and will inherit culture and knowledge to the descendent teams. (Figure 5-41)

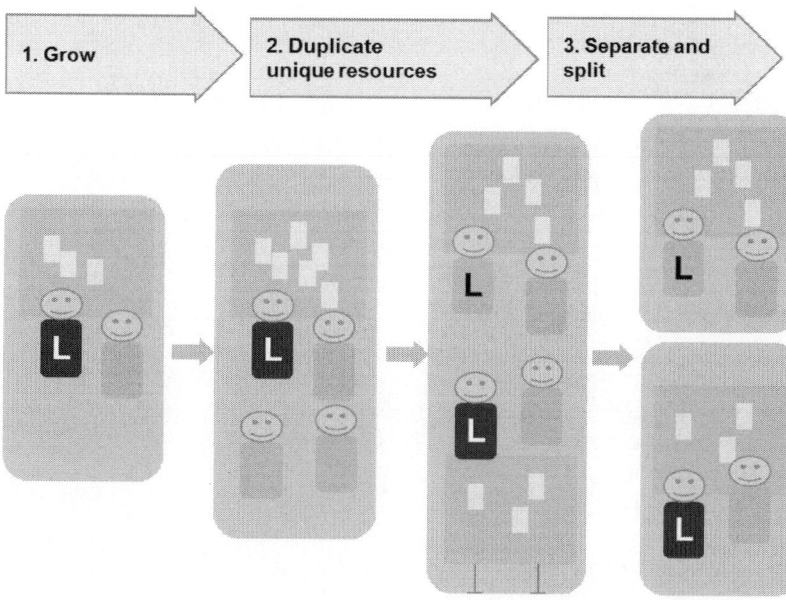

Figure 5-41: Simplified team cycle. The team has to duplicate all resources, before it can split itself into two descendant teams.

[282] Section 8 shows how this works in project business in a greenfield and a brownfield environment.
[283] This can be applied to Viable Teams and to Large Viable Organizations as well. A Large Viable Organization, which consists of several Viable Teams, can be split es well; maybe to address two different markets. In such a case both new organizations need to have all functions.

5.5.2 Summary

5.5.2.1 Insights

Biology teaches us that cells, except LUCA[284], do never start from scratch. Cells are always created by splitting an already existing cell. Each cell has a lifecycle, where it duplicates all resources, before it separates them internally, to be ready for a cell division.

The most important point first: Viable teams also do not start from scratch. They are descendants of other teams that split. They inherit well educated team members and a good team culture from their ancestor team.

The split of a team happens if it becomes too large. As preparation for the split the team has to replicate all needed resources like team roles and equipment. Then it assigns the resources and responsibilities to sub-teams that can work more independently.

In a final step the sub-teams can be separated from each other and form independent teams.

5.5.2.2 Requirements

A Viable Team (VT) shall implement the following requirements (Table 19):

ID	Requirements	Comment
5.5.1	The VT shall be able to detect the need to split	Needs sensitive leaders how recognize the need for a split in advance.
5.5.2	The VT shall be able to replicate all resources	Needs education of new team members.
5.5.3	The VT shall be able to assign all resources to the descending teams, such that both teams have all needed resources	Needs fair behavior. The original team lead looses 50% of the team members.
5.5.4	The VT shall be able to split itself	

Table 19: List of requirements related to lifecycle

5.5.2.3 Existing Solutions

Forming-Storming-Norming-Performing

The life cycle of teams is subject in many publications. Bruce Tuckman introduced the forming-storming-norming-performing model as a life cycle model for teams.[285] The model explains the specific challenges in every phase. He extended the model later with the adjourning phase. [286]

This classical lifecycle assumes that a team starts from scratch. Viable Teams are always created by growing and splitting existing teams. This split of teams adds an additional phase before forming that covers all activities to duplicate resources and finally to separate two independent teams.

---

[284] The Last Universal Common Ancestor is the name of cell number one

[285] See (Tuckman, 1965)

[286] See (Tuckman, et al., 1977)

5.5.3 *Annotations*

About Cell Cycle

The cell cycle is much more complex than described in the translation. One important step is the replication of the genetic material with DNA-polymerases.[287]

For our translation this step is not so important. If the process knowledge is already written down, then it normally can be accessed by both teams. The problem with the split of teams is more the competences of the employees.

Limitations of Growth/Apoptosis

In section 5.6.5 we describe the mechanisms that control the growth of cells or trigger the programmed cell death (Apoptosis).[288] Both mechanisms are important for multicellular organism.

---

[287] See (Alberts, et al., 2015 pp. 237-298) for the details of DNA replication
[288] See (Alberts, et al., 2015 pp. 1021-1034) for the programmed cell death.

5.6 Adaptation

Business

In the previous sections we have seen that an organization needs to have mechanisms to import and protect everything that is needed for its best working conditions.

How can it guarantee that this environment will remain stable? For the moment our organization depends on the factor markets and must hope that all production factors are always available outside. All these markets change. Labor market has different requirements than ten years previously. If an organization wants to attract top talents, they have to do people-work differently than ten years before. Customer markets change even faster. There is high competition. Finance markets are also changing.

The basic problem here is that it is not sufficient to have only one perfect mechanism for growth and protection. The organization needs to have different mechanisms for different situations, and it needs processes to decide for the right one. These decisions are based on gathered information about the inside and outside worlds and select the best of a set of possible reactions.

So, there is a demand to adapt the organization to a changing world. This adaption has to be early enough to avoid problems.

Biology

Every life form has the same challenge and has perception, decision and action mechanisms.

The perception mechanism collects information of the inside and outside worlds that correlate with something important. In most cases it is about food, threats and reproduction. A decision process has to integrate the perceptions and set priorities to create a defined action. Action mechanisms change something to adapt to the situation adequately.[289]

These mechanisms are available in very simple organisms, even in bacteria.

For most organisms these are permanent processes and with that they can react agilely to their environment.

---

[289] See (Alberts, et al., 2015 pp. 813-888)

Translations

On the next pages you can find the following translations (Table 20):

	Translation	Biology	Business
1	Perceive, Decide, Act	Control activities of the cell based on signals	Control the organization based on information
2	Control metabolism	Regulate metabolic pathways	Kanban
3	Control creation of catalysts	Gene expression	Control creation of the best working conditions
4	Change form and position in environment	Change form of cell and move	Change own form and position in market
5	Control Lifecycle	Control growth, cell division and cell death	Control growth of team, team split and shutdown of activities
6	Signal Pathways	Signal pathways	Decisions
7	Perceive outside world	Receptors	Understand markets and stakeholders and future developments

Table 20: Translation of adaptation mechanisms

5.6.1 *Perceive, Decide, Act*

Biology

Cells have many mechanisms to create their own constituents and to protect them. These mechanisms must be extremely effective and efficient; thus, they are highly optimized to a specific environment. However, the environment changes very often. Even the self-creation mechanisms themselves contribute to a change in environment. A cell consumes available food to be able to grow. After a while the amount of freely available food decreases. The solution is adaptation. The cell has to perceive and react to the changing world (Figure 5-42). It has to collect information and influence its own mechanisms accordingly.

Cells adapt to a changing world by perceiving themselves and their environment and selecting one of several possible actions.

All this is possible because proteins can change their form if certain signal molecules attach to them. If these proteins are enzymes, this can enable or disable their ability to catalyze reactions. If they are structural proteins, this can lead to a change of the overall form of the cell.

This feature is used in various situations. It can be used to change metabolism, change the gene expression or influence the lifecycle. The signals which change the forms of effector proteins, can come from inside the cell, or from outside[290].

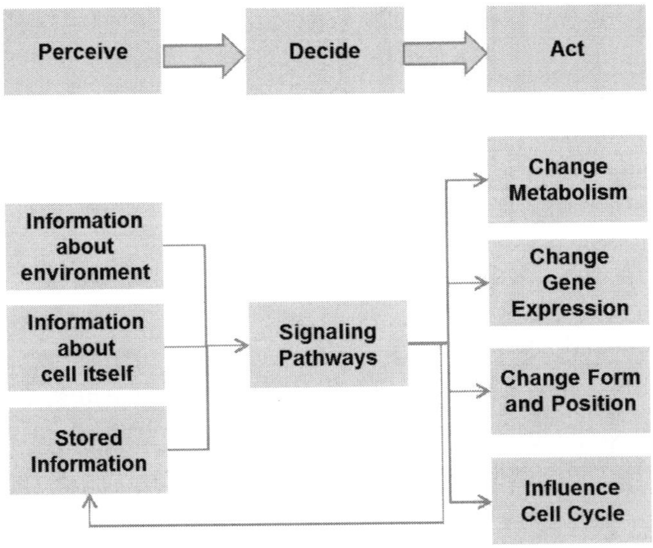

Figure 5-42: Perceive, decide and act in organisms

In the following sections we will describe the purpose of the actions in more detail.

[290] See (Alberts, et al., 2015 pp. 813-888) for the principles of cell signaling

Business

Every organization has the challenge that the world is always changing. This applies to large organization and to single teams like project teams. Customer demand is changing. New competitors come up with completely new products. Governments change laws. Suppliers have problems to deliver and employees enter or leave the organization.

Organizations have to adapt to these changes (Figure 5-43).

Organizations should adapt to a changing world by perceiving themselves and their environment and selecting one of several possible actions.

Adaptation needs the ability to control processes and structures based on decisions. Every process should be started or stopped by decisions to reduce or intensify them. Workplaces should be created or removed by decisions. The structure of the organization and the balancing of resources is based on decisions, and the growth of the organization as well.

The decisions are made based on information that comes from inside the organization (e.g. work-in-progress, productivity) or outside (e.g. customer demand, employee satisfaction). This is called evidence-based decision making.[291]

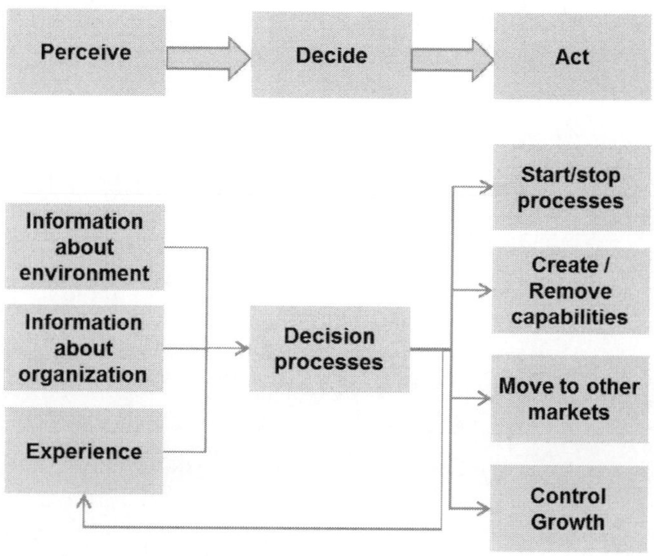

Figure 5-43: Perceive, decide and act in organizations

The overall picture shows that there are three main requirements: collect information, make good evidence-based decisions and be able to implement the decisions quickly.[292,293] Let us start with the implementation of decisions and see how this is connected to the idea of best working conditions.

[291] See (ISO 9000, 2015)

[292] The perception of the inside and outside world can be seen as the purpose of system 1 in the Viable System Model of Stafford Beer. See (Beer, 1995) and the literature review in section 2.7

[293] From a perspective of Beer's Viable System Model, the actions influence the operations of system 1 (Beer, 1990). These operations are not detailed in VSM. Viable Business shows that operations are about the functions of living organisms: self-creation, self-protection, self-replication.

5.6.2 Regulation of Pathways

Biology

A very important use case for adaptation is the control of metabolic pathways. Cells have the problem that they need to have all amino acids and nucleotides in the right mix. If they produce too much of one molecule type, this would require space for the intermediates and the product, which are both not needed. This reduces the relative concentration of all other molecules. Thus, a good idea is to stop a pathway if there is no demand for the product.

Cells are able to switch the catalytic behavior of enzymes to produce molecules on demand.[294]

Many pathways are regulated at their first reaction as follows (Figure 5-44). The enzyme for the first reaction has a specific form, which allows that the product of the pathway attaches to it. This attachment slightly changes the form of the enzyme, especially its active site. This inhibits the catalytic activity. The reaction does not produce any more intermediates and products.

The connection between enzyme and product is not permanent. The connections are always created and removed. It is in equilibrium. If the concentration of the product goes down, then many more connections are removed than created, and the enzyme can continue to catalyze its reaction and create more products.

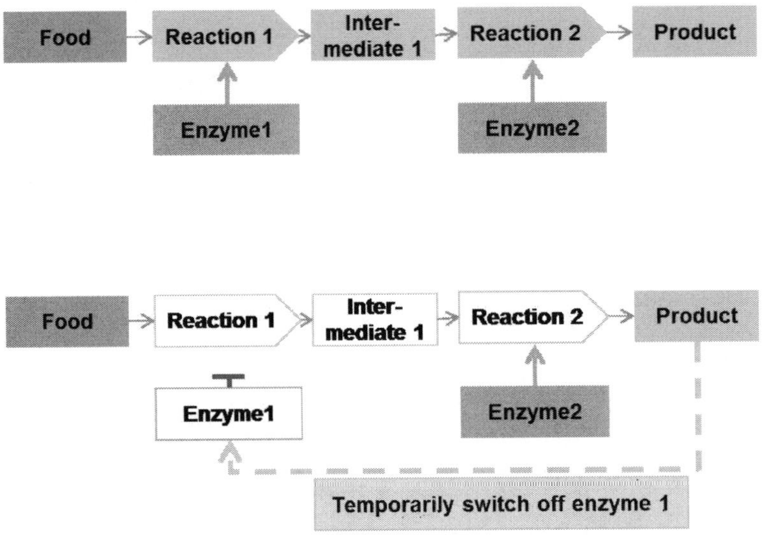

Figure 5-44: Regulation of pathways by allosteric inhibition. Products inhibit enzyme activity to stop production. The regulation of pathways allows everything to be in the right mix. The concept also explains why a cell mainly consists of end products rather than intermediates.

[294]See (Berg, et al., 2015 p. 444) or (Alberts, et al., 2015) for allosteric control of enzymes

Business

Organizations want to produce exactly what is needed to avoid stocks that need space and bind resources. Lean production uses pull systems like Kanban to regulate process chains. Agile software development uses the same idea to reduce work-in-progress.

Organizations should be able to switch their production processes on demand.

Business processes also have to be able to produce on demand (Figure 5-45). This can be done by a decision to start or stop. The decision influences the best working condition, which includes the "Would, Could, Should." If an employee gets the decision not to produce, then the "Should" is missing, even if the employee "Could" do it and "Would" like to do it. This decision can be based on the product of the entire process chain. If there are already enough products available, then the first process can be stopped. This is easy to implement if the employee at the first process knows the number of products. If the products cannot be seen directly because the products are at a different location, this can be done with a Kanban system. Kanban cards are used as signals to indicate that the first process should continue to produce. [295]

In a production team kanban cards are used to indicate the material supply team to replenish material. In a software team kanban cards are used to indicate whether the product owner can ask for new features. In both cases the kanban boards reduce the work in progress.

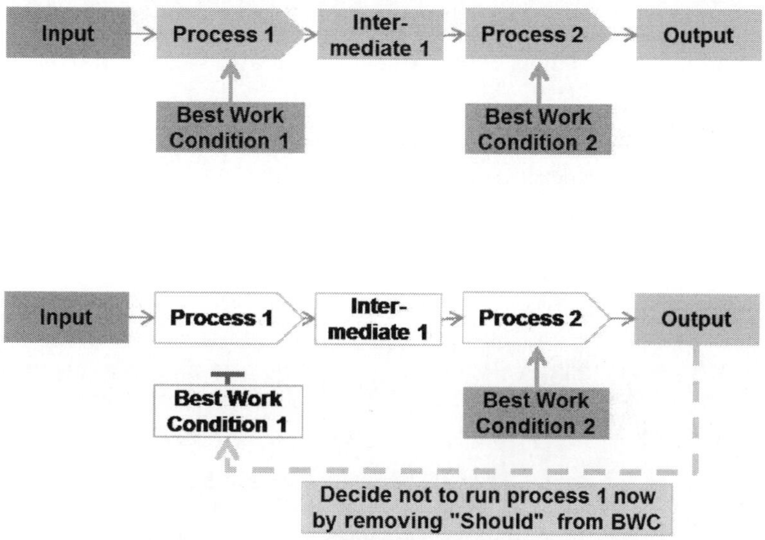

Figure 5-45: Kanban system. Production processes are only running if there is a need for more products.

One benefit of the pull systems is the reduction of needed storage space to increase concentration of production factors. Another benefit is to avoid investing too much working time in producing something that is not needed yet. Employees can concentrate on that what is needed now. The next section shows a second strategy that goes in the same direction and increases concentration even more.

---

[295] See (Gorecki, et al., 2013 S. 258)

5.6.3 *Regulation of Gene Expression*

Biology

It is a good concept to switch off enzymes if the product is not needed. Sometimes the products are not needed for a long time. The products and the intermediates of metabolic pathways are very small compared to the enzymes. So, it would reduce even more space if the enzymes themselves could be removed. Since enzymes are always destroyed after some time, it only needs to inhibit their creation (Figure 5-46).

Cells are able to control their gene expression in order to create proteins (enzymes) only when they are needed.[296]

New enzymes are always created at the ribosome based on their corresponding messenger RNA (mRNA), which is a copy of a specific gene on the DNA.[297] If a cell is able to inhibit the transcription of the gene, then there is no mRNA, and no enzyme. This mechanism is called regulation of gene expression. Enzymes are created only when they are needed. Gene expression contributes to the high concentration of the other enzymes which are really needed at the moment.

Gene expression is realized with proteins that cover sections of the DNA molecule. This coverage inhibits the RNA-polymerase to read the genes and create RNA for new proteins. In other words: Covering sections of the DNA allows to switch-off the protection of proteins and thus to influence metabolism. Gene expression can be influenced by signal molecules. Thus, the gene expression is not fixed. It can be controlled by signals which reflect information about the inside and outside world.[298]

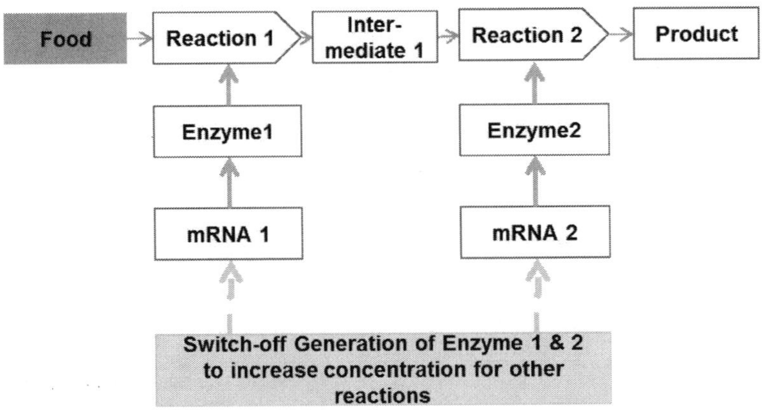

Figure 5-46: Gene expression

Gene expression is also the reason that all cell types in the human body have different sets of proteins even if they have the same DNA. Every cell uses a subset of genes only.[299]

---

[296] See (Alberts, et al., 2015 pp. 369-438) for a description of gene expression
[297] Standard translation as described in (Alberts, et al., 2015 pp. 333-362)
[298] See (Alberts, et al., 2015 p. 372)
[299] See (Alberts, et al., 2015 pp. 369-371)

Business

Workplaces require space and bind capital. They should be established only if they are really needed.

Organizations should control the creation and removal of workplaces based on demand.

In a factory this is quite clear: every workplace requires space. If the workplace is not needed for a longer time, then the organization can decide to deconstruct it (Figure 5-47). This frees space and allows all other workplaces to be brought closer together. In other words, it increases concentration. If the workplace is necessary at a later time, then it can be set up again because all necessary information for its build-up is available.

In an engineering environment the workplaces mainly consist of IT-tools. If they are not used for a longer time, they can be uninstalled, and licenses can be used otherwise. This example shows that a second purpose is to make resources available.

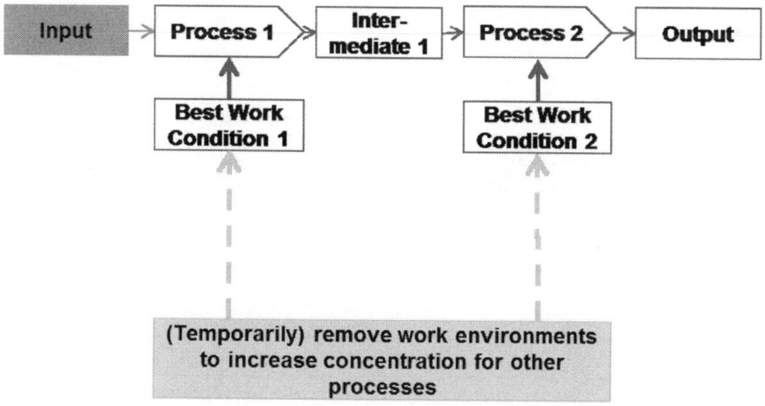

Figure 5-47: Creation or removal of work environments for processes

The translation of gene expression needs to make the term of mRNA more precise. It should be understood not only as the information on how to create a protein. Instead, it also includes the order to do it. In business this can be understood as the order to create a workplace.

The information how to do that is already available in the know-how of the organization. The decision to create the workplace (gene expression) creates the order to the work preparation team.

Influencing the processes directly or influencing the best working conditions of processes allows to control the process networks. However, this only works internally. Other mechanisms influence the outside world as well.

5.6.4 *Change of Form, Movement*

Biology

Regulation of metabolic pathways and gene expression allow increasing concentration inside. They only need information from inside the cell. But the speed of growth depends on the availability of the food outside. A cell has problems to influence the world outside directly. If it creates enzymes and releases them to the outside, their concentration would be low and decrease continuously.

Another possibility for the cell is to move itself to where the food is (Figure 5-48).

Cells can change their form to change their position relative to the environment

Proteins are not only used as enzymes but also as structural proteins that are used in the cytoskeleton or in the flagella. The cytoskeleton allows the cell to maintain a specific form. This is a prerequisite for a controlled change of form. [300]

The flagellum consists of proteins that are connected to a specific form. This form can change if signal molecules attach to it. The change of form creates a movement which drives the cell to the opposite direction.

The benefit of the movement is the chance to get more food and fewer threats at another location. Movement contributes to self-creation and self-protection.

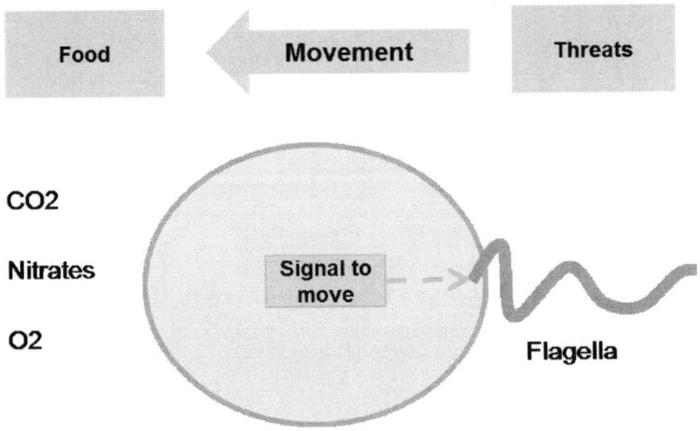

Figure 5-48: Movement with flagella

The probability to move to a better location is much higher, if the cell perceives information about food and threats. With that information it can create the signals to move more precisely and move in the best possible direction.

[300] See (Sadava, et al., 2011 S. 109)

Business

The usage of pull systems and the control about creating workplaces helps the organization to increase the concentration of production factors inside, but it does not ensure that it gets enough resources from outside. All the resources come from specific markets.

If the organization is not able to get resources from a market, even if it is able to produce the best working conditions inside, then it needs to move to another market (Figure 5-49).

Organizations should be able to move relative to the environment

[301]For customer markets it is clear: if the current product cannot be sold, then the organization does not earn money. The reason can be internally: the product does not meet expectations of customers or is too expensive. Here the organization has to improve processes internally.

Sometimes the organization has to find new customers in new markets. This is like relocation in the markets. The relocation can be physical if the organization wants to get customers in a different country, it can start a regional organization. The relocation can also focus on portfolio elements; the organization starts a new business unit for new products and services.

In both cases it moves the organization to an environment where it has more potential customers than before. The movement requires creating new structures and processes and assign resources to them.

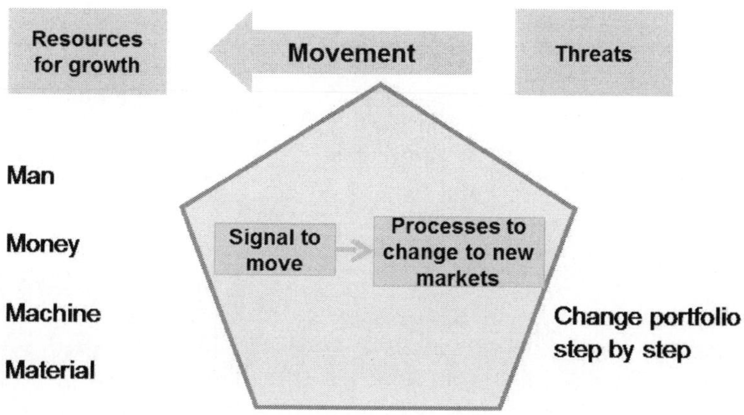

Figure 5-49: Movement to new markets

This movement to new markets is more successful if the organization has information about the outside world. It has to know where potential customers are. It also has to know about the threats (e.g. if it wants to move to a different country).

---

[301] The translation of customer relationship is described in section 5.7

5.6.5 Control of Lifecycle

Biology

If the cell can create high concentrations inside and move to an environment that provides enough food, it can grow. This growth is the purpose of life.

In multicellular organisms this growth has to be limited. Here the size of organs needs to be in a specific balance, so it is not allowed for the cells to grow as they want. The growth needs to be regulated (Figure 5-50). It should be as fast as possible, but without loosing the overall balance. Adult multicellular organisms stop their growth when they've reached the optimum proportions.

Cells are able to control their growth and their replication.

The growth of cells is regulated by signals that come from outside the cell. If a cell does not get the signals to survive, it undergoes a programmed cell death and decomposes itself to release its components for further usage.[302]

If it gets this signal to survive, it continues to live at the same size without growing. It only creates itself to replace damaged components. With the additional signal to grow the cell is allowed to grow and to replicate itself. Other signals indicate the cell that they should specialize to a certain cell type.

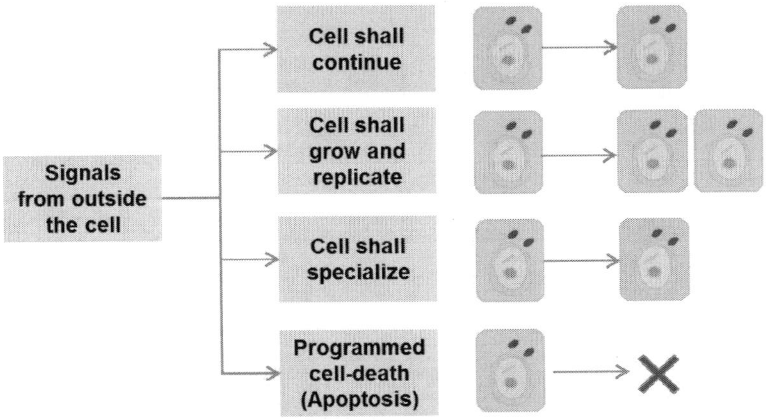

Figure 5-50: Regulation of growth and lifecycle in a cell.

This control of lifecycle is not needed by unicellular organisms. But it is prerequisite for all complex multicellular organisms, which need to keep the proportions of their organs.

---

[302] See (Alberts, et al., 2015 pp. 1021-1034) for programmed cell death

Business

For a small startup the sky is the limit. There is no need to self-destruct the one and only team, to specialize or to stop growing, but for large organizations this is necessary. They often have the problem that some central departments grow faster than they should, whereas others grow too slowly. Large organizations need a precise balance.

This requires that the lifecycle of teams has to be controlled (Figure 5-51).

Organizations should control their growth and their replication.

First of all, it must be possible to stop teams. Some team types like project teams[303] are created only for the duration of a project. At the end of the project they have to release their resources. In other cases, teams have no expiration date, but the organization wants them to stop activities.

In all other cases the teams shall continue to work. They are allowed to grow only if they get an external decision to do so. The specialization is also allowed only if there is a decision to do so.

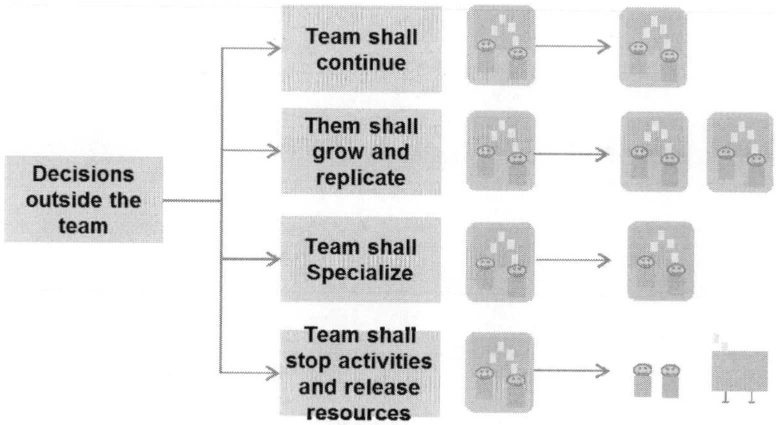

Figure 5-51: Regulation of growth and lifecycle for teams. Teams should be able to control their lifecycle

These mechanisms are used during the growth of a large organization. It starts with one team which grows and replicates. Later the teams specialize. During the entire journey it needs a control of the lifecycles to keep the entire organization balanced.

Prepared with the mechanisms that influence the processes, the form and position and even the lifecycle, let us see how all this is connected to the information that has been perceived from inside and outside.

---

303 Section 8 explains the different team types and their lifecycle in detail

5.6.6 Signal Pathways

Biology

We have discussed four different ways how adaptation helps the cell to improve self-creation and self-protection: Regulation of pathways, gene expression, movement and control of lifecycle. All of them are based on signals that trigger the actions (Figure 5-52). Where do the signals come from?

Cells are able to combine several input signals in signaling pathways to create output signals.

Every molecule of a cell can be used as a signal. If it binds to an enzyme, this enzyme can change its form and change its catalytic behavior. We have discussed that with the regulation of metabolic pathways.

Sometimes several of these signals are used, but not to switch an enzyme of a metabolic pathway. They can also switch an enzyme that produces other signals. This uses the same mechanisms, but it creates something new. It allows many signals to be combined logically to produce new ones. It is like a neural network where many inputs are combined to trigger an output.[304]

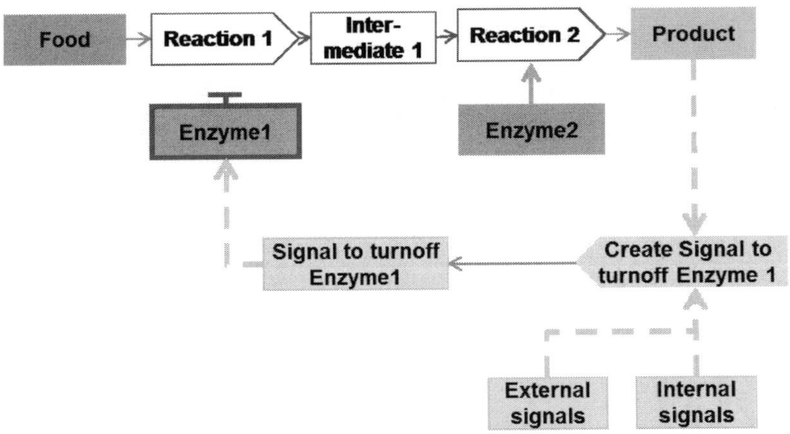

Figure 5-52: Signal pathways. Cell can combine several input signals to trigger actions.

In our example this would allow the control of pathways based on combined information.

[304] See (Alberts, et al., 2015 pp. 816-817)

Business

We have seen various ways to adjust structures and processes. All of them start with a decision. Decisions are made internally by management (Figure 5-53). Where do they come from?

Organizations should be able to combine information sources in decision chains to make decisions.[305, 306]

A typical example is the decision of production. It not only depends on the number of products that are already available. A decision about the process should also consider whether demand is increasing or decreasing in the future, or whether sufficient resources will be available in the future. Thus, there is a need to combine all this information to make the decision.

The same situation applies to the other decisions of the organization. An entry to a new market is based on many internal and external factors. It needs a lot of information that has to be combined.

Very often the decisions go stepwise. A first decision to enter to a new market is the basis for a second decision about the new organizational structure, which is the basis for a third decision.

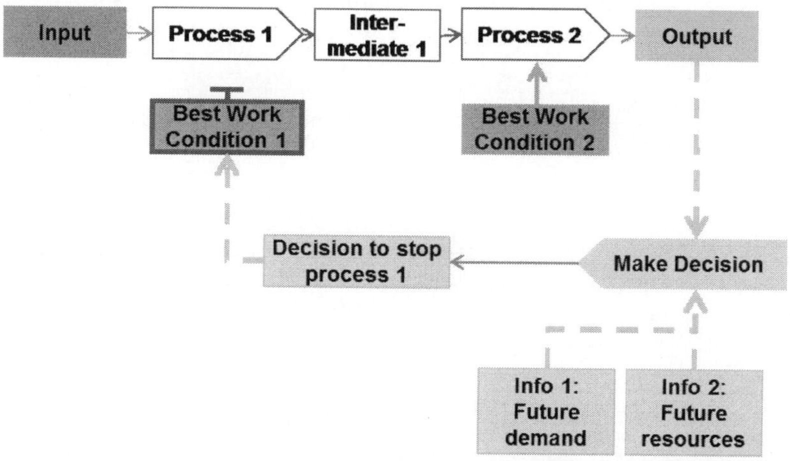

Figure 5-53: System analysis. Organizations combine several isolated information sources to one big picture before they trigger actions.

Good decisions are based on good information. An organization needs to have a clear picture about itself, combines the information and makes the decision. Information about stakeholders and markets are important for many decisions. What can we learn from biology about the perception of the outside world?

---

305 Evidence-based decision making is described in (ISO 9000, 2015). Decisions should be based on data, facts and numbers, rather than on opinions.

306 Luhmann's organization theory is also based on chains of decisions. Each decision reduces the variety of the system and triggers other decisions. See (Luhmann, et al., 2017) and (Luhmann, 2011)

5.6.7 Perception of the Outside World

Biology

Some of the signals come from the outside. Which signals are needed? How are they imported?

Cells are able to create signals based on perceived information from the outside world.

The cell has receptor proteins in the cell membrane that react to specific molecules of the outside world (Figure 5-54). If the molecule binds to it, the receptor creates a specific signal molecule for the inside. This is called the second messenger. A cell can only react to the outside world if it has the correct receptors.[307]

Very important are those receptors that identify molecules indicating food or threats. Since a cell needs various food components, it can have several specialized receptors.

The receptors not only copy the signals, they can also amplify the signal by creating a higher concentration of signals inside. Good examples are the photoreceptors in the eyes, which amplify the signals. The signal can be combined with others and trigger the necessary actions.[308]

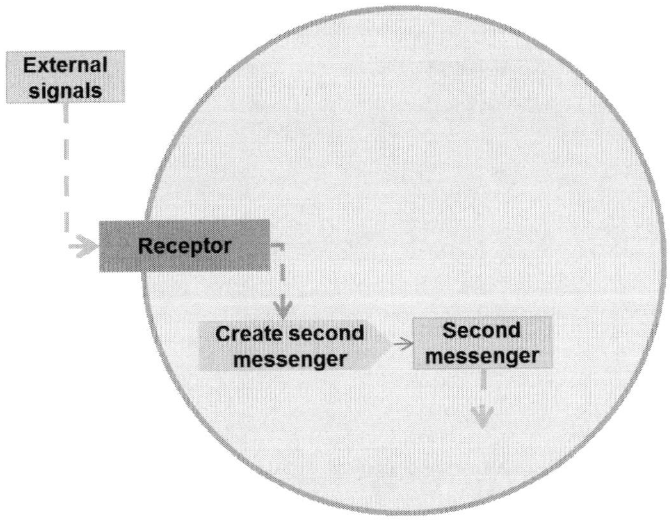

Figure 5-54: Receptors and second messengers. Receptors create a second messenger to make information available inside

[307] See (Alberts, et al., 2015 pp. 815-816)
[308] See (Alberts, et al., 2015 pp. 819-822)

Business

Organizations need to have a clear picture of the environment (Figure 5-55). They need to understand the opportunities and threats of the markets they are in. They need to understand their stakeholders. They also need to observe potential threats from the environment. This information has to be made available.

Organizations should be able to collect information about the outside world and make them available internally for decisions and actions

The collection of information is done by various processes. Information about employee satisfaction comes from an engagement survey. Information about new tools can be found during conferences or visits to a supplier.

All this information has to be made available internally to get a precise, integrated picture of the outside world that is in focus. Such a picture can then be used to make good decisions.[309]

Some information is especially important. The organization has stakeholders in several markets which provide resources. It is important to get information about the stakeholder needs and its satisfaction with what the organization gives to fulfill them. This knowledge allows internal processes to be aligned to the stakeholder needs.

Additionally, the organization has to understand the trends in the markets. This is necessary to find new stakeholders in the future.

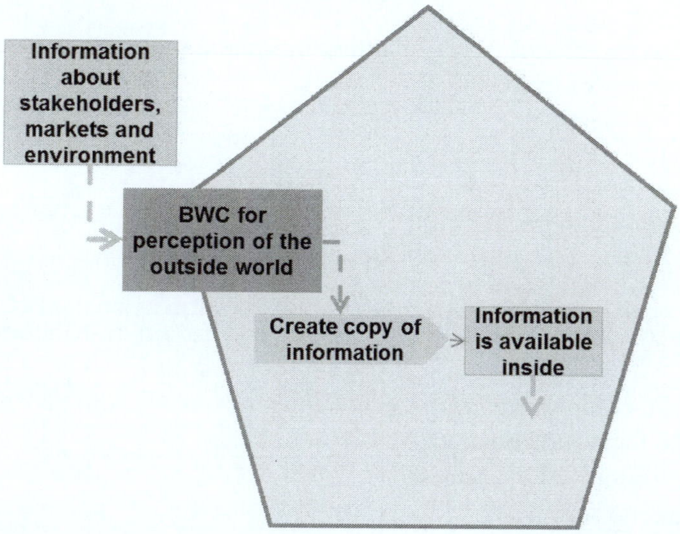

Figure 5-55: Perception of the outside world. Perception processes have to collect information about the environment and make them available inside.

[309] This perception of the outside world leads to an absorption of variety. The organization can only perceive a subset of the possible information. Beer describes the perception in system 1 of his Viable System Model. See (Beer, 1990).

5.6.8 Summary

5.6.8.1 Insights

Biology teaches us that cells are able to adapt to their environment. They are able to perceive their inner state and the relevant outside world and combine this information to trigger an action. Cells can influence their metabolism, they can change their gene expression, change form and position, and they are able to control their lifecycle.

The growth of the organization depends on its ability to create and protect itself. Adaptation has the purpose to adjust the creation and protection mechanism according to the situation.

Adaptation can be done by changing the regulating processes, by creating or removing workplaces, by movement relative to the environment or by controlling growth. In all cases the actions are triggered by decisions.

These decisions are made based on information that is collected about the organization and its environment. Performance indicators provide relevant internal information about the organization. Stakeholder satisfaction surveys and information about markets give a picture of the outside world.

5.6.8.2 Requirements

A Viable Team (VT) shall implement the following requirements (Table 21):

ID	Requirements	Comment
5.6.1	The VT is able to control the execution of processes based on decisions	Requires discipline in execution
5.6.2	The VT is able to control the creation of removal of work environments	
5.6.3	The VT is able to move to a different environment	e.g. Change the purpose of the team to get more contracts with customers
5.6.4	The VT is able to control its lifecycle	Growth, replication, specialization or self-destruction.
5.6.5	The VT makes decisions based on numbers, data, facts and knowledge	Evidence-based decision making
5.6.6	The VT collects information about its own processes and resources	
5.6.7	The VT collects information about the outside world	
5.6.8	The VT stores information for later usage	Learning

Table 21: List of requirements related to adaptation

5.6.8.3 Existing Solutions

The following methods might help to implement the Viable Business:

Retrospective

SCRUM uses the retrospective as part of the sprint meeting. In the retrospective the team members speak about their current working conditions and make proposals to improve them. The result of the retrospective is input for the next sprint planning.[310]

Scorecard, Metrics and Key Performance Indicators, Dashboard

Performance indicators give a good quantitative picture of the organization. A balanced scorecard can help to present this information to many employees. This helps to align decisions to this information and to motivate people to reach the goals[311].

In Chapter 1 we will discuss a very specific form of the scorecard. The functional overview diagram of Viable Business contains a top-level overview of all organizational functions. The performance indicators of the scorecard can also be located in this picture. This gives a clear relationship between function and indicators, which simplifies the analysis of the system.

Maturity Assessments

Assessments and audits compare process maturity with a reference model and provide insights about strength and weaknesses of processes. Typical examples are the ISO 9001 quality audit, the ISO 9004 self-assessment or the CMMI assessments. They provide information about the organization that can be used for improvements [312]

Pull Systems, Kanban, WIP-Limit

Pull systems create products only if there is a demand. A typical implementation is the Kanban system. It is used in Lean production and in Agile Software Development [313]

5.6.9 Annotations

Other Aspects That Have Not Been Translated

One important aspect that has not been discussed is the ability to learn. Even *E. coli* bacteria are able to make predictions about future conditions. Tagkapoulos tested this behavior in the lab by changing temperature from 25° to 37° followed by a time-delayed change of oxygen concentration from 20% to zero. The experiment showed that bacteria learned to change the metabolism to anaerobe, even before the oxygen concentration goes down. This ability to predict a situation and to act in advance gives them an evolutionary advantage. This learning is based on an internal representation of the learned dependencies between temperature change and oxygen change. [314]

Business also needs the ability to predict a situation. This allows starting countermeasures early enough. Base of the prediction is an internal representation of the cause / effect relationship.

---

[310] See (Schwaber, 2004)
[311] See (Kaplan, 2001)
[312] See (ISO 9001, 2015) (ISO 9004, 2015) (SEI, 2018)
[313] See (The Productivity Press Development Team, 2002a) (Butterhof, 2017)
[314] See (Tagkapoulos, et al., 2008)

5.7 Interactions with Other Viable Systems

Business

In section 5.4 we have seen the importance of the boundary. The main purpose is to increase the concentration of production factors by limiting, restricting or generating flow between inside and outside. This high concentration guarantees the best possible work environment inside.

Now the question is: where do the competences and the tools come from? The obvious answer is that the organization gets them by hiring employees or buying them from suppliers.

Before we study that in detail, we should have a look at some other possibilities that have to be considered. Let us compare it with the benchmarking partner.

Biology

All cells need resources from outside to grow. They need material and energy. There are several possibilities known:

1. Autotroph
2. Heterotroph
3. Mutually beneficial symbiosis
4. Parasitism

Autotroph

In the simplest case the cell is an autotroph and creates everything by itself based on inorganic material. Such a cell can survive even if there is no other cell around. [315]

In a business environment this would mean that an organization uses only self-made tools and material. It has to do everything from scratch. This model can be named the Robinson Crusoe model. It is the perfect choice if someone gets lost on an unknown island and want to start a business.

Heterotroph

Some cells try to get their resources by eating other cells. Those other cells have protection mechanisms to avoid that. Thus, it needs some additional mechanisms for heterotroph organisms to overcome the protection.[316]

In business this is known as mergers and acquisitions (M&A). An organization buys another one with the intention to get all its resources for growth. This nutrition model is widely used when a market consolidates, and the larger players buy the smaller ones. The organization does not need to hire people because they are already hired.

---

[315] See (Sadava, et al., 2011 S. 1411)
[316] See (Sadava, et al., 2011 S. 1411)

Mutually Beneficial Symbiosis

Cells are also able to collaborate with others. In a symbiosis they exchange resources with other cells. The symbiosis not only happens between cells of different species. The same thing happens between the cells in a multicellular organism.

In this section we will focus on this type of collaboration to get resources from outside.[317]

Parasitism

A last one should be mentioned here as well. If cells mutually exchange resources, there is always the risk that parasites take the resources without giving something in return.[318]

In a business environment every stakeholder group can have parasites that use the benefits of the organization without giving something back. On the other hand, the organization itself can also be parasitic regarding stakeholders (especially regarding environment).

Translations

Here we focus only on the most important interaction with others: The mutually beneficial symbiosis. This explains the relation to all kind of stakeholders and the integration of stakeholders and the value creation processes in our overall picture. (Table 22)

	Translation	Biology	Business
1	Symbiosis	Symbiosis with another organism (mutually beneficial)	Interaction with stakeholders in a win-win situation

Table 22: Translation of collaborations

[317] See (Sadava, et al., 2011 S. 1564;1573-1579)
[318] See (Sadava, et al., 2011 S. 860)

5.7.1 Mutually Beneficial Symbiosis

Biology

Some raw factors needed by the cell are only available in the environment because they are produced by other organisms. So, it makes sense that the cell moves itself to the producer because the highest concentration is there. Another solution is to attract the other organism by producing something that is needed by the other cell. The cell gives something to the other cell and receives something in return. This mechanism is called mutually beneficial symbiosis and is beneficial for both cells (Figure 5-56).[319]

Cells can live in mutually beneficial symbiosis with other cells.

A typical example in nature is the export of carbohydrates to attract organisms that provide nitrates. This exchange has the advantage that the cell can start to specialize. If nitrates are available in the environment, then it can reduce its own production, and focus even more on the production of carbohydrates. The carbohydrates that are not needed can then be shared with the other cell. The specialization allows both cells to increase concentration for the production of one substance.

The symbiosis may also introduce new catalyzed reactions which are not needed for self-creation. These reactions work indirectly. The cells give something in order to be able to get resources that they need. Symbiosis is a give and take.

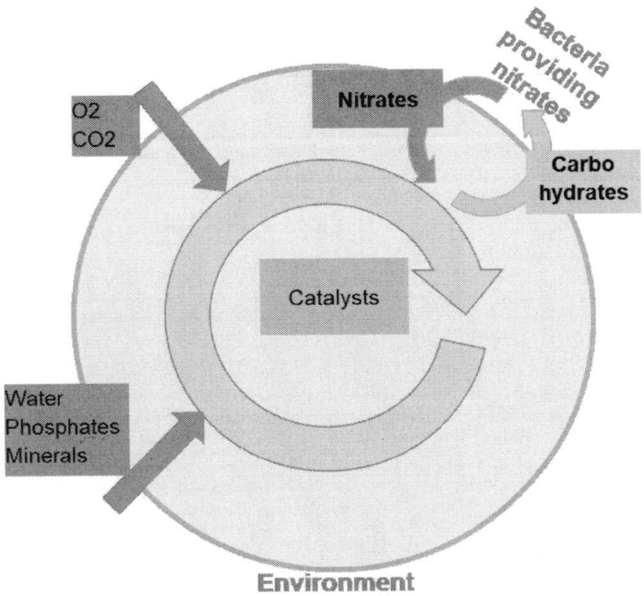

Figure 5-56: Symbiosis

This give, and take is the same mechanism that can be found in multicellular organisms. Cells can be specialized for one task, because the other cells take care of the other tasks (see section 6.1)

[319] See (Sadava, et al., 2011 S. 1564) for a description of symbiosis

Business

Most resources are not available for free. Labor is not for free. Material and tools are also not freely available. They come from two stakeholder groups: employees and suppliers. The organization has to give something to attract them and close contracts (Figure 5-57).

Organizations live in symbiosis with their stakeholders (give and take).

Let us follow our process network backwards to identify the main stakeholder groups. First of all, we need employees and suppliers only because they provide the resources for the creation of the best working conditions. These two stakeholder groups are directly linked to our system of self-creation. We need money to pay employees and suppliers.

This introduces the customers as the third stakeholder group. The organization is able to create the best working conditions for all processes. With these efficient and effective processes, it can also create value for customers. The customers in return pay for this value. A large portion of that money is needed to get the people, tools and material for customer value creation, but some money can be used to reinvest in better work preparation.

The investors allow the organization to pay salaries before the customer pays for the product. In return the investor expects to get interest from the organization. All these interactions are based on the economic system provided by society, which expects to get taxes.

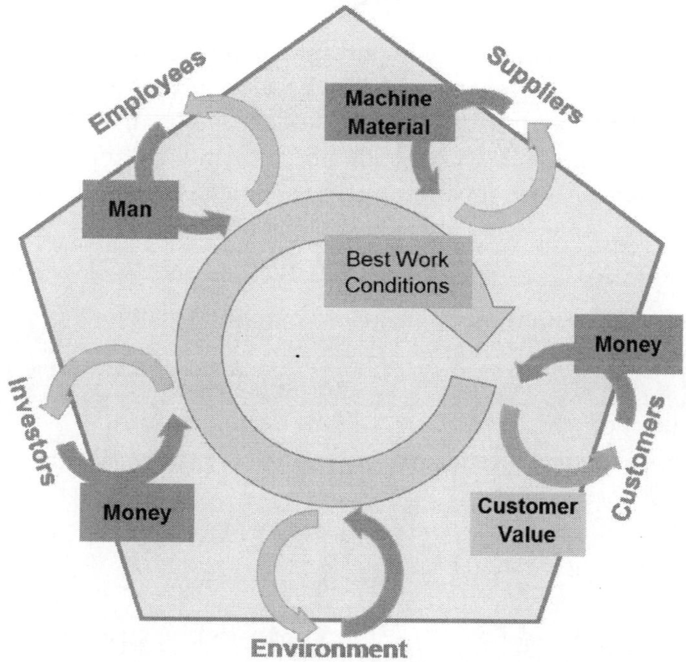

Figure 5-57: Collaboration with stakeholders

Employees are stakeholders that are strongly connected to the organization. They are like obligate endosymbionts. Employees provide the motivation and energy to drive processes, the have the competences to run processes, and they have the ability to perceive the system. On the other hand, the organization provides the full environment for employees during working hours, like canteen, restrooms or coffee machine. This creates a strong mutual dependency.

5.7.2 Summary

5.7.2.1 Insights

Biology teaches us that organisms have a benefit when they create a mutually beneficial symbiosis with other organisms. They give something and receive something. The symbiosis allows both organisms to be more focuses on some processes.

Organizations interact with all their stakeholders in a win-win situation (symbiosis). Typical stakeholder groups are investors, employees, suppliers, customers and the environment (society).

All these stakeholders are important because they provide the needed resources. The stakeholders also want something from the organization what they need. This leads to a give and take that is mutually beneficial.

Practically this is implemented by contracts and agreements that create rights and obligations. These interactions lead to some new process chains: For every resource that is needed from outside, the organization has to find and attract many potential partners. Then it needs processes to select the best possible partner and create a formal contract or an informal agreement. Additionally, the organization has to deliver what the other party needs and has to ensure that it gets the needed resources. The collaboration with others is usually very beneficial. The organization focuses only on a few numbers of processes, in return it gets everything it needs.

The symbiosis with employees is especially important. Employees provide the energy that drives all processes in the organization, and they provide the necessary competences. In return the receive a good work environment and a salary. This is similar to the mitochondrion, which is an endosymbiont of the eukaryotic cell. Mitochondria provide the ATP that is required to drive processes in the surrounding cell; in return they get the food and the best environment.

5.7.2.2 Requirements

A Viable Team (VT) shall implement the following requirements (Table 23):

ID	Requirements	Comment
5.7.1	The VT shall have processes to identify other stakeholders that can provide necessary resources	Example: marketing
5.7.2	The VT shall have processes to identify the needs of these organizations	Example: marketing
5.7.3	The VT shall have processes to create mutually beneficial contracts	Example: acquire customers
5.7.4	The VT shall have processes to create value for other organizations	Example: create customer value
5.7.5	The VT shall have processes to get resources from other organizations	Example: expediting, claim management

Table 23: List of requirements related to collaborations

5.7.2.3 Existing Solutions

The following methods might help to implement the Viable Business:

Stakeholder Analysis

Stakeholders are symbionts of the organization. They provide resources and they expect something. A stakeholder analysis describes this relation.[320]

Market Segmentation and Value Propositions

Value propositions express what the organization wants to provide to a stakeholder group (profit, customer value, and good workplace). In most cases the main statement refers to the value proposition for the main stakeholder group, the customers. Since most employees work for customer value creation, the mission can help to create an identity.

In the light of the system dynamic of Viable Organizations we see that everything is clearly connected. In Chapter 1 we will propose describing the give and take with all stakeholder groups in the (extended) vision statement. This allows these vision statements to be located in the functional overview diagram, which describes the system dynamics.

Customer Orientation

Lean management very often refers to customer orientation[321]. The agile manifesto also has customer satisfaction in its first statement.[322] The customer decides what value is. Since support teams only work indirectly for customers, this concept is also used for other teams which are treated as internal customers.

In our concept we explicitly see all stakeholders as customers of the organization. The organization needs to have at least some processes that are aligned to fulfill their needs.

Employees and Organizational Behavior

There is rich literature about the relation between organization and employee.[323] Robbins and Judge describe topics like emotions, motivation, working in teams, leadership and culture in detail. All that is very important! The only reason that it is not in the focus of this work: The focus on employees would distract from the overall system dynamics – the core of this book.

5.7.3 Annotations

Other Aspects That Have Not Been Translated

Parasitism has not been discussed here even if it is important for the organization to protect itself (e.g. with contract/claim management) against parasitic stakeholders.[324]

---

[320] See (ISO 9000, 2015) for stakeholder relations
[321] See (Liker, 2006) (Liker, et al., 2007)
[322] See (Agile Manifesto, 2018)
[323] See (Robbins, et al., 2017)
[324] See (Sadava, et al., 2011 S. 860) for parasitism

5.8 Compartments

Business

Processes are more effective and efficient if all inputs are available in high concentrations. We need the right people, the right information as inputs and the right tools. This is the reason for excellent planning and work preparation.

Complex teams face the challenge that a high number of processes are necessary. However, this leads to a high number of required competences and tools and best practice descriptions. Every team member has to learn many different processes and must repeat the activities with a lower frequency. This leads to a lower efficiency.

The countermeasure for this problem is to build substructures of the team. A substructure is a perfect environment for a group of processes where all inputs and catalysts for a process are available in high concentrations.

A well-designed and staffed workplace in a factory is such an environment. A reserved time for doing some processes could be one. A shop-floor meeting for daily planning is another one. Well-organized meetings provide the best environment for the contained processes. That is their purpose. That explains nearly all meeting rules: increase concentration of production factors for a process.

This leads to the question: Which substructures are most important?

Biology

During billions of years of evolution some bacteria developed more complex behavior, they also had more chemical reactions to manage. Therefore, they also needed to have different enzymes. Unfortunately, these additional enzymes required space. This reduced the concentration for each of the enzymes, thus it slowed down metabolism and slowed down self-creation and growth.

Bacteria also have a second problem: if they produce digestive enzymes and export them to the outside, they have a low concentration outside, because diffusion distributes them in all directions. It is also problematic to produce more of them, because it disturbs the metabolism inside.

Eukaryotic cells found a solution by separating the interior into different compartments, the organelles. Each organelle is an environment for a group of related processes. The cell sorts its molecules and brings them to the right environment. This creation of organelles as substructures was one of the most important steps in evolution. It not only allowed the cell to grow, it also allowed cells to create molecules in high concentration and deliver them to the outside. This ability is the base for multicellular life.[325]

Today these eukaryotic cells are the basis for all multicellular life.[326]

---

[325] See (Alberts, et al., 2015 pp. 641-694) for a description of intracellular compartments
[326] See (Alberts, et al., 2015 p. 30)

Every cell can teach us which mechanisms are necessary to maintain such substructures. We will see that the main organelles of a cell have a clear purpose and design that can be translated to a business world.

Translations

We start with some general mechanisms of substructures and explain how they increase concentration of production factors and speed of processes Table 24().

Then we look to the organelles that can be found in every animal or plant cell. This includes the cytosol, nucleus, nucleolus, mitochondrion and endoplasmic reticulum. Some other organelles like the Golgi apparatus and vesicles and mechanisms like endocytosis or organelles will be discussed in the annotations.

	Translation	Biology	Business
1	Compartments		
2	Compartments for self-creation	Cytosol	Time / room for self-organization
3	Compartments for export	Endoplasmic reticulum	Time / room for value creation
4	Compartments for creation of blueprints	Nucleus	Time / room for learning
5	Compartments for creation of the primary auto-catalyst	Nucleolus	Time / room for learning the ability of self-organization
6	Compartments for endosymbionts that renew energy	Mitochondrion, chloroplasts	Time / room for regeneration of motivation

Table 24: Translation of substructures

There are many possibilities to structure the environment for business processes. The assignment of areas to processes is quite obvious and can be seen in every factory. Thus, we will focus on the assignment of time slots instead. This can already be done by a single person, and it is still needed when the team grows, and roles are introduced.

5.8.1 Compartments

Biology

Eukaryotic cells are 1000 times larger than bacteria. They also have a more complex behavior which demands more enzymes. The high number of different reactions and enzymes decreases the relative concentration and reduces speed. Eukaryotic cells have a solution for this problem: organelles help to increase concentration by providing separated process spaces (Figure 5-58).[327]

Organelles are compartments of a cell that are surrounded by membranes.

Examples for organelles are the nucleus, the endoplasmic reticulum, the mitochondrion, the Golgi apparatus and the transport vesicles. The separation of the organelles makes the cytosol to an additional specialized compartment.

The important point is the process focus. Substructures like organelles have one purpose: They increase concentration of reactants and catalysts. This accelerates reaction speed and thus the growth rate of the cell.[328]

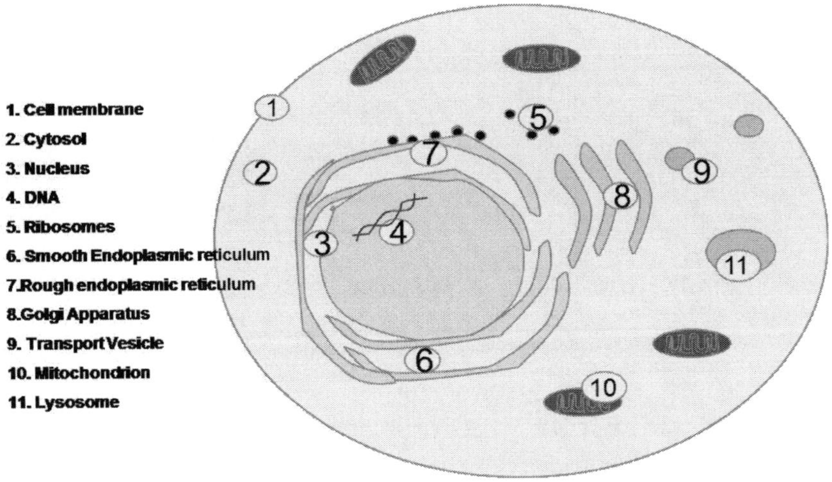

Figure 5-58: Organelles of the eukaryotic cell

There is a second important point. All organelles except the cytosol and the nucleus are treated as "outside". This allows things to be done for the outside in high concentrations and in a very controlled environment. The cytosol and the nucleus are the "inside" where the cell creates itself.[329]

---

[327] See (Alberts, et al., 2015 pp. 23-39) for a description of the eukaryotic cell
[328] See (Alberts, et al., 2015 pp. 641-694)
[329] See (Alberts, et al., 2015 pp. 641-694)

Business

If teams have too many processes, they need to have many competences and tools as well. This decreases concentration. Substructures solve this problem by providing environments with a high concentration of production factors (Figure 5-59).

Organizations can create specialized environments for their processes.

This happens not only in the physical but also in the social world.

In a physical world the team can assign processes to dedicated areas in the office or factory. Everything has its place. Lean methods 5S and zoning help that every object has its best place.[330]

Additionally, the team can create roles and assign specific responsibilities. And it can create time slots where specific processes have to be done regular meetings with a defined agenda.

All these substructures can help to increase the concentration of production factors. Every object has its location. Every process has its location and its time. This helps to accelerate all the business processes.

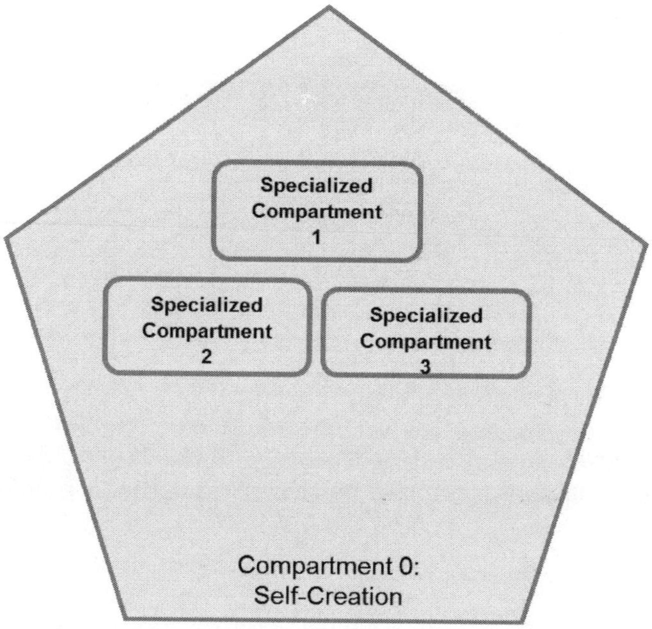

Figure 5-59: Substructures of a team

There are many possibilities to structure the processes. The assignment of areas[331] is quite obvious. On the next pages we will focus on the assignment of time slots. This can already be done by a single person, and it is still needed when the team grows, and roles are introduced.

Let us now go through the environment, starting with the most important one.

---

[330] See (Gorecki, et al., 2013)

[331] Also called zoning. Areas are reserved for processes or for storage

5.8.2 Compartment for Self-Creation

Biology

The main compartment of the cell is the cytosol. It contains all chemical reactions that are needed to create catalysts. [332]The created catalysts dramatically accelerate these reactions. Thus, the cytosol can be understood as the main compartment, where the "self" is located (Figure 5-60).

Cytosol is the compartment where the self-creation of the cell happens.

Ribosomes are located in the cytosol. They create enzymes – the catalysts for all reactions.

In a eukaryotic cell the DNA is located in a different organelle, the nucleus. This allows the cytosol to focus on the creation of catalysts. It starts with the raw material that comes from outside and the genetic information that comes from the nucleus and ends with enzymes.

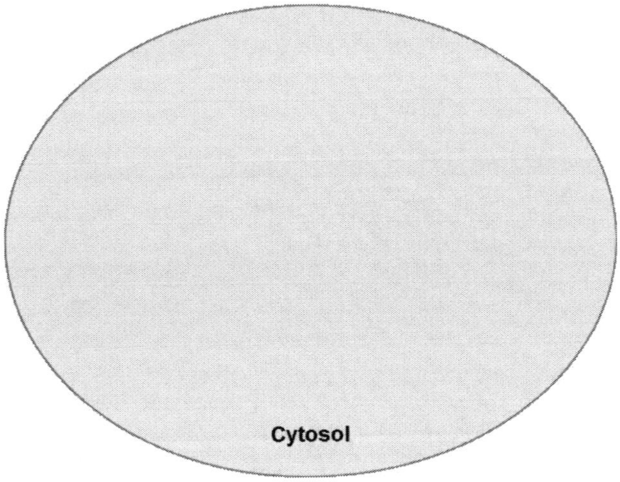

Figure 5-60: Cytosol

The specialized environment of the cytosol increases the efficiency in catalyst production, so the cytosol can easily produce the catalysts for the processes in all other organelles.[333]

---

[332] See (Alberts, et al., 2015 p. 642)
[333] See (Alberts, et al., 2015 p. 642)

Business

The best working conditions are the catalyst for business processes. Everything that is needed to provide the best working conditions is important and needs the highest possible concentration of production factors and mental focus. Thus, it makes sense to explicitly separate planning and work preparation from all other activities (like learning, value creation, etc.) (Figure 5-61).

An organization needs an environment for work preparation.

Self-organization techniques, like "Getting Things Done" by David Allen, focus on providing the best working condition for tasks.[334] Agile sprints and standup meetings have the same purpose for an entire team.[335] The purpose of work preparation is to ensure that all preconditions of a task are available at the beginning of the task. This requires know-how about the specific process.

Time	Mo	Tu	We	Th	Fr
08:00					
09:00	Self-Organization: Planning + Work Preparation				
10:00					
11:00					
12:00					
13:00					
14:00					
15:00					
16:00					

Figure 5-61: Environment for planning and work preparation

The planning and work preparation are the core processes, and the last mile of work preparation is always the self-organization of people. How good are you in self-organization? Just ask yourself at every task whether you had the best working conditions. If not (which is nearly always the case), then you can ask what has to be improved in self-organization. This question leads backwards through teamwork to the design of the entire organization.

If the environment for work preparation is necessary for the self-creation processes: Where is the environment for customer value creation processes?

---

[334] See (Allen, 2013)

[335] See (Schwaber, 2004)

5.8.3 Compartment for Export

Biology

Cells not only create catalysts. They also produce something for the outside. In case of a pancreas cell this might be insulin that is created for other cells. In case of a single-cellular organism this might be digestive enzymes. The creation of these molecules is not done in the cytosol but in the endoplasmic reticulum (Figure 5-62). [336]

The endoplasmic reticulum (ER) is used to produce something for the outside world

This allows something to be created at a high concentration without disturbing the ability to create catalysts. A good example are digestive enzymes that are used for the digestion of other cells. They are needed in a high concentration and would destroy the cell itself if they would be produced in the cytosol. Another example is the production of hormones in high concentration.

The endoplasmic reticulum is not responsible for creating enzymes as catalysts. All catalysts are created by the ribosomes in the cytosol (or more precisely: they are bound outside at the membrane of the ER).[337]

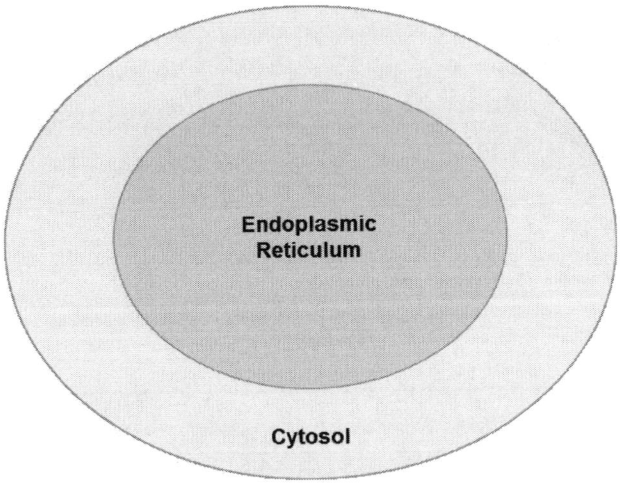

Figure 5-62: Endoplasmic reticulum. This organelle is the environment for all processes that create something for the outside

This separation of ER and cytosol allows the cell to do two things: the ER gets all its necessary catalysts for free and it can focus on its own processes. The cytosol can focus on the creation of catalysts. It makes no big difference if more enzymes are made, because the difference is only described in the genetic information; the required amino-acids are the same.

[336] See (Alberts, et al., 2015 pp. 669-691)
[337] See (Alberts, et al., 2015 p. 642)

Business

Now we have seen that self-organization, training and motivation are important factors that have to be organized. All these factors allow the best working conditions for processes to be provided.

Let us talk about the processes that create something for others. These processes require other inputs than self-organization, so it is better to create a separate environment (Figure 5-63).

The creation of customer value processes has to be separated from self-organization

This allows high value to be created without undermining the ability of self-organization

Customer value creation can expect to get the best working conditions for every task. Catalysts are created during work preparation.

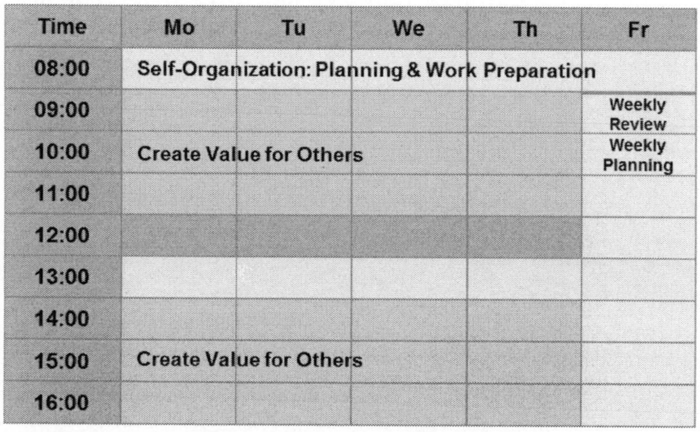

Figure 5-63: Environment for the creation of stakeholder value. Customer value creation has to be separated from work preparation to avoid loss of concentration in self-organization.

This split is the reason behind the 7 mudas and the 5S in Lean. Customer processes always have to be perfectly prepared. This leads to a lot of work preparation tasks. But Work preparation is a standard process that is also very much standardized and highly efficient. It is no problem to create the best working conditions for some customer processes. [338]

The processes to create the best working conditions are always the same: use the work instruction with the best practice to identify all the needed resources, ensure that they are all available at the right time, and use the step-by-step explanations which describe what to do.

The endoplasmic reticulum may be not so famous as another organelle. One of the most prominent organelles in cells is the nucleus that contains the DNA. How can this be translated to business?

[338] See (Gorecki, et al., 2013)for a description of 5S and the 7 mudas.

5.8.4 Compartment for Creating Blueprints

Biology

Animals and plants consist of cells that have a nucleus (Figure 5-64). It is the inheritance of 2 billion years of evolution from the first eukaryotic cell. Obviously, the nucleus is very important.[339]

The nucleus is the organelle with the best environment for storing, copying and post-processing of genetic information.

The nucleus hosts the DNA and all chemical reactions that are needed to transcript and post-process messenger RNA.

All material and energy needed to produce new DNA or RNA comes from outside. In the nucleus the DNA is read, and pre-mRNA is produced, then it is post-processed to its mRNA. Only completely processed mRNA molecule can leave the nucleus and enter the cytosol.[340]

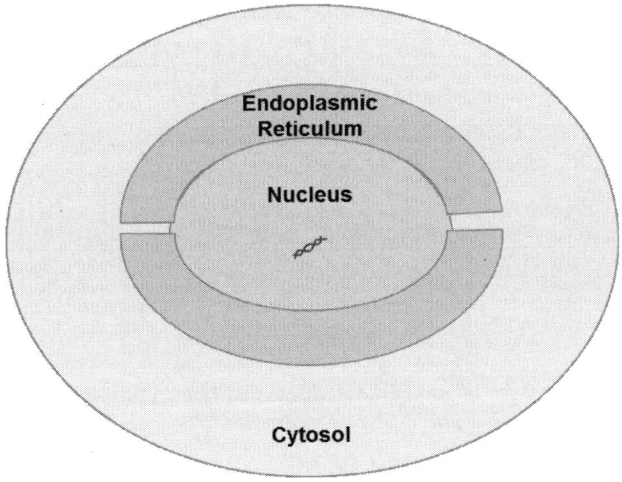

Figure 5-64: Nucleus

Bacteria have a DNA but no nucleus. Due to a non-optimal environment they are not able to post-process the mRNA. They also have no such complex gene regulations like the eukaryotes.[341]

---

[339] See (Alberts, et al., 2015 p. 642)
[340] See (Alberts, et al., 2015 pp. 649-658)
[341] See (Alberts, et al., 2015 p. 315)

Business

What does it mean for a business environment? The DNA of a team is the documented knowledge and transcription is the ability to share it with new team members. All this is critical. If know-how gets lost, then the ability to provide the best working conditions also gets lost.

Thus, it makes sense to provide a protected environment that allows to document knowledge and to transfer it with high concentration: the training environment (Figure 5-65).

Organizations need an environment for creating and sharing know-how.[342]

Examples**:** Training center with trainer, possibilities to apply new knowledge

Knowledge management contains everything from storing know-how to making sure that the right know-how is available for the right person.

It starts with the selection of which know-how is important in the current situation. There must also be a post-processing that adapts the important know-how to a specific situation. This selected know-how can be transmitted from trainer to trainee. At the end the trainer ensures that the trainee is able to apply the knowledge.

Time	Mo	Tu	We	Th	Fr
08:00	Self-Organization: Planning & Work Preparation				
09:00					Weekly Review
10:00	Create Value for Others				Weekly Planning
11:00					
12:00					
13:00					
14:00					
15:00	Create Value for Others				Learn
16:00					

Figure 5-65: Environment for learning

If there is no good training environment, then the training results are poor, and the know-how level is low. This leads to poor work preparation processes and finally to problems with value creation.

Some processes are more critical than others. Especially the work preparation needs excellent competences. This is reflected by an even more specialized environment.

[342] (Training within Industry Program, 2009a) and (Liker, et al., 2008) both discuss the importance of a dedicated training environment where the employee is focused on learning instead of creating customer value

5.8.5 Compartment for Creating the Primary Auto-catalyst

Biology

We have seen that the cytosol is the location where the ribosomes create the enzymes. The nucleus provides the mRNA that contains the blueprint for the enzymes. The ribosomes themselves also come from the nucleus. They consist of two major parts that are created in the nucleolus, a special region within the nucleus (Figure 5-66).[343]

Nucleolus is the location in the nucleus where the two parts of the ribosomes are created.

The ribosomes are some of the most critical parts of a cell. Even cells as different as human cells or archaea in the hydrothermal vent share ribosomes as the factories for proteins. Even after 4 billion years of evolution the ribosomes have only minor changes.[344]

This criticality might explain why the creation of ribosomes happens in the protected area of the nucleolus. This ensures a high quality of the auto-catalyst.

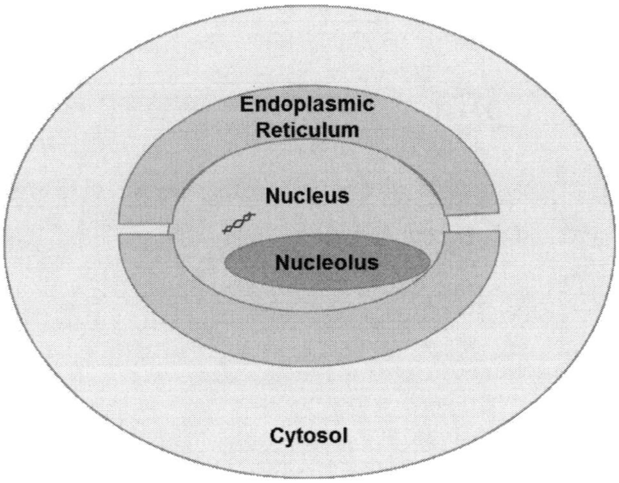

Figure 5-66: Nucleolus. Nucleolus is the location where ribosomes are created

[343] See (Alberts, et al., 2015 pp. 329-331)
[344] See (Alberts, et al., 2015 p. 6)

Business

The auto-catalyst of the business is the best working condition for planning and work preparation. This is more than the task list, the calendar and the file structure. It also requires the competence of creating the best working conditions.

And there must be a location where this competence comes from (Figure 5-67).

The environment to get the competence of self-organization provides the most important competence.

The ability of self-organization is the prerequisite for the usage of all other know-how that can be trained. It allows converting knowledge and skills to the best working conditions.

In large organizations there is a broad distribution in the ability of self-organization, which leads to very different levels of productivity.

A key to success is to strengthen this ability by understanding it as the most basic ability that has to be trained. This training needs an explicit environment. The trainer has to ensure that the trainee has not only understood self-organization or applied it. The trainer must also work as a coach until high self-organization has become a strong habit.

Time	Mo	Tu	We	Th	Fr
08:00	Self-Organization: Planning & Work Preparation				
09:00					Weekly Review
10:00	Create Value for Others				Weekly Planning
11:00					
12:00					
13:00					
14:00					Learn Work Preparation
15:00	Create Value for Others				Learn
16:00					

Figure 5-67: Environment to create competence of self-organization

Since it takes a long time to develop a high capability, it is not the job of a two-day training but a life-long job for a good leader to develop the self-organization capabilities of all team members.

5.8.6 *Compartments for Endosymbionts*

Biology

Many chemical reactions depend on the availability of energy. Without energy a cell cannot survive. In a eukaryotic cell a large part of the energy is created in mitochondria (Figure 5-68).[345]

Mitochondria provide the energy for all the chemical reactions in form of adenosine-triphosphate (ATP).[346]

The main aspect here is that the creation of ATP as an energy carrier requires many chemical reactions, and it makes sense to have a specialized organelle that provides highest concentration for these reactions.

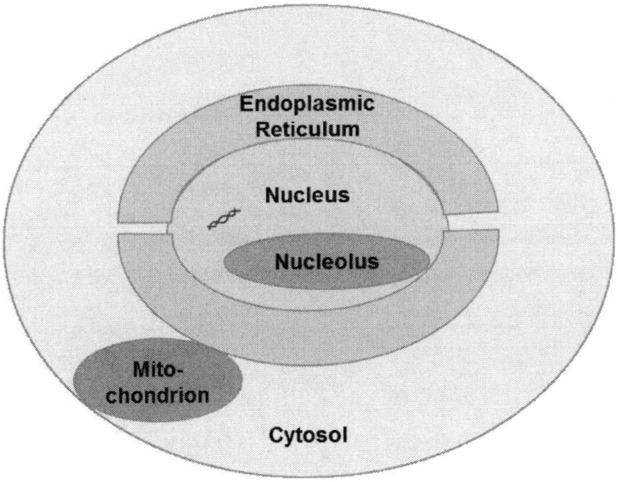

Figure 5-68: Mitochondria as the location where energy is renewed

A mitochondrion is an endosymbiont. Originally it was a bacterium that entered a eukaryotic cell and was specialized to the task of creating energy in the form of ATP. In return it gets all its food by the enclosing eukaryote.[347]

As a former bacterium the mitochondrion has its own DNA and its own ribosomes.[348] So, it is a good example that the DNA of the cell does not know everything. Sometimes the knowledge is decentralized.

---

345 See (Alberts, et al., 2015 pp. 753-761)
346 See (Alberts, et al., 2015 pp. 774-777)
347 See (Alberts, et al., 2015 p. 25)
348 See (Alberts, et al., 2015 p. 25)

Business

Every business process requires energy. Every notebook and every mobile phone require energy. People require energy too. Some workers in factories need muscle power, but in most cases, it is more about mental energy, about motivation. This is very often a critical factor.

Motivation is the energy that drives us forward. As in a cell there is the question where all the motivation comes from (Figure 5-69).

Organizations need a special environment for the processes that renew motivation.

Motivation can be seen as the mental energy that drives people to do processes

High motivation is also the result of processes – maybe of processes that are not fully understood. In an organization many activities are to create motivation. This includes having a shared vision to provide sense, the creation of team spirit or a reward system. In an organization it is important to understand the individual motivations of people and to use them. At the end all these processes contribute to the motivation as the driver for processes.

The creation of motivation is a very complex thing. Motivation and demotivation happen during the day. However, it makes sense to spend additional room for the replenishment of energy and motivation. Team events very often have this purpose, beside the purpose to transmit information.

Time	Mo	Tu	We	Th	Fr
08:00	Daily Planning & Work Preparation				
09:00					Weekly Review
10:00	Create Value for Others				Weekly Planning
11:00					Renew Energy
12:00					
13:00	Work Preparation for Afternoon				
14:00					Learn Work Preparation
15:00	Create Value for Others				Learn
16:00					

Figure 5-69: Environment for the creation of motivation

We should also be aware of a subplot in biology: the cell does not know everything about the mitochondrion, but it provides the best environment to get energy in return. Maybe it is the same in business. The roots of motivation are hidden, but it is clear that good conditions are needed to (re-)create this motivation.

5.8.7 Summary

5.8.7.1 Insights

Biology teaches us that multicellular organisms are based on eukaryotic cells, which have organelles. The organelles provide dedicated environment for a set of processes. Organelles allow to sort enzymes to the right location and to increase concentration and thus speed.

Substructures help to increase the concentration of resources. The higher concentration will result in a higher efficiency and effectiveness. Substructures require resources to be stable. Thus, it is always a tradeoff that has to be chosen carefully.

Substructures can be created in the physical world (locations and defined time periods), in the psychological world (skills of persons) and in the social world (roles and responsibilities). An ideal substructure considers all dimensions.

A good example for a substructure is a jour-fix meeting. The purpose of the meeting is to provide the best environment for some processes that have to be done regularly by a group of persons. This can be the weekly planning process, for example.

Now it makes sense to make the meeting at the same defined location (e.g. Obeya) and to ensure that the participants have the right motivation and skills. It is also important that the participants are empowered to decide about planning. If people are not empowered, or have no skills or motivation, or come too late, or sit in the wrong room, then the processes in the meeting have no best working conditions.

The purpose of all meeting rules is the creation of high concentration of success factors. They describe behavior that ensures that all needed factors are available. Everybody has to be punctual and prepared. Agenda is clear. Team members are empowered to decide. They also describe behavior that is not accepted, because it reduces concentration. Mobile phones have to be turned off. Non-contributing participants have to leave. New agenda points, often a by-product of a discussion, are not allowed. Meeting agendas describe the passive protection, import of needed factors and export of unneeded factors, for an important substructure: the meeting.[349]

This example shows that a missing dimension of the substructure results in a missing production factor for the process and makes planning ineffective and inefficient.

5.8.7.2 Requirements

A Viable Team (VT) shall implement the following requirements (Table 25):

ID	Requirements	Comment
5.8.1	The VT shall be able to create more specific environments for processes	These environments need to have the same mechanisms as the outer boundery: Passive protection, import of everything that is needed, export of everything unneeded, and storage.

[349] The mechanisms of concentration management are already described in section 5.4.

5.8.2	The VT shall have a specific environment for work preparation	
5.8.3	The VT shall have a specific environment for stakeholder related processes	
5.8.4	The VT shall have a specific environment for learning	Know-how is critical input for work preparation and has to be created in high quality
5.8.5	The VT shall have a specific environment for learning self-organization	
5.8.6	The VT shall have a specific environment to provide energy	Motivation of team members is not for free. It needs processes (team events, target setting...) to renew this energy.

Table 25: List or requirements related to substructures

5.8.7.3 Existing Solutions

There are many possibilities to structure a team according to the dimensions

1. Time
2. Location
3. Roles and Responsibilities

Here are only a few examples:

Lean/Agile: Visualization on Kanban boards, Obeya

The Obeya is a dedicated location to visualize information and to conduct meetings. The Obeya has several boards that clearly structure the information. Obeya is a location with highly structured information visualization[350].

Agile: Sprint meeting, dedicated roles

The sprint meeting provides the best environment for review, retrospective and planning processes. It also has defined roles like the product owner and the SCRUM master which have clearly defined responsibilities.

The ranked backlog that is used in Agile also creates substructures. It separates the next task from all the others. [351]

Getting things done: Plan

This method not only has a defined planning environment, but in addition all next steps are assigned to an environment (“At home”, “At work”). This association creates perfect environments for the activities.[352]

---

[350] See (Lean Enterprise Institute, 2018)
[351] See (Schwaber, 2004)
[352] See (Allen, 2013)

5.8.8 *Annotations*

Other Aspects That Have Not Been Translated

There are some more organelles that have not been discussed so far:[353]

1. Golgi apparatus
2. Vesicle
3. Endosomes
4. Lysosomes

They are needed to fully understand the main process routes of the endomembrane system:

1. Exocytosis: Export something from inside to the outside
2. Endocytosis: Import something from outside to the inside.
3. Transcytosis: Import something on one side of the cell and export it on the other side.

Here is a short description of these important mechanisms. In later editions of this book they can be described in more detail.

Exocytosis, Golgi Apparatus and Vesicle

We have already learned that the endoplasmic reticulum is used to produce something for the outside. Very often this creation has to be done in defined steps. The Golgi apparatus helps to perform these steps in the right sequence. It consists of a stack of subunits, called the cisternae. Each cisterna is responsible for some chemical reactions. The material flows through the cisternae and can be modified in a defined sequence, before it can be stored in vesicles and transported to the outside. The export at the cell membrane can be triggered by a signal. This allows the cell to create a high amount of material that waits for the right moment to be exported. [354]

The translation to business is simple: sometimes a sequence of workplaces is needed where material goes from one workplace to the next. Material can only pass to the next process if it has high quality. At the end the product can be stored in a staging area until it gets the signal to be sent to the outside.

Endocytosis, Endosomes and Lysosomes

In order to grow, cells have to import material to the cytosol. This process has to be controlled because everything from outside can also be a threat for the cell.

Endocytosis is the process of such a controlled import. Everything that comes from outside enters the endosomes instead of the cytosol. Endosomes are organelles that have a lower pH-value and are able to destruct some of the complex structures that have been imported. Everything that is more complex to destruct enters the lysosome which has an even lower pH-value and some hydrolase enzymes that are able to destruct complex macromolecules and make their constituents available for the cell.

---

[353] See (Alberts, et al., 2015 p. 642) for a description of organelles
[354] See (Alberts, et al., 2015 pp. 741-750)

Lysosomes are also used to destroy molecules that are no longer used and to recycle their constituents. [355]

This ability is also needed in business. Destruction means to get high value for the team and to destroy complex structures that are not needed or a threat.

If the team has strong shared beliefs and a good culture, it is always difficult to include new employees. So, it is important to learn which new beliefs are good for the team and which are not. Sometimes the team leader has to destroy wrong beliefs like "If you have a problem – never tell it to your boss." So, the purpose is to get excellent new team members – but without the wrong "shared belief." This work on wrong beliefs cannot be done during normal work. It also needs a special environment.

Transcytosis

Transcytosis imports something on one side of the cell and exports it on the other side. This is only used by cells that are part of a multicellular organism. A typical example is the small intestine cell that imports carbohydrates, amino acids and fatty acids from the lumen and exports them on the other side to the cardiovascular system. [356]

In business this is typical for all activities that have to do with external stakeholders. A customer project team gets resources from the customer (payment), but this payment has to be forwarded to the resource management teams that manage finances.

[355] See (Alberts, et al., 2015 pp. 730-737)
[356] See (Alberts, et al., 2015 p. 737)

6 Multilevel Viable Systems

Business

Small one-team organizations need to implement many mechanisms to have sustained growth and to be able to adapt to a changing environment.

Large organizations have the same goals and challenges as small ones. The main difference is that all necessary processes are distributed over many teams. All these teams have to work together internally and with their neighbor teams to reach the overall goal and to react to challenges.

Large organizations often have problems with this collaboration. Teams are focused on targets that are not clearly aligned. The communication between teams is slow. Teams do not treat their neighbor teams as internal customers. All these problems lead to an inefficiency of the entire organization, which in return has problems to provide the best working conditions for each of the teams. It is a vicious circle: problems of the teams lead to problems of the entire organization, which create additional problems on the team level.

How is it possible that large plants or animals manage it that hundred trillion of cells work perfectly together, when an organization already has problems with some hundreds of teams?

Biology

Evolution has solved all these problems in the last billions of years. It created various architectures of multicellular organisms, where cells work perfectly together.

Multicellular life can have many possible forms. A tree has a completely different structure from an animal, but all multicellular organisms have some mechanisms that are shared, because they all have the goal to grow and they face the same basic challenges. What are these basic challenges? Every multicellular organism has to ensure that the contained cells get all the resources they need. They also have to synchronize the adaptation to the outside world. Cells cannot react isolated from each other. They must share their information and coordinate their actions.

The challenges of the multicellular organism are not new. Even the first cell wanted to create, protect, adapt and replicate itself. Unfortunately, the split to several daughter cells introduced new problems which require new solutions. Let us start with the first cell, follow the replication and study the typical problems of ontogenesis. Then we will come step-by-step to the needed organ systems and the collaboration of cells in tissues.

This knowledge can then be used to design a Large Viable Organization that consists of Viable Teams.

C. Dachs, *Viable Project Business*, Contributions to Management Science, https://doi.org/10.1007/978-3-030-62904-5_6

6.1 Collaboration & Specialization

Business

Collaboration already starts inside a team. Every team member has different strengths and weaknesses. It is beneficial if everybody can focus on the strengths and let the weaknesses be compensated by others. This is the reason for different roles in a team: everybody can work with higher productivity.

The team is successful and grows until it reaches a team size where the communication between team members gets more and more difficult. There are some team members working closer together than others. This can be understood as a sub-team which has strong communication inside and is connected to other sub-teams. If the team splits into two teams, the split is mainly based on the sub-teams that existed before. This leads to descending teams which are also specialized. The direction of specialization can be in various dimensions (functional specialization, specialization on products or markets).

This specialization of a team allows focusing only on selected processes. The high focus results in a high efficiency of the team. It also has a disadvantage: by specializing the teams can lose the contact to some stakeholders. Then the teams do not get all needed resources from their environment anymore. However, resources are part of their best working condition. If they are missing, the processes are not catalyzed anymore and slow down, the team becomes less productive.

The structure of the large organization has to solve this problem. Every team gets all needed resources – from other teams. Thus, the purpose of all teams is to serve either a stakeholder or other teams. This is also true for teams which create customer value. They do that because the customer pays for it, and all other teams need that money to pay employees or suppliers. At the end every team contributes to the success of other teams.

Let us think about the best working conditions in reality. Does every team in a large organization have all resources it needs? In many cases: No! This is a big problem for many large organizations. The teams very often have a lack of resources.

Biological organisms obviously have no problems with that. Let us look how this is solved in detail.

Biology

Cells can continuously grow until they have reached a maximum size and have to split themselves. If the cells have special membrane proteins, they build cell junctions and connect to each other. Every cell adapts to its environment. Since the siblings are directly connected, they are permanent neighbors.[357]

We have seen that cells can live in symbiosis with other cells. They only have to exchange some of their resources. If one cell is a little bit better at producing a certain molecule and gives it to the other cell, then the other cell does not have to produce it and can focus more on other molecules and share them. This mechanism leads to a positive feedback loop. It amplifies small differences in

---

[357] See (Alberts, et al., 2015 pp. 1145-1216) for the development of multicellular organisms.

the metabolisms and let the cells specialize. The ongoing specialization increases the concentration in every cell and increases speed.[358]

All the specialization is only possible if resources can be exchanged. A cell can focus only on one resource if all the other resources are freely available.

Some of the cells in a multicellular organism do not contribute directly to this exchange of resources. In the human body, only the respiration system and digestive system contribute to the import of resources. All other organ systems are only consumers of resources. They contribute to the organism by realizing other functions like protection or adaptation. [359]

These other specialized organs can only evolve if the provision of resources is organized. The other organs must bring additional benefits that are higher than the disadvantage of being an additional resource consumer.

The understanding of the purpose of all the organs in multi-cellular organisms can help business to design well-balanced sub-organizations. However, before we can focus on the organs, we need to understand the basics of collaboration between cells.

Translations

Let us focus on the basics of specialization and the mechanisms that ensure that every Simple Viable System has all resources it needs (Table 26).

	Translation	Biology	Business
1	Junctions	Cell junctions keep cells permanently together	Connections between teams have to be defined
2	Specialization	Specialization to cell types	Specialization to team types
3	Bilateral Exchange of Resources	Exchange of resources from cell to cell	Direct exchange of resources between teams
4	Multilateral Exchange of Resources	Central provision of resources for all, extracellular matrix	Central resource management

Table 26: Translation of basic mechanisms in multicellular organisms

The starting point is the question: why do cells stay together after replication? And more philosophically: how many organisms do we have after the split? Two or three? What is the relation between the whole and its parts?

[358] See (Alberts, et al., 2015 p. 1153)
[359] See (Alberts, et al., 2015 p. 1147)

6.1.1 Junctions

Biology

Multicellular organisms evolved because some membrane proteins kept the cells together after replication. This had the evolutionary advantage that a multicellular organism is bigger and thus a worse target for predators than two smaller single cells.

Cell junctions can keep cells together after replication.[360]

Multicellular organisms start as a single cell that splits and stays connected by cell-cell-junctions or the extracellular matrix (Figure 6-1).[361] Repetition of replication leads to multicellular organisms. They have no differentiated tissues yet. Every cell gets food from the environment for itself. Thus, the form of the multicellular organism must keep all cells at the boundary to the environment.

Very simple multicellular organisms are connected in a line, a plane or as a hollow sphere. The connection is beneficial if the additional protection in the group is higher than the new competition about food between the cells.

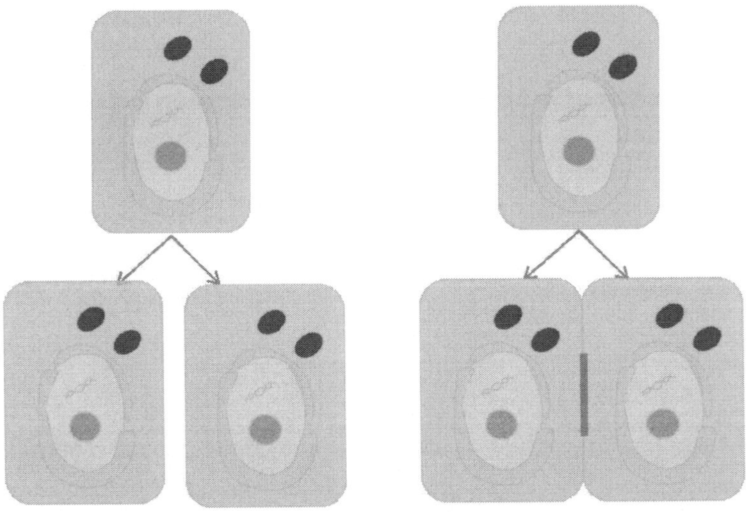

Figure 6-1: Cell junctions connecting siblings. On the left-hand side, cells are not connected. On the right-hand side, they are connected with a cell-junction and form a multicellular organism.

Multicellularity evolved several times independently.[362] It always happens if a cell creates membrane proteins which keep the siblings connected after replication. Animals evolved from single-cellular choanocytes. These single-cellular organisms connected together to the first sponges, the oldest forms of animals.[363] Plants have developed multi-cellularity independently.

---

[360] See (Alberts, et al., 2015 p. 1035)
[361] See (Alberts, et al., 2015 p. 1035)
[362] See (Alberts, et al., 2015 p. 880)
[363] See (Sadava, et al., 2011 S. 764)

Business

If a small organization continuously grows, it exceeds the size of a team; therefore, it has to replicate and create two teams. They are not completely independent. Instead they are both parts of the organization as a legal entity.

Defined interfaces keep Viable Teams connected after replication.

A small organization grows and creates two teams. Every team has a subset of the responsibilities and a subset of the resources and knowledge of the organization.

If the replication happens several times, the organization has 2, 4, 8, 16 or more teams which stay connected in the same organization. They are connected because they are part of the same legal entity, and there are no responsibilities that got lost during the splits (Figure 6-2).

One example for the model can be a consulting company, which grows and hires new consultants and keeps everything connected. The advantage can be the shared branding.
Another example can be software startups that want to be completely decentralized.

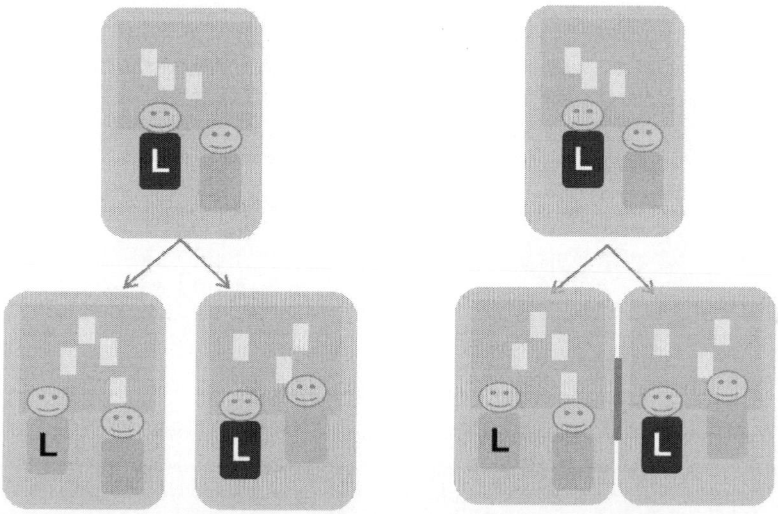

Figure 6-2: Connection of teams. On the left-hand side the teams are disconnected. On the right-hand side they are connected. This leads to a Large Viable Organization, that consists of two Viable Teams.

The biggest benefit for staying connected is to allow specialization. Every team relies on the fact that neighbor teams are permanently in its environment. What are the mechanisms of specialization?

6.1.2 Specialization

Biology

Connected cells are better protected from predators. This is already a benefit. Additionally, the junctions are the prerequisite for an additional opportunity. If cells are connected, they are permanent parts of their environment. It allows collaboration between them that enables them to specialize (Figure 6-3).

Cells can specialize after replication[364]

The reasons for specialization can be

1. The components of the parent cell were not distributed equally during replication.
   There is a small difference in the speed of processes[365]
2. The environment for the daughter cells is different
   There is a difference in getting raw material from outside
3. Daughter cells get signals from neighbors that trigger specialization[366]

These internal and external differences might be small, but they open an opportunity. Cells can focus more and more on what they are good at, if they can exchange resources with their neighbor cells. This is a mutually beneficial symbiosis.

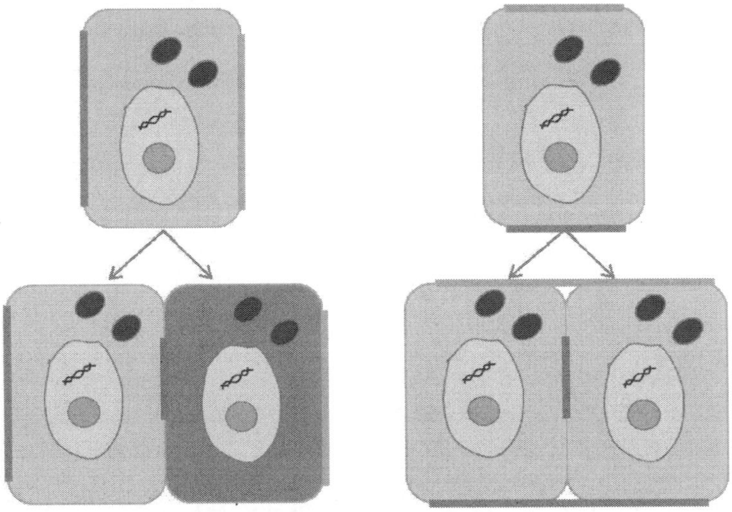

Figure 6-3: Split direction. The red and blue line indicate a difference in the environment. On the left-hand side both cells have a different environment after the split. On the right-hand side, they have the same environment. The colors red and blue represents different environments of the cells.

The specialization requires the ability to exchange resources. Before we discuss the different forms of resource exchange, let us have a look at the translation to business.

[364] See (Alberts, et al., 2015 p. 1145)
[365] See (Alberts, et al., 2015 p. 1153)
[366] See (Alberts, et al., 2015 p. 1150) for inductive signaling

Business

In most large organizations many specialized teams work together. Where does this specialization come from? Let us go back to the first team. When it has its first replication and creates two teams, these teams can specialize (Figure 6-4).

Viable Teams can specialize after replication

The reasons for specialization can be

1. The resources such as know-how were not equally balanced between the teams after replication. Each of the teams has its specific strengths and weaknesses
2. Not every team has connection to all stakeholders.

A team has several collaborations with stakeholders. If a descendent team does not have all of them, it can specialize to a subset of resources and must get the missing resources from their siblings. The specialization requires a stable exchange of resources between the teams.

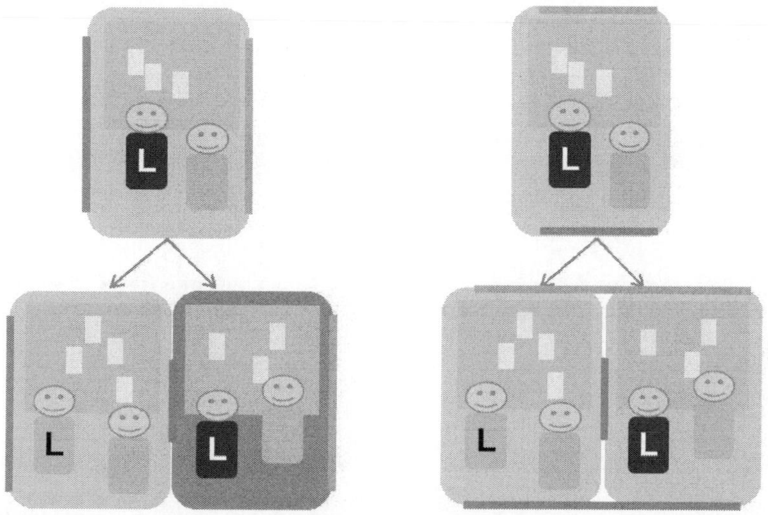

Figure 6-4: Split direction of teams. The split direction can lead to different environment for Viable Teams. The colors red and blue represents different environments of the teams. These environments can be the relationships to specific stakeholder groups like investors or customers.

The specialization leads to defined team types. The first team started as a generalist team that was fully responsible for the entire organization. After a split to different team types, every descendant team depends on each other.

The splits are also the reason for target cascading. The descending teams have different goals than the original first team. The goals of the Large Viable Organization have to be broken down to the teams.

Let us now focus on the exchange of resources between the teams.

6.1.3 Bilateral Exchange of Resources

Biology

Cells that are connected are mutually a part of their mutual environment. Thus, they adapt to this environment and form a kind of symbiosis (Figure 6-5).

Cells mutually exchange resources to allow specialization

The specialization is only possible if cells are able to share their resources. If cell A is better at creating a resource and then shares it, cell B can adapt to the situation and reduces its own processes to create that resource. As a result, cell B has a higher focus on other resources, which can be shared with cell A. Even small imbalances between the metabolisms are amplified by mutual adaptation to the imbalance.[367]

Higher specialization allows sharing resources with others. Sharing of resources allows a higher specialization. This is a reinforcing loop that amplifies even small differences between cells.

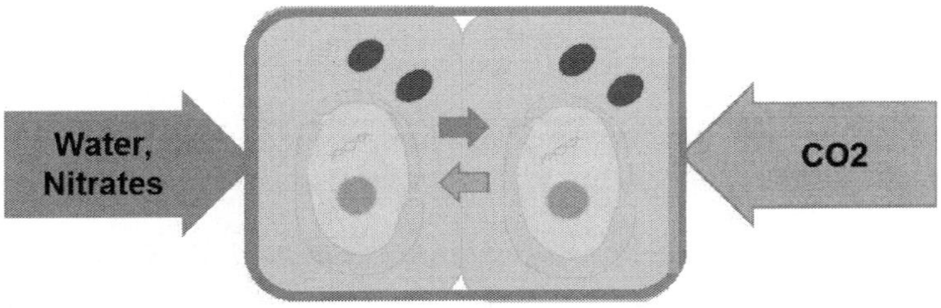

Figure 6-5: Specialized cells exchanging resources. One cell focuses on import of nitrates; the other cell creates carbohydrates. Both cells need both factors, thus there is an exchange of resources.

For the specialization only one thing is important: how reliable is the supply of resources. It makes no difference whether the resources come from the same or a different species. Only if the needed resources are freely available for a longer time, can the cell adapt to the situation, switch off the enzymes or regulate the gene expression.

Specialization is also possible if cells are not physically connected. The physical connection has the advantage that it allows a better exchange. Nothing can get lost during exchange. There are especially no losses caused by diffusion or by parasites. Close collaboration is obviously an evolutionary advantage.

Symbiosis with cells of other species works the same way. The exchange of resources between a cell and its endosymbionts, like the mitochondrion, is very safe, because nothing can get lost.

---

[367] See (Alberts, et al., 2015 p. 1173) for the principles of asymmetric cell dicision

Business

Teams that split can specialize to different organizational functions. An exchange of resources allows every team to have all resources it needs. From a perspective of a single team, it forms very long-term and stable stakeholder relations with the neighbor teams (Figure 6-6).

Teams need to mutually exchange resources to allow high specialization

Even if people in a startup have the same education and knowledge, they have different strengths and weaknesses. It is normal that everybody focuses on the task he is good at. Other people have other benefits and can focus on their strengths. The mechanism of specialization continues even if a team is split. In such a case it becomes more explicit, that both teams depend on work that is done in the neighboring team. They have to exchange resources.

Figure 6-6: Specialized team exchanging resources. Two teams specialize on procurement and HR topics

This specialization is similar to the collaboration with stakeholders. Large organizations can have very strong relationships, because the neighboring teams are permanently in the environment. This is the difference to a market where collaborations are weaker.

The bilateral exchange of resources requires having a directly connected partner for every resource. How can this work in a larger setup?

6.1.4 Multilateral Exchange of Resources

Biology

The direct exchange of resources between cells allows specialization which increases concentration and finally the growth rate. It also has a limitation: it does not allow a high number of specialized cell types.

Advanced multicellular organisms have mechanisms to share many resources

Cells enclose an interstitial space and make their resources available there (Figure 6-7).

The interstitial space allows many specialized cells to contribute. Every cell gets all resources from the interstitial space in high concentration and can focus on only one factor. This allows a high specialization and high productivity of every cell. Some of the cells do not contribute to the provision of food. Instead they contribute to other aspects like protection or adaptation.

Animals, for example, consists of three germ layers, the ectoderm, endoderm and the mesoderm. The mesoderm cells are inside the organism and have no contact to the environment. It is the job of the endoderm cells to provide all resources to them. This transfer of resources is not done by direct cell-cell connections. Instead, all endoderm cells have to give something to all mesoderm cells. In vertebrates the endoderm develops to the gut tube, and the mesoderm develops muscles and the cardiovascular system which help to distribute all the resources.[368]

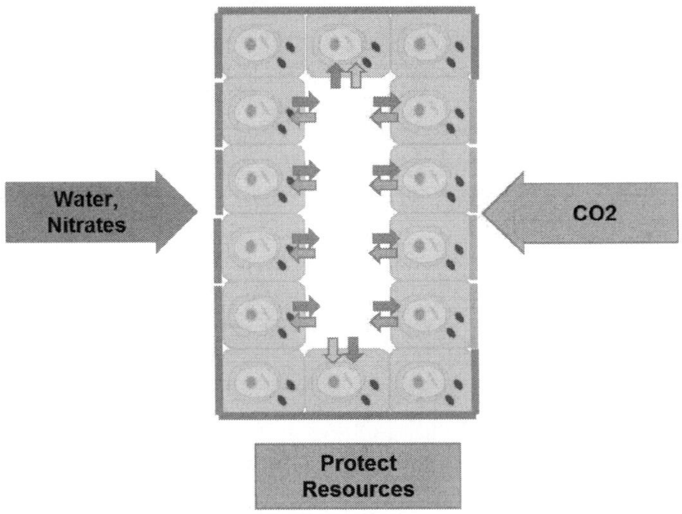

Figure 6-7: Interstitial space. The interstitial space allows sharing multiple resources and further specialization of cell types.

At the end this specialization leads to various organ systems that focus on the realization of functions which already existed on a cell level.

[368] See (Alberts, et al., 2015 p. 1147)

Business

Even a small organization has many stakeholder groups like investors, employees, suppliers and customers. If the organization grows, then the stakeholder relations are in the responsibility of specialized teams.

Large Viable Organizations have to manage that every team gets all needed resources.[369]

Every team relies on the resources that come from other teams (Figure 6-8).

Resources contain everything that is needed by the teams to do their work. Every team can specialize to one task because it has the best working condition that is provided by the entire organization.

Some of the teams are not responsible for importing resources. Instead they contribute to protection or adaptation. One example for teams that protect something is the security service with guards or cybersecurity. Examples of teams that collect information about organization and markets, make decisions and implement them are strategy teams or communication teams. All these teams do not create resources on their own, thus they are resource consumers.

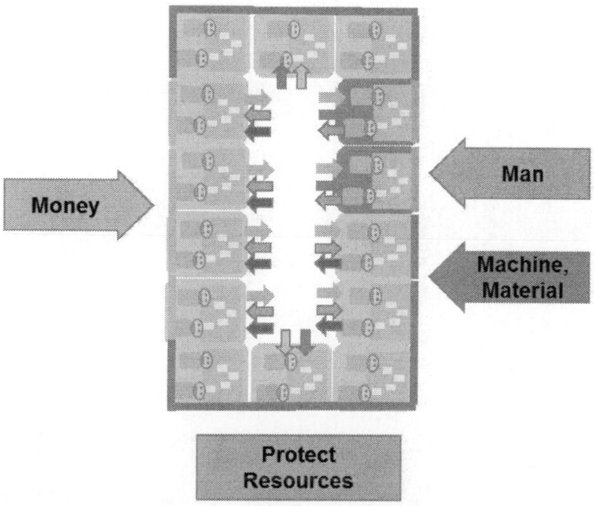

Figure 6-8: Central resource management. Central resource management allows sharing multiple resources and further specialization of team types

This specialization leads to specialized sub-organizations. Every team depends on the ability of the entire organization to import and distribute resources.

[369] See (Krajewski, et al., 2016 S. 457-502) for resource planning

6.1.5 Summary

6.1.5.1 Insights

Biology shows us, that multicellular organisms are based on the splitting of a single stem cell. The descendant cells stay connected by cell junctions and start to specialize to different cell types, which focus on the production of a smaller set of resources or functions (e.g. protection).

The translations showed the main mechanism. Teams have to split if they become too large. The split can lead to teams that are quite independent of each other. Independency has some advantages: The teams can act autonomously. They do not need additional coordination.

It also misses an opportunity: teams that are connected can also exchange resources and start to specialize. The higher specialization leads to higher productivity and faster growth. The specialization is only possible if resources are shared between teams with high reliability.

The specialization of teams leads to sub-organizations which are responsible for a sub-set of functions. On the lowest level the functions are realized by a group of teams or a single team. The group of teams is more than its parts. It needs clear collaborations between the involved teams to realize the function.

6.1.5.2 Requirements

A Large Viable Organization (LVO) shall implement the following requirements (Table 27):

ID	Requirements	Comment
6.1.1	A LVO shall start its lifecycle as a single Viable Team, which splits itself to connected Viable Teams.	This is the **ontogenesis** of LVOs.
6.1.2	A LVO shall ensure the best working conditions for each contained VT even after splitting teams	This describes the **homeostasis** of LVOs
6.1.3	A LVO shall allow a high specialization of contained Teams	
6.1.4	A LVO shall be described by an organization chart that clearly maps functions to teams.	This is the anatomy of the LVO that realizes the physiology.
6.1.5	A LVO shall be based on team types	

Table 27: List of requirements related to stable collaborations

6.1.5.3 Existing Solutions

The following methods can help to implement the requirements

Self-Organizing Teams

Laloux describes the healthcare organization Buutzorg. It is based on self-organizing teams. Buutzorg has around 9000 employees that work in teams of 10-12 members. The central governance only needs 28 employees [370]

This example shows that even large organizations can consist of many independent teams. Every team is self-organizing. In the case of Buutzorg there are no complex dependencies between the teams necessary, because every team has its own stakeholders. Thus, it is an excellent example for self-organization, but not for complex collaborations that are needed in project business.

Holacracy

Holacracy designs organizations with circles and roles. The top-level circle is called anchor circle. At the beginning it contains all roles. If the organization grows it can create sub-circles[371]. This mechanism is similar to our approach. The anchor circle that represents the legal entity correlates to the multicellular organism as a whole. With our approach we can discuss the collaborations in more detail. We understand the LVO as a symbiosis of many Viable Teams which depend on a reliable supply of resources. Additionally, we have team types that realize functions.

Team Types in Project Business

Team types are the templates to construct groups of teams with defined collaborations. They allow designing organizations with well-understood building blocks.

PMI and IPMA already use four pre-defined team types: the project, the program, the portfolio and the P3Office. We can use them as building blocks when we define our Viable Project Business.

---

[370] See (Laloux, 2017)
[371] See (Robertson, 2014)

6.2 Subsystems

Business

In the previous section we discussed how teams can specialize and contribute to the overall organization. At the beginning everything is about the exchange of resources. The large organization has to ensure that every team has the best working conditions. This allows a further specialization of teams, which then can better contribute to the organization as a whole.

This specialization leads to a high dependency between the teams. The large organization has to ensure that every team has sufficient resources. This requires complex coordination of the teams to ensure that the mix of resources matches the demand.

How can this be realized?

Biology

Before we discuss how organ systems and organs contribute to the organism, we should remember the four main functions that are needed to realize sustained growth:

1. Create self
2. Protect self
3. Adapt self
4. Replicate self

How are these functions realized?

For undifferentiated organisms it is sufficient if every cell takes care of itself. The four main functions of the cell lead to growth and replication of cells. Thus, the multicellular organism adds more and more cells to itself. All of the cells protect themselves. Adaptation to the environment has to be done by every cell individually.

This strategy has its limits if the cells are specialized.

Cells depend on a highly reliable supply of resources. Some organs are responsible for bringing the food to every cell and removing the waste. Such a system allows feeding other organs that do not focus on resources.

Since the cells are all connected, it is possible to reduce the threats for all cells by a centralized protection system. Threats come from the outside. If the multicellular organism protects its outer boundary, then all cells inside are protected. This allows every cell to reduce its self-protection and to focus even more on the provision of resources.

With the increased specialization the work of cells has to be synchronized. Cells that are specialized in the production of one resource, do not know that the entire organism has a lack of a different one. A central adaptation system can solve that problem by sending signals to the cells that indicate how they can contribute best to the overall organism. Hormones are signal molecules that are created inside a cell and exported to transmit information to other cells.

The specialization of cells has the side effect that most cells of an organism cannot survive if they are alone. Multicellular organisms have specialized replication systems that produce unspecialized stem cells and support them with food and protection until they are able to survive on their own.

Translations

In the following translations we show that the main functions of cells can be found again at a higher level: as the organ systems of multicellular organisms (Table 28). This is due to the specialization of cells according to the functions that are already implemented on the cell level.

For a better understanding we use the human body as an example of the different organ systems, even if it is clear that a tree has completely different organ systems that fulfill the same purpose. On the business side there are no standard sub-organizations that can be used, but it is possible to give some typical examples.

	Translation	Biology	Business
1	High-level structures	Organ systems	Suborganizations
2	Low-level structures	Tissues	Groups of teams
3	Examples for specialized subsystems	Examples for human organs	Examples for sub-organizations and their functions

Table 28:Translation of subsystems

6.2.1 High-Level Structures

Biology

Large multicellular organisms are not only a collection of cells. They have structures between the entire multicellular organism and the contained cells (Figure 6-9).

The anatomy of multicellular organisms can be described on various levels.

The highest level is the entire organism. This organism needs to implement several functions. These functions are realized by organ systems that consist of organs.

Each of the organs has a specific function to implement. The main work is done by the functional tissue of the organ that consists of cells of a specific cell type. Every organ has some other tissues that stabilize the form of the organ or connect the organ to others.[372]

Tissues consist of cells that work close together. Their cell types are specialized to realize a particular function, and their connections and communications influence the form and function of the tissue.[373]

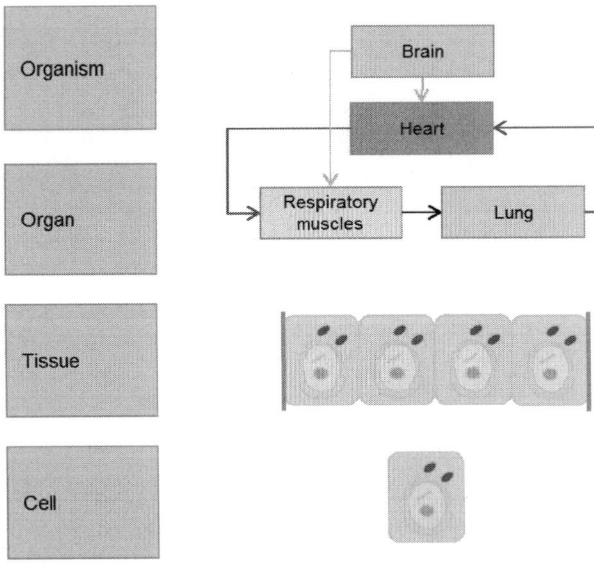

Figure 6-9: Structural levels in organisms

Even if multicellular organisms have trillion of cells, they only have a couple of hundred of cell types. Humans have around 200 cell types.[374] Plants have even fewer cell types than animals. The cell types allow understanding form and function at the tissue level.

[372] See (Sadava, et al., 2011 S. 1104-1106)
[373] See (Sadava, et al., 2011 S. 1104-1106)
[374] See (Alberts, et al., 2015 p. 1217)

Business

Large organizations are composed of teams (Figure 6-10). As in the multicellular organism, it is difficult to understand them by considering only two levels.

Large organizations need to be understood on various levels.

On the highest level the entire organization wants to have sustained growth. This requires some functions to be implemented by specialized sub-organizations. Every sub-organization is focused only on a subset of functions. As in biology this mapping between functions and sub-organizations is not 1:1, instead there might be an m:n relation.

Sub-organizations can break the job down to functions, which can be realized by a single team or a group of teams. The group of teams is designed to jointly realize one or more functions.

We can use this model to compose large organizations based on Viable Teams. The understanding of the organizational functions allows specialized teams to be described.

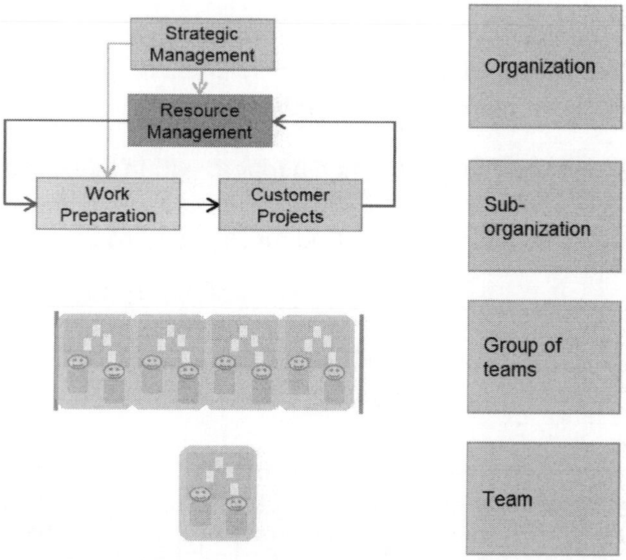

Figure 6-10: Structural levels in organizations

As in biology a group of teams jointly realizes a particular function. If a single function has to be realized by a group of teams, it needs to be defined how all the teams have to work together. The necessary collaborations have to do with the function that has to be realized.

This collaboration between neighboring teams is one of the typical problems in large organizations. Let us look at this level in detail.

6.2.2 *Low-Level Structures*

Biology

All cells of a multicellular organism are descendants of the first cell. The cells can specialize, but they do not need to. In such a case there are many cells with a common ancestor doing exactly the same. This leads to the concept of cell types and tissues (Figure 6-11).

Cells of a tissue have the same cell type and contribute to the same function.

Let us go back to the first cell that was the common ancestor of the tissue. The entire function was fulfilled by only one cell. When the entire organism grows, this cell is no longer sufficient to fulfill the function alone. Thus, it also has to grow and to replicate itself to cells of the same cell type. Now the descendant cells need the same resources from other cells, they have the same functions, and are embedded in the same environment. The tissue is more than the cells. It needs a specific overall anatomy and physiology that allows the contained cells to make their contributions. This anatomy and physiology are implemented by the cell type.[375]

One example is the epithelial tissue of the skin. Cells have to be close together without any gaps. They are fed by cells inside and have the shared goal to protect the outer boundary of the organism. This task requires many cells that are connected closely together.[376]

Another example is muscle tissue. The contraction happens within every muscle cell. But the tissue as a whole has to connect from sinew to sinew without any gaps. The connection must be very strong due to the high forces. The muscle cells need mechanisms to contract at exactly the same time.[377]

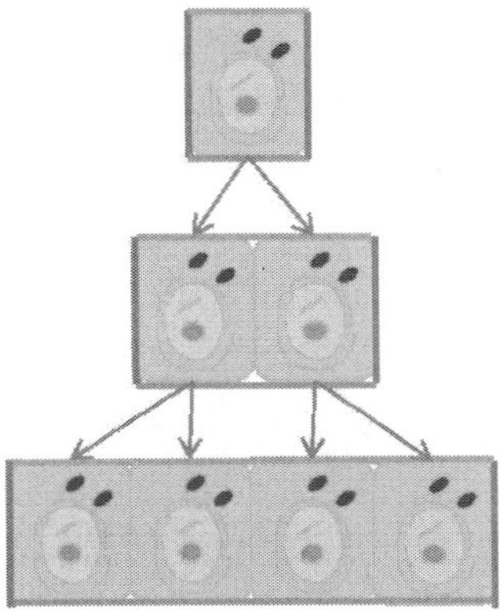

Figure 6-11: Tissues. Many cells of the same cell type work together to realize one function.

Tissues fulfill a specific function of the organization. Thus, the number of cells in a tissue depends on how much each cell can contribute to this overall demand. The number of cells determines the size of the tissue relative to others.

---

[375] See (Sadava, et al., 2011 S. 1104-1106)
[376] See (Sadava, et al., 2011 S. 1104)
[377] See (Sadava, et al., 2011 S. 1106)

Business

If a large organization grows, the newly created teams get more and more specific. The differences between the teams that split in later phases get smaller. At the end there are some teams which are quite similar. We refer to them as a group of teams (Figure 6-12). The target of all contained teams is derived from the target of their common ancestor team before it split.

Teams in a group of teams have the same team type and contribute to the same function.

One example is the continuous improvement team that is responsible for the entire end-to-end process.[378] This team solves problems whenever they come. It collects issues, finds the route cause and changes the processes to eliminate them. If one team is no longer sufficient, then the team splits. After the split every team is only responsible for a part of the end-to-end process. The group of teams has to ensure that there are no gaps between them. Now the new challenge is that the symptom and the cause of problems are distributed between teams. They need specific collaborations to ensure that even these problems can be solved. One team has to change processes to create a benefit in a different team. This example shows that it is problematic to give individual targets to teams. Instead the group of teams has a shared target, and every team has specific contributions to it.

The team type has to reflect that. The split of responsibilities and resources and the typical collaborations have to be designed at the team type. This allows the team type to be used independently of the exact number of contained teams. The puzzle pieces always work together.

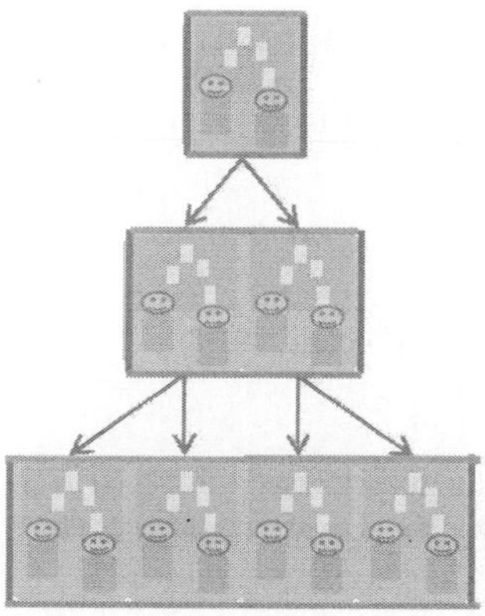

Figure 6-12: Groups of teams. Many teams of the same team type have to work together to realize one function.

Every organization requires a good balance between the resources. If it requires more work to get one resource than another, then it needs more teams to realize that. In reality most teams are focused on creating customer value in order to get the resource money, whereas there are fewer teams involved in the contact with suppliers or investors.

[378] The team types will be explained in section 8

6.2.3 *Specialization of Subsystems*

Biology

At the end every specialized cell only focuses on a small set of functions. It is part of a specialized organ system. Plants and animals must realize the same functions, but they have different environments and thus different strategies to implement the functions with a set of organs.

The example of the human body shows that every organ system fulfills one or more functions that stems from the requirements of living cells (Table 29).

	Functions	Human Organ System	Comment
	Create Self		
1	Import	Digestive system[379] Respiratory system[380]	
2	Distribute	Cardiovascular system[381]	Distribution is a multi-purpose system that is used for the transport of food, waste and signals.
3	Export	Urinary system, [382]Respiratory system	Can also be seen as part of protection.
	Protect Self		
4	Protect Outer Boundary	Integumentary system	
5	Protect Inside	Immune system[383]	
	Adapt Self		
6	Perceive, Decide	Nervous system, Endocrine system[384]	
7	Act	Skeleton Muscle system[385]	
	Replicate Self		
8	Replicate	Replication system[386]	

Table 29: List of human organ systems

Maybe the most difficult translation is the muscle system and the skeleton. They are needed to move from one location to the other. Both organ systems are pure resource consumers, but they help as actors to bring the organism to a location with more food and fewer threats.

[379] See (Sadava, et al., 2011 S. 1410-1440) for the digestive system
[380] See (Sadava, et al., 2011 S. 1354-1379) for the respiratory system
[381] See (Sadava, et al., 2011 S. 1380-1408) for the cardiovascular system
[382] See (Sadava, et al., 2011 S. 1442-1469) for the urinary system
[383] See (Sadava, et al., 2011 S. 1151-1185) for the immune system
[384] See (Sadava, et al., 2011 S. 1248-1329) for the nervous system and the sensory systems
[385] See (Sadava, et al., 2011 S. 1330-1353) for the skeleton and muscle system
[386] See (Sadava, et al., 2011 S. 1186-1218) for the replication system

Business

Large organizations can have various structures. The form depends on the business type and the size of the organization, but the functions of all organizations are the same. If an organization wants to have sustainable success, it has to implement the basic functions of viability. Every team contributes to that by realizing one of the functions with high effectivity and efficiency. With the right organizational architecture this creates the best working conditions for all teams (Table 30).

	Subsystem	Mechanisms	Comment
	Create Self		
1	Import	HR (Resource: Man) Procurement (R: Material) Finance (R: Money) CRM/SCM (R: Money)	All sub-organizations that import resources (This includes all symbiotic activities like customer value creation)
2	Distribute	Resource management	Ensure that every team has all necessary resources
3	Export	Terminate contracts, sell or recycle inventory, clean everything.	Remove everything that is unneeded
	Protect Self		
4	Protect outside	Guards, Cybersecurity experts Contract managers, Lawyers	Protect at the boundary to the outside
5	Protect inside	Health & safety Data backups	Protect inside the system. Protect the system from its self-created problems.
	Adapt Self[387]		
6	Perceive Decide	Business intelligence, top management	Perceive inside / outside world and make decisions
7	Act	Teams that help to move the organizations to new markets.	Allow the organization to move to new markets
	Replicate Self		
8	Replicate	Business excellence	Copy the architecture of the system to others (e.g. to other business units or to suppliers)

Table 30: List of suborganizations

The teams that realize movement to new markets are only resource consumers. They do everything that is needed to get started in a new growing market. This allows the organization to move all internal resources (budget, people) to new markets and reduce the resources in shrinking ones. Without the ability to move, it does not help to perceive change or to make decisions. The organization would starve due to a lack of resources.

---

[387] Adaptation in a Large Viable Organization follows the logic of the Viable System Model, which explains how complex systems can be controlled using 5 systems. See (Beer, 1990) and section 2.7

6.2.4 Summary

6.2.4.1 Insights

Biology teaches us that large organisms consist of organs that fulfill a function. Organs are composed of tissues of cells with the same cell type.

The translation of organ systems shows that large organizations need various team types that have to work together.

The primary challenge is to ensure that every team gets all resources needed to work. This is the job of resource management which realizes the distribution. Resources come from the stakeholders. We need teams that create symbiotic relations with stakeholders, produce something they need and get the needed resources. These activities are similar to those we have discussed on the team level.

Some teams have to protect the organization at the outer boundary. This includes many security activities (e.g. security guard) which are independent from stakeholders. Additionally, it needs internal mechanisms to protect assets (e.g. data backup).

The adaptation of large organizations is an additional challenge. The highly specialized teams cannot react to the changes of the environment without coordination. Large organization need to integrate all information into one big picture, make decisions and derive actions that can be implemented by all the teams.

Finally, the replication requires unspecialized teams that can replicate the organization from scratch.

6.2.4.2 Requirements

A Large Viable Organization (LVO) shall implement the following requirements (Table 31):

	Requirements	Comment
6.2.1	A LVO shall contain Viable Teams that are responsible for importing resources	Digestive system Respiratory system
6.2.2	A LVO shall contain Viable Teams that are responsible for exporting everything that is not needed	Urinary system
6.2.3	A LVO shall contain Viable Teams that are responsible for distributing resources to and from all Viable Teams	Cardiovascular system Lymphatic system
6.2.4	A LVO shall contain Viable Teams that are responsible for protecting the LVO at the outside	Integumentary system
6.2.5	A LVO shall contain Viable Teams that are responsible for protecting the LVO inside	Immune system

6.2.6	A LVO shall contain Viable Teams that are responsible for collecting and processing information	Nervous system Endocrine system
6.2.7	A LVO shall contain Viable Teams that are responsible for maintaining or changing form	Muscular system skeleton
6.2.8	A LVO shall contain Viable Teams that are responsible for replicating the entire LVO	Reproductive system

Table 31: List of requirements related to large viable systems

The Large Viable Organization, i.e. a project business, starts top-down with the top management team. All of the requirements have to be realized by this team, simply because it is the only one. If the top management team splits, it hands over the responsibility to the other team. If it does not, then it is still responsible for the requirements.

This perspective is similar to the anchor circle of Holacracy. At the end the anchor circle is responsible for the entire legal entity, whether it has been delegated or not.[388]

6.2.4.3 Existing Solutions

Most descriptions about methods and principles are very generic. Thus, they do not apply to the specific subsystems that are mentioned here. The human nervous system is studied very well and already copied by neuro-cybernetics.

Viable System Model, Cybernetics

Beer describes his Viable System Model as copy of the human nervous system. His focus is on the relation between the parts of the organism that work independently and their central control [389] [390](see section 2.5).

Beer's Viable System Model can be used to model the adaptation system of the Large Viable Organization.

Hoshin Kanri

The lean method of Hoshin Kanri is about target definition and policy deployment. It aligns the activities of all teams to the overall target of the organization [391,392,393] Hoshin Kanri is helpful to align target of teams to the targets of sub-organizations and finally the overall goal. Hoshin Kanri does not ask why it is needed.

Our approach focuses on the reasons for the need of collaborations and target alignment. Everything is necessary because of the growth of the organization and the split of teams. One

---

[388] See (Robertson, 2014) for Holacrazy
[389] See (Beer, 1995)
[390] See (Beer, 1990)
[391] See (Colletti, 2013)
[392] See (Jackson, 2006)
[393] (Kudernatsch, 2013)

example is the function to improve processes. The cause of problems is very often at a different location than its symptom. If cause and effect are located in different teams, there must be a special collaboration between some teams. It also creates the needs to align the targets.

6.2.5 Annotations

Classification of Subsystems by the Main Functions

Export should be part of the protection system. It does not contribute anything to creation. Instead it removes unneeded or dangerous objects. Practically, the export function can be done by the same organs that realize the import (example: respiratory system).

The distribution system can be used for various purposes: food, waste, immune system and hormones. Thus, it is used as a multi-purpose organ system for distribution.

Human Medicine: Anatomy and Physiology

For an understanding how teams how to work together in detail, it is interesting to understand the anatomy and physiology on a detail level. Rohen explains the functional histology of humans starting with cells and typical cell types, before the typical tissue types are introduced. Interesting point are also the different designs of organ tissues (lung tissue should have maximum surface, muscles should be able to handle strong forces) that result from their purpose. This work gives an excellent understanding about the micro-anatomy.[394]

Physiology focuses on the dynamics of the human organism, especially the question how the homeostasis is realized. The books of Hall or Speckman et al. both explain the metabolism of the human body to understand why every cell has a perfect environment.[395]

Human Medicine: Pathology, Inner Medicine and Pharmacology

Literature about pathology and special medicine is interesting for three reasons. First of all, the general pathology describes typical problems that can be found in every tissue. In our translation this has a direct equivalent in problems that can occur in every team of the organization. Secondly, the specific medicine disciplines have a systematic approach to identify and therapy diseases. If an organization is like an organism, then business consultants are like physicians, and they could use the same systematic approach.

Wilkinsens Clinical medicine is a good starting point to understand, how physicians structure their knowledge about diseases systematically. This can be used for a transfer of the medical approach to business consulting.[396]

Books about pathology describe diseases that could occur everywhere (e.g. tumors). For our purposes it is sufficient to have an overview about the variety of problems that could occur. Nennstiel gives a brief introduction to pathology that answers this question.[397]

---

[394] See (Rohen, et al., 2000)
[395] See (Hall, 2016) (Speckmann, et al., 2013)
[396] See (Wilkinson, et al., 2017)
[397] See (Nennstiel, 2013)

Rent-Polster et al describe the field of inner medicine, which includes all human organs having to do with importing resources, distributing them or exporting waste. This book not only contains details descriptions of the anatomy and physiology of inner organs. It also describes the typical symptoms and signs of diseases, the diagnosis of diseases with clinical tests and typical therapies.[398]

Pharmacology is interesting, since it separates pharmacodynamics from pharmacokinetics. The first question is why a drug helps for a specific problem. The second question is how the drug comes to the right location.[399] This difference is well known in business consulting, where many ways can be taken to implement a specific improvement. This scope is typically covered by change management. The change aspects have to be considered in parallel to the question how the specific improvement helps to increase organizational fitness.

---

[398] See (Renz-Polster, et al., 2012)
[399] See (Kuse, et al., 2009)

7 Viable Business

7.1 Introduction

Here we want to design the **Viable Business Management System (VBMS)**. It is a very abstract description of a management system, which can be used by many organizations. It is neither focused on a specific organizational size, nor on a specific business type like project business. For that reason, it contains only the basic elements, and it needs to be detailed by other designs.

A more specific description will be given in the Viable Project Business Management System (VPBMS), which focuses on the specific teamplay in project business. It will be explained in Chapter 8.

What is special to the Viable Business Management System? It is not designed by using the current best practice in the business world; instead it utilizes the system dynamics of life.

In chapters 4, 5 and 1 we have analyzed the system dynamics of life and collected many requirements for Viable Organizations. What are these requirements about? Some of these requirements are about forms, others are about functions.

1. Forms are about the structure of the organization. Examples for forms are roles, teams, documents or regular meetings. In a Viable Organization each form contributes to the acceleration of one or more processes.
2. Functions are about the dynamics of the organization. Most[400] functions are realized by business processes. In a Viable Organization the final purpose of all the functions is to create the forms.

This logic is typical for healthy biological organisms: [401]

Structures accelerate Dynamics, Dynamics create Structures.

Both perspectives influence each other positively. This is the reinforcing loop that leads to growth. The challenge is to identify all necessary functions and forms on a top-level perspective. If one critical function or form is missing, then the entire reinforcing loop won't work.

Before we can start to discuss the functions and forms in detail, we should have a look at the terminology used in business. We need to understand basic terms like "organization", "management system" or "sustained success".

For our work we refer to the terminology, which is defined in the ISO9000 standard and used in ISO9001 and ISO9004.[402] This terminology is already well-known in quality management departments and very often used in formal process descriptions.

---

[400] Some functions like "keep in position" do not necessarily require a process, since nothing changes or moves, they can be realized by forms. (e.g. cytoskeleton keeps organelles in place)
[401] See section 5.1 for the discussion of autocatalysis as the base for growth.
[402] ISO 9000 contains the glossary for quality management systems which is used in ISO 9001-9004. ISO 9004 contains a recommendation for advanced quality management systems which goes beyond the minimum requirements for ISO 9001. This recommendation works with the same stakeholder groups as VBMS.

C. Dachs, *Viable Project Business*, Contributions to Management Science, https://doi.org/10.1007/978-3-030-62904-5_7

Usage of the known terms helps to implement Viable Business in real organizations. Additionally, the exact definitions give some additional requirements, that have to be considered by the Viable Business Management System.

7.1.1 Organization

ISO 9000 defines the **organization** as a "person or group of people that has its own functions with responsibilities, authorities and relationships to achieve its objectives." (ISO 9000, 2015 p. 27)

This definition already shows several aspects that have to be considered by the Viable Business Management System. Organizations have

1. Person or group of people
2. Objectives
3. Functions
4. Responsibilities and authorities
5. Relationships

Person or Group of People

Persons form the organization in order to achieve an objective. This definition is very general. It applies to a large company as well as to a private soccer team.

In both cases the organization has a context: These persons also exist outside the organization. They are only part of the organization, because they are interested in the objective ("earn money" or "have fun"), and they contribute ("invest money", "invest time"). Persons are stakeholders of the organization.

In the Viable Business Management System, we will strictly separate the organization from the persons. This allows to analyze their interactions with the organization in more detail.

Objectives[403]

The ISO 9000 standard defines **success** as the "achievement of an objective" [404]and **sustained success** as "success over a period of time"[405] . This success over a period of time requires that the organization also needs to exist over the same period of time.

Most organizations want to have sustained success. This requires that "live long and prosper" is at least one of several objectives. In the Viable Business Management System, we will be even more strict and focus only on the objective to exist over a period of time, or even better: to continuously grow. This focus is not a restriction; instead it highlights what is common to all organizations that want to have sustained success.

[403] ISO 9000 defines objectives as "results to be achieved". (ISO 9000, 2015 p. 42)
[404] See (ISO 9000, 2015 p. 43)
[405] See (ISO 9000, 2015 p. 43).

Most organizations describe their purpose in the mission statement and their objectives in the vision statement. Thus, the Viable Business Management System has to show, how this can be done for the (very generic) objective of growth.

Functions

The definition shows that the organization has functions to achieve the objective. The functions cover the dynamic aspect of the organization. Which functions are needed to realize sustained success?

In section 4.2 we discussed four main functions that are needed to realize the objective:

1. Create self (to achieve gross growth)
2. Protect self (to minimize losses)
3. Adapt self (to adapt all functions to a changing environment)
4. Replicate self (to grow beyond a maximum size)

In Chapter 5 we showed mechanisms how to implement the four main functions and we derived some requirements.

Responsibilities and Authorities

There is more than one function to be realized in order to achieve the objectives. If the organization consists of a single person, then this person can do one task after the other.

If the organization consists of a group of people, then it can execute more than one function in parallel. A group member has the authority to use resources in order to fulfill the responsibility to execute a task.

If the team uses roles, this can increase efficiency of each team member. This kind of specialization will not be covered by the Viable Business Management System but by the Viable Project Business Management System, because specialization depends on business type and organizational size[406].

Relationships

People realizing the responsibilities and having authorities need to collaborate. There must be relationships between them. Typical organizations also have relationships to interested parties in the outside world, the stakeholders.

The Viable Business Management System describes the relationships between the organization and its stakeholders, but not the internal collaborations. Internal relationships between teams will be described in the more specific Viable Business Project Management System. Prepared with an understanding of the elements of an organization, we can now focus on the management system.

[406] We already compared humans and trees. They have to realize the same functions but use very different structures to implement them. Different business types are like different types of animals (or plants).

7.1.2 Management System

The ISO 9000 standard defines a management system as a "set of interrelated or interacting elements of an organization to establish policies and objectives, and processes to achieve those objectives"[407].

The definition contains two parts:

1. Set of interrelated or interacting elements of an organization to establish policies and objectives
2. And processes to achieve those objectives.

Set of interrelating or interacting elements[408] of an organization to establish policies[409] and objectives

One of the purposes of the management system is to establish objectives and policies, whereby the policies shall support the achievement of these objectives.

The most important part is "set of [...] elements of an organization to". It shows that policies and objectives are defined by the organization itself, and that it needs a set of elements to do that. The management system of an organization contains everything that creates the objectives.

... and processes[410] to achieve the objectives

As a second part the management system includes all processes that are needed to achieve all the self-created objectives. This is typically the more difficult part.

7.1.3 Viable Business Management System

What is now special about the Viable Business Management System? It is its objective and its focus on system dynamics.

The Viable Business Management System focuses on only one main objective: "live long and prosper". Thus, the VBMS has to contain all elements to refine and detail this main objective, derive the necessary policies, and it has to contain all processes to achieve the objective.

VBMS details and realizes the main objective with a defined system dynamic which has been derived from biological organisms. The aspects of the organization like objectives, functions, responsibilities and authorities or relationships are derived from this system dynamics.

---

[407] See (ISO 9000, 2015 p. 36)

[408] ISO 9000 defines a **system** as "Set of interrelated or interacting elements". (ISO 9000, 2015 p. 36), thus the entire statement could be simplified to "Organization's system to establish policies and objectives".
Like other definitions of the term system, it refers to interrelated elements or objects that are not further defined. IEC 81346-1, for example, defines a system as a "set of interrelated objects in a defined context considered as a whole and separated from the environment" (IEC, 2009 S. 11). IEC has the focus more on the difference to the environment.

[409] ISO 9000 defines policy as "intentions and direction of an organization as formally expressed by its top management" (ISO 9000, 2015 p. 38)

[410] ISO 9000 defines a process as a "set of interrelated or interacting activities that use inputs to deliver an intended result" (ISO 9000, 2015 p. 33). Here we can see the definition of "system" applied for activities.

In the following subsections we discuss the various aspects of organizations and management systems that are required by the ISO 9000 definitions. This will show that VBMS fulfills all requirements.

7.2 System Dynamics

This section gives a short introduction to the fundamental part of the system dynamics. It explains the reinforcing loops that allow the system to grow.

In section 7.5 we will detail the system dynamics in a functional model and explain which functions need to be implemented. In section 7.6 we will discuss, how these functions can be realized by a single person, a team or a large organization.

7.2.1 Best Working Conditions for Each Process

The dynamics are realized by business processes. Even small organizations have to execute many different processes like "create customer value", "get contract", "purchase tool", "pay invoice". Each of these business processes need to have best working conditions to be effective and efficient. Let us refer to them as "**each process**" to indicate that we refer to each individual process of the organization.

Each process needs some input factors to produce the output.[411,412] This is especially true for the creation of customer value. We can categorize the factors with the fishbone diagram (Figure 7-1).[413]

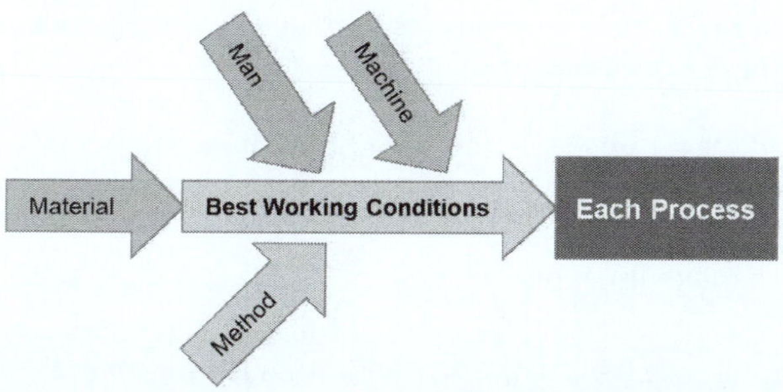

Figure 7-1: Best working conditions for Each Process. Best working conditions contain information, which can be copies, and resources that have to be allocated. Resources can be divided in human resources and other resources.

The fishbone includes man, machine, material, method (and maybe some other factors that are needed) to make a process effective and efficient. Material is everything that will be consumed by the process or loses its value. Man, and machine are the two parts of the catalyst that accelerates the process. Method is the information how man and machine have to interact with the material.

[411] ISO 9000 defines a process as "set of interrelated or interacting activities that use inputs to deliver an intended result" (ISO 9000, 2015 p. 33). The term result has been used to avoid circular dependencies between process and output.

[412] See section 5.1 for the translation of biocatalysts to best working conditions.

[413] Also known as 4M (man, machine, material, method) or Ishikawa Diagram. Literature also shows 8M-fishbones that additionally include factors money, management, milieu and measurement. (Bicheno, et al., 2009)

7.2.2 Resources

Factors man, machine and material always have their origin outside of the organization (Figure 7-2). We *do not* make new employees, we hire them. We also *do not* create all material and tools from scratch.

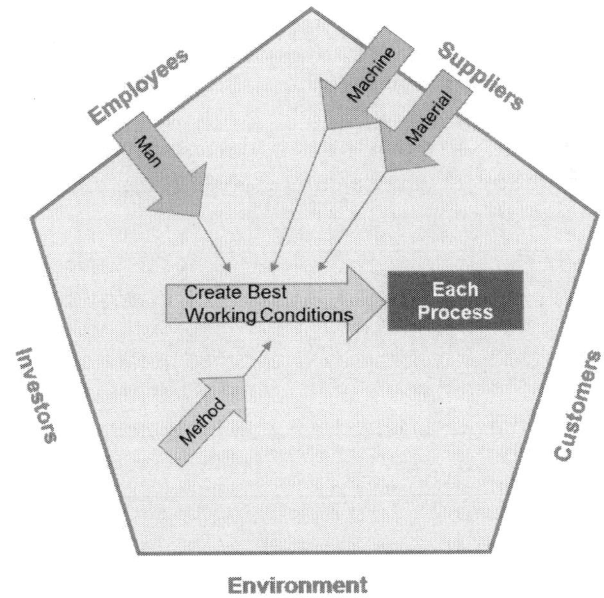

Figure 7-2: Import and development of resources from outside. Only information can be created inside. All other factors come from outside and have to be developed.

If we follow the resources on their way, we can see some major steps:

1. Import resource from outside [414]
2. Develop resource to make it usable [415]
3. Protect resource [416]
4. Allocate resource to create best working conditions for a process[417]

Import

First of all, the organization has to get the raw factors from outside. This can be done by hiring people with a standard qualification and knowledge, or by buying standard tools like furniture, computers or standard software.

Develop

Development contains all processes that are necessary to make the imported resources usable in order to run processes. For the factor man this contains all people development activities to create

[414] Import of resources is the translation of membrane proteins. It is described in section 5.4.3
[415] Development of resources is the translation of metabolic (anabolic) pathways. It is described in section 5.3
[416] Protection is the translation of the cell membrane, which keeps self-created molecules together. See 5.4.1
[417] Allocation mechanisms (complex transportation and sortation) are realized in eukaryotes to bring molecules from one organelle to the next (see section5.8), but they are not necessary in bacteria, where diffusion processes are sufficient to bring everything to the right place.

competences (social skills, knowledge about the product...) that are needed to run a process. Development of tools contains all activities that are needed to configure the bought tools to a working tool environment.

Each process has its specific method. These methods also have to be developed by the organization.

Protect

Since development costs a lot of time and energy, it is also necessary to protect all the resources. It is good if people stay at the organization after they have been trained. Tools have to be protected too and data has to be protected with data backups and cybersecurity. Protection mechanisms do not create anything new, but they help to reduce the losses. If people leave the company, we have to hire other people and train them from scratch. If data gets lost and no backup is available, then a lot of time and energy is needed to create it again. Thus, protection contributes to growth by avoiding unneeded replacement activities.

Allocate

The last step is to allocate the resources for a process. The resources have to be at the right time at the right location. Together with the necessary method and other information, the resources are part of the best working condition.

This network shows the pathways of all resources from their origin outside the organization to the point where they are a part of the best working condition of a process.

7.2.3 Reinforcement Loop

The result of the entire process network is best working conditions for each business process. It creates catalysts for all the processes it consists of and is thus an auto-catalytic system. [418]

This creates a reinforcing loop which is the base of the entire system dynamics.

The best working conditions accelerate all business processes. The business processes lead to best working conditions.

How can this reinforcing loop work at the maximum speed?

7.2.4 Balancing Loop

The speed of growth depends on the slowest process in this network or on the most limited resource from outside. Thus, the organization needs adaptation mechanisms to identify the internal and external bottlenecks and eliminate them.

The identification of the bottleneck requires information about the internal processes and availability of external resources.

---

[418] Many biological organisms (like archaea in the hydrothermal vent) are organized that way. They do not need other lifeforms as partners. They have an inorganic dietary and transform all the inputs to catalysts. In other words: They are the catalyst which accelerated the creation of itself.

The elimination can then be done by balancing the resources such that all processes work at the same speed. This increases the speed at the bottleneck, while decreasing speed at non-bottlenecks, and increases the growth rate of the entire organization.

This concept is a balancing loop which tries to keep the entire system in balance.

7.2.5 Stakeholders

Let us now focus on the external bottleneck of growth: How to find all the employees and suppliers that provide manpower, tools and material for free? Especially, if the only target is to create best working conditions for getting even more resources.

Unfortunately, most organizations do *not* get all these resources for free. Employees want to earn money, Suppliers want to be paid. Thus, the organization has to give something beneficial to its stakeholders in exchange for the needed resources (Figure 7-3). [419]

In other words: The organization needs money.

Fortunately, this is no problem: The organization already knows how to create best working conditions for all kind of processes. Thus, it can use this ability to create value for a third stakeholder group, the customers, who provides the money that is needed to pay employees and suppliers.

The most important point here is not to use all resources for the creation of customer value. Of cause, it is important to create customer value to earn money, because the organization needs to pay employees and suppliers. The organization also has needs resources to improve the relationships with its other stakeholders. Employees not only want to have money. They want to have a great workplace. It needs processes and resources to provide that. The same is true for suppliers.

All this is important, but it is at the periphery of the system dynamics. Stakeholder relationships are symbiotic relationships of the Viable Organization to others with the purpose to get the resources for the self-creation processes.

Thus, at the end it is about the resources the organization can use to create best working conditions for all the processes. This includes processes of work preparation, resource development, resource management, development of know-how and adaptation.

---

[419] See section 5.7 for the translation of the symbiosis to stakeholder relations.

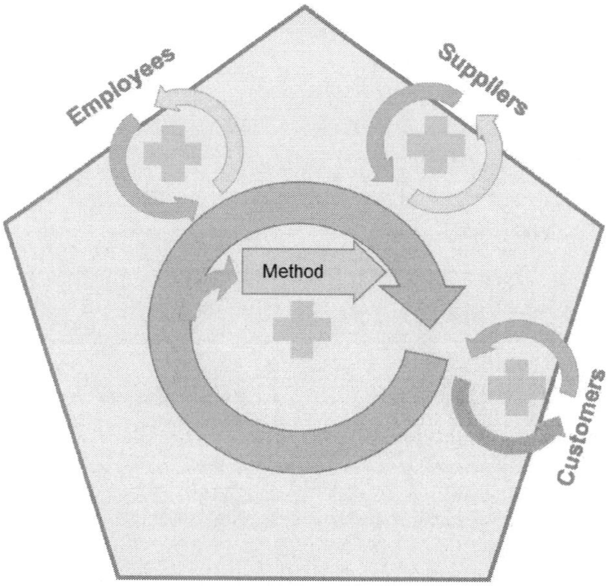

Figure 7-3: Give and take with stakeholders. The organization has to give something to employees and suppliers to receive something for them.

These processes create the catalysts, which accelerate each process and make the organization stronger and stronger. This strong organization with its effective and efficient processes is then able to create value for all the stakeholders.

This is important for the organization and all the stakeholders. Self-creation ensures existence and is the base for all sustained success and for the sustained ability to create value for others.

There is only one problem: where is the starting point of this loop. Either the customer has to pay in advance, or the employees and suppliers have to wait for the money. This problem can be solved by the investor who starts the loops with an investment.

The storyline is as follows (Figure 7-4):

1. The investor has trust in the organization, its target and the team, and invests money.
2. This money can be used to pay employees and suppliers.
3. In return the organization gets the workforce of the employees and the machines and material from the suppliers.
4. The resources can be used to create best working conditions to create customer value
5. The customer pays for the product
6. The organization is able to pay taxes to society, which provides the environment needed to do business.
7. The organization is also able to pay interests to its investors.
8. The cycle can start again.

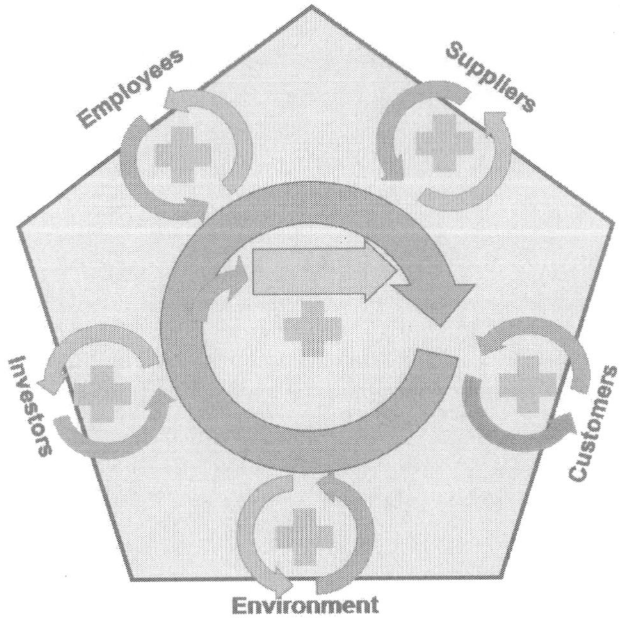

Figure 7-4: Typical reinforcing loops with main stakeholder groups. The organization maintains a symbiosis with each stakeholder group. Everything is connected by the inner reinforcing loop of self-creation. If one of the loops is weaker than the others, then it limits the entire system.

Balancing Stakeholder Relationships

If the organization wants to grow continuously, then its relationships to stakeholders also have to grow continuously and create mutual trust. The organization needs more resources from outside, which requires providing more value to the stakeholders. The reinforcing loop of growth is connected with the reinforcing loops with the stakeholders.

Since we need all our resources in the right mix, there is no stakeholder group that is more important than the others. Viable Business is not about "investor first" or "customer first" or "employee first". Viable Business requires all resources from the stakeholders in a balanced way.

Of course, the markets are asymmetric. It might be easier to find a new supplier than a new customer, but this is only a matter of proportions. Most work may be related on the creation of customer value, and only a small portion on providing value for suppliers. But it has to be ensured that also the supplier collaboration is excellent. If we have problems with suppliers, then this leads to quality problems or high material cost and influences our customer value. If the customer is not satisfied, this influences investors and employees as well.

Give ...

All the stakeholders are important for the organization. It is important to establish a good relationship. The organization has to give something, what the stakeholders really need.

For a good value proposition, it is important to understand the stakeholder groups and their needs and expectations. A good practice is a market segmentation for each stakeholder group and a detailed stakeholder analysis.

The more precisely we know what the stakeholders really want, the more effectively we can deliver that in high quality at low cost. This is not only true for the customer relationship, but also for the other relationships (Figure 7-5).

This value proposition is important for us, because it shows the value of our organization from the viewpoint of the stakeholders. It shows our stakeholders why they should be interested in giving us something back what we need: resources.

These resources are needed to provide all the value we give and allow the organization to grow.

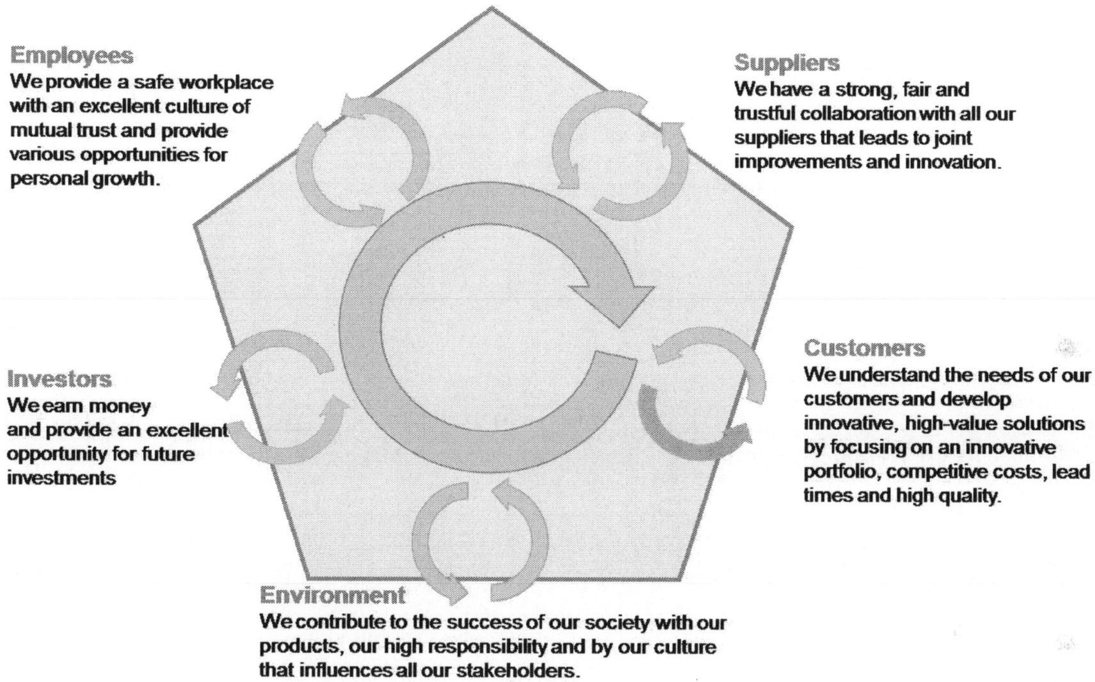

Figure 7-5: Example for a value proposition. It shows what the organization intends to give to each stakeholder.

… and Take …

As with the value we give to the stakeholders, it is good to understand precisely which resources are needed from outside (Figure 7-6). In the case of money, it is easier than with the workforce. What do we really require from our employees? The answer depends on the roles the employees have. We have to make a segmentation of roles that have different characteristics. In HR this is often done by defining career paths and identifying critical soft-skills that are needed to fulfill roles.

For a high-level value proposition, it is sufficient to focus on the basic characteristics (discipline, motivation, willingness to learn, team spirit…) that are required by many roles.

The organization need to have processes to ensure that it gets all the resources with these characteristics.

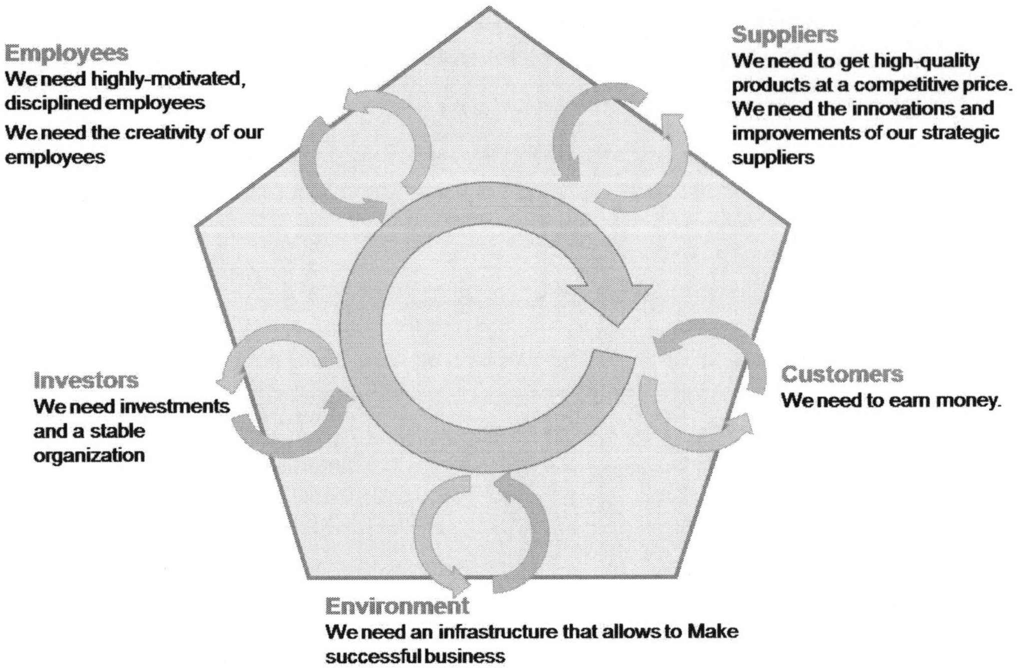

Figure 7-6: Example for resources from stakeholders. The organization has the symbiosis with its stakeholders only, because it needs several resources from outside. The organization has to understand what it needs, to be able to communicate it.

The resource from outside are the defined starting point for the development activities which have already been discussed in section 7.2.2. All the developed resources are then needed by the resource allocation to create best working conditions.

... to Create Best Working Conditions for Each Process

All the resources have to be allocated at the right time to the right point. This requires know-how about processes. This know-how is the differentiator of the organization. The best working conditions require fewer resources to produce results (Figure 7-7).

Thus, it is important for the organization to develop, use and improve processes and knowledge. They are the reason for high productivity and for good value propositions to the stakeholders.

Figure 7-7: Best working conditions. The purpose of all internal functions is to create best working conditions now and in the future.

7.2.6 Summary

The system dynamics consist of several connected reinforcing loops (central system, relationships to stakeholders), which are controlled by balancing loops (adaptation), that determine and eliminate bottlenecks with the objective to maximize growth and to “live long and prosper”.

Reinforcement Loop of Self-Creation

Each process can be very effective and efficient if it has the best working conditions. A network of processes can provide all the needed resources and thus creates best working conditions for each of its processes.

Balancing Loop to Identify and Eliminate Bottlenecks (Adaptation)

The speed of the process network depends on the most-limiting process or resource. The adaptation mechanisms try to identify and eliminate this bottleneck. This balances the system such that each process works at the right speed.

Reinforcement Loops of Give and Take with Stakeholders (Symbiosis)

Since the resources are not for free, the organization needs to create value for stakeholders, which in return provide the needed resources for growth. All the relationships to stakeholder group are connected via resources with the central system and thus indirectly connected with all the other stakeholder relations.

Prepared with these general considerations we can discuss form and function of the organization. Let us have a first general view on the “self” of the organization, before we study the main function “create self” that is necessary to realize the goal “live long and prosper”-

7.3 Organization

7.3.1 Organization and Stakeholders

The discussion of system dynamics showed that the organization has to interact with stakeholders like investors, employees, suppliers, customers and the environment. This perspective has the direct consequence that the organization is not a "person or a group of people..." as defined by ISO 9000[420]; instead, it interacts with them. This is the main difference: The Viable Business exists as a subject independently from its stakeholders.

In large organizations it is clear that the organization is independent from the person who executes the role of the CEO. The organization would persist, even if this important person leaves. The same is true for all other roles in the organization. The organization also exists independently from the investors. It only needs the investment as a resource but does not depend on a specific person who invests.

If a small company has only one entrepreneur who is the only investor and employee, we can use the same model. Of course, the organization would fail, if the entrepreneur would stop working, but it is helpful to separate the organization from the role of the employee and investor as well. This separation allows to model the give and take between the small company and the entrepreneur.

Let as focus on the organization as person-independent entity.

7.3.2 Organization as a Legal Entity

A second aspect of the system dynamics is that the organization is able to give something to others. It also receives resources from others to build its own structures. The organization has property and is able to receive rights from or transfer rights to others.

In other words, we need to understand an organization as a legal entity and thus carrier of rights and obligations. This ability to have property and rights, especially the right to make contracts with other parties, clearly separates the organization from its context. [421] There is a clear distinction between inside and outside.

As a legal entity the organization consists of everything it owns.

7.3.3 Self

In section 4.3 we already discussed the relationship between form and function. This was necessary to understand the term "self" in the functions "create self", "protect self", "adapt self" and "replicate self".

The "self" of a cell are all the molecules the cell consists of, and they are located in the three-dimensional physical space. This gives the self a form and a position relative to others. This form

---

[420] See (ISO 9000, 2015)

[421] See section 5.4 for the discussion of the system boundary

and the system boundary are important to understand the concepts of flow, concentration and their relation to process speed and growth rate. Furthermore, it is necessary to understand protection mechanisms that protect the cell at the outer boundary, and also the inner protection that has to identify, what does not belong to the cell.

In the translation we discussed the business objects in the conceptual space. Organizations have a variety of objects, like tools, contracts, shared beliefs, values or emotions that are not only located in the physical space. We already discussed the space of social constructs and the psychological space as the conceptual space where business objects are located. Like in biology we need to understand the form and position of the system, to understand and manage flow, concentration, process speed and growth.

Thus, the questions are: What is the space, the organization lives in? Where is the organization located? What is the form of the organization in this space? Can form or position be changed by the organization itself? How can something enter or leave the system boundary?

The organization as a legal entity allows us to connect all the other spaces to a multi-dimensional conceptual space consisting of subspaces:

1. Space of physical objects (3-dimensional space)
2. Space of subjects (psychological space) incl. motivation, know-how, emotions
3. Space of shared beliefs (social constructs) incl. concepts of legal entities, properties and rights
4. Space of information

The connection of the objects starts with the legal entity that has properties and rights and is able to owe something to other legal entities. The existence of legal entities and property and rights is obviously a social construct.

Since property and rights can be linked to physical objects, there is induced relation to the physical space. An organization can own real estate or at least rent an office, it has a right to use a part of the physical world exclusively: the company's site. This gives it a defined form and position in the three-dimensional world as well. All owned movable physical objects have a form and position as well. It is easy to decide whether they are at the company site or not. On the other hand, it is easy to identify all physical objects at the company site which are not owned. High concentration in the physical world means to bring all owned objects together and remove all not-owned objects.

If the organization has relationships to employees, then it has the right to use the workforce. These relationships connect the organization to the psychological space of employees, who carry motivation, culture, know-how and skills or personal networks. People belonging to the organization can be understood as inside, all others are outside.

A fourth subspace might be the space of information. The organization can store information and protects its access, such that the information can only be used by itself. The availability of information defines a clear boundary.

All these spaces are connected by the legal entity. If process speed depends on the concentration of success factors for a process, then it requires to create high concentration in all of the subspaces. This is similar to the purpose of the cell membrane and the membrane proteins.

Analogue to eukaryotic cells, it makes sense to create substructures in order to increase concentration internally. Substructures can be created in all the subspaces.

The perspective of an organization as a legal entity is the starting point for all internal substructures. The organization can assign the authorities and responsibilities to roles and assign it to persons. It can create teams and define collaborations with them. It can assign resources to the teams. It can observe itself and create an inventory of all its structures. It can create contracts with other legal entities. All that has the starting point with the organization as a legal entity and the ability to own and owe something.

Of course, these considerations have not answered the question about the "self". They just explained in which space it is located, how this space is structured.

As already discussed in section 5.1 the self is an auto-catalyst that is able to create itself in this space. This means, that the self is all the business objects that are needed to accelerate the processes to move or create these business objects.

Before we analyze these processes, let us focus on the objective of the organization.

7.4 Objectives

7.4.1 Mission Statement

In the previous sections we discussed that the purpose of organizations is to have sustained success or "live long and prosper". Such a description of the purpose is helpful to understand the system dynamics, but it is not helpful to create an identity. Every organization would have the same purpose.[422]

The mission statement describes the purpose of the organization from a higher perspective[423]. It shows that the entire organization fulfills a defined function in the society and highlights the specific contribution that differs from other organizations.

In many cases, this specific purpose is to create value for the main stakeholder group, the customers. Sometimes the text describes the products; in other cases, it describes the value of the product for the customer. The mission statement can be used to communicate the identity and purpose of the organization to the stakeholders.

In Viable Business this mission statement can be derived from the value proposition for the customers.

---

[422] A pancreas cell has the ultimate goal to stay alive. This is the same for all cells in the human body. On the other hand, the pancreas cell has a purpose for the entire body: it produces insulin. Here the mission statement of the cell would be "We create insulin for the human body", and not "We want to be alive".

[423] ISO 9000:2015 defines mission as "organization's purpose for existing as expressed by top management". (ISO 9000, 2015 p. 38)

7.4.2 Vision Statement

The vision statement is focused on the future. A long-term vision statement describes and details the ideal value proposition for stakeholders in the future.[424] Thus, it should be similar to the value propositions of the ideal system dynamics.

For the purpose of communication, it makes sense to create a more compact version of the intended value propositions and highlight the most important aspects.

Important aspects are especially those, where the organization identifies a need for change. This allows to use the vision statement as a compact description, where the organization wants to go to, and allows to derive strategic measures.

7.4.3 Principles and Values

Most organizations have guiding principles and values.[425] Principles describe rules that are considered to be true in general, independent from the question, whether we like them or not. Many principles have their origin in the system dynamic of organizations and describe an important cause – effect relationship, that is so evident that the effect can be omitted in the written statement.

Examples for Principles are

> "We respect people"
> "We focus on creation of customer value"
> "We always use the current best practice until we find a better one"
> "We continuously improve processes"

These principles can be applied by every employee at every process. In our example the principles help to align a process to the customer, by providing exactly that what is necessary. Usage of the best practice ensures to get a high efficiency, and continuous improvement increases both efficiency and effectiveness. Respect people contributes to good relationships between employees and thus to a good collaboration.

Since not everybody follows the principles at all times, they can be understood as an objective for the "How", whereas the vision statement is the objective for the "What" and the mission statement describes the "Why".

7.5 Functions

In section 7.2 we have seen the overall system dynamics that are needed for growth. Here we want to create a functional structure of the organization. Which organizational functions are needed to implement the system dynamics?

---

[424] ISO 9000 defines the vision as "aspiration of what an organization would like to become as expressed by top management". (ISO 9000, 2015 p. 38)

[425] Terms "Value" and "Principle" are very often used ambiguously. Here we follow Covey (Covey, 1989), who uses the term "Principle" for rules that are true in general, whereas "Values" are preferences.

7.5.1 *Create Best Working Conditions for Each Process*

Starting point is to create best working conditions for each process of the organization (Figure 7-8). Processes are only efficient if all production factors are perfectly synchronized.

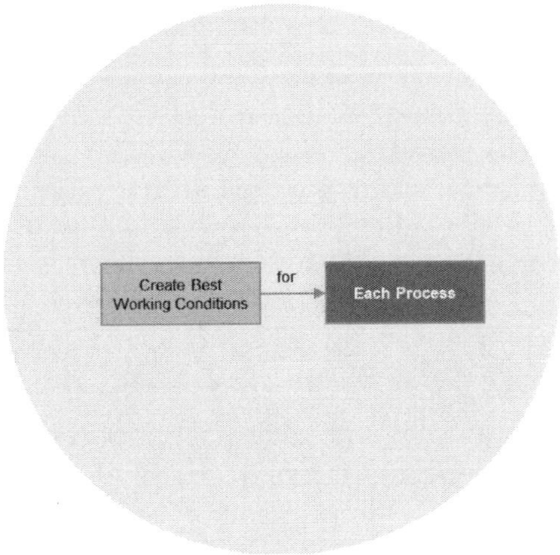

Figure 7-8: Create Best Working Conditions. The core function is to create best working conditions for each process.

Best Working Conditions

There are three types of factors:

1. **Material:** Physical material will be consumed during the process execution. If it is information that is needed only one-time, it will not be consumed, but is loses its value.
2. **Man, Machine:** They will not be consumed, but the resources need to be allocated for the duration of the process. The process consumes man-hours and machine-hours.
3. **Method:** It will not be consumed, since it is pure information. It can be reused.

Factors man and machine are both parts of the catalyst that accelerates the process without being consumed. The method is the description, how to build the catalyst. It explains how man and machine must be organized to transform the material to the product. How does the catalyst work in detail?

The duration of a process from start to end includes value creating activities plus the time it takes for the transport of material and movement of people and tools. Since all transports and movements don't contribute to the process result, they are considered to be waste (muda)[426] and have to be eliminated. Especially lean methods like 5S[427] try to ensure that all factors are placed best point. This reduces search, movement and transport times and thus accelerates processes. If all factors are available at the best point, then there is less waste, than if one factor is missing. When lean methods analyze processes and search for Muda, then they want to change the process and create even better working conditions. They improve the catalyst.

---

426 Lean production tries to identify the 7 wastes (mudas) in production (overproduction, transport, storage, movement, waiting, over-processing, defects). See (Ohno, 1988 p. 152)

427 5S is a lean method to bring all needed factors to the best possible position. It includes removing unneeded tools and bringing the right tools to the best position. (Bicheno, et al., 2009)

Non-production Environments

In a production environment the fishbone concept is already well-known. But it is not limited to production processes. Other trades like engineering or business administration have processes, which convert inputs (material) to outputs. These processes are performed by people (man) using tools (machine) and follow a method. In these examples, processes can be accelerated too, if everything is best prepared, before the process starts.

Create Best Working Condition

The process can be realized in three steps:

1. Get the competence to execute the process.
   This includes the knowledge, which resources (material and tools) are needed and how to organize them.
2. Get all resources.
3. Position all resources at the best possible place

If this is done, then the process should be much faster than an unprepared process, where know-how or resources are missing or at the wrong place and thus have to be moved.

The process "Create best working condition" itself is generic. The method consists of the 3 steps. The input for "Create best working condition" is the competence and all the resources for the process to be prepared. The catalyst is the person which has the competence to consequently follow these steps.

All this should be no problem. The real problem very often is, that know-how or resources are not available. Where do all the factors come from?

7.5.2 Create Competences, Develop Documented Knowledge

The function "Create Competences" is responsible for the knowledge and skills that are needed to execute processes. Competences are the result of learning. A typical method for creation of competences is TWI (Training within industry), where trainees learn knowledge and skills, supported by a trainer.[428]

The function "Develop documented knowledge" creates written knowledge (best practices or other information) that can be used directly or for learning. Documented knowledge is more stable than the individual competences of people and can be seen as the DNA of the organization.[429]

---

[428] Competences are the translation of RNA. They can be both catalytic (ribozymes) or used to store information how to build other catalysts (mRNA encodes enzymes).

[429] Documented knowledge is more stable then competences of experts. It can be used to train people to the same best practice and be used for continuous improvement. This is similar to DNA which allows several RNA-copies and is the subject of evolution.

7.5.3 Provide Resources

The function "Provide Resources" needs to provide all resources (people, tools) for a process (Figure 7-9). This includes all activities to allocate all resources at the right time to the right processes, but not the creation/development of resources.

Resource management is based on a clear inventory of all available resources and the knowledge about current and potential resource suppliers and consumers. This enables the resource manager to document which resources are used and which are available and can thus be allocated. If all consumers and suppliers of resources can give forecasts, it is possible to understand future bottlenecks.

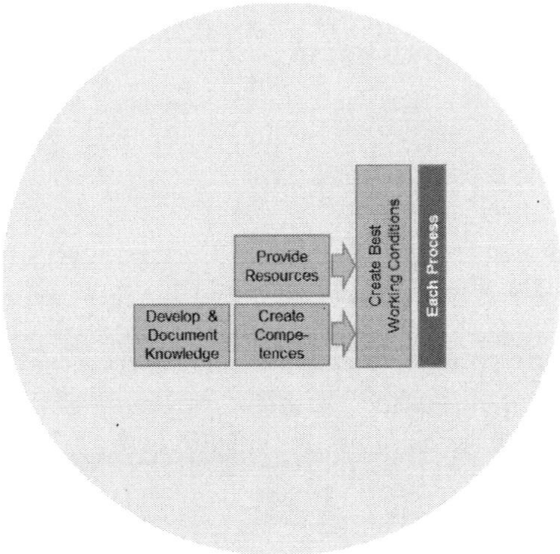

Figure 7-9: Provide Documented Knowledge and Competences. These functions provide the documented know-how and the competences of people.

So far, we have defined 4 functions. Provide Resources realizes the scheduling process which has to ensure that the other 3 functions and itself get sufficient resources (man, machine). It has to ensure that the generic function "Create Best Working Conditions" is always done before execution. This can be done with Kanban-boards using the PDCA logic (Plan, Do, Check, Act). Create Best Working Condition is the process that is executed in the plan/prepare phase, to make the do-phase efficient.

Create Competences is responsible to provide the know-how for all 4 functions. The competence is already needed to identify all the needed resources.

Now our system is able to work, if all necessary resources are available from outside. Unfortunately, this is nearly never the case. Available resources from outside are different from that what is needed inside. What can close the gap?

7.5.4 Import and Develop Resources

It needs several resources to create best working conditions for a process. Typically, these resources are not freely available. Instead, they have to be imported and developed (Figure 7-10)[430].

A simple example is a process, which requires some software tools. Even if the computer and the software is available, it requires some work to setup the computer, create logins, install and configure the software. The physical workplace consists of a desk, chair and other tools, which come from outside, but the organization needs to have processes to put everything at the right place.

All that requires processes, which require best working conditions as well. These processes take something what is available outside, import it and transform it until it is usable as a resource for a specific process.

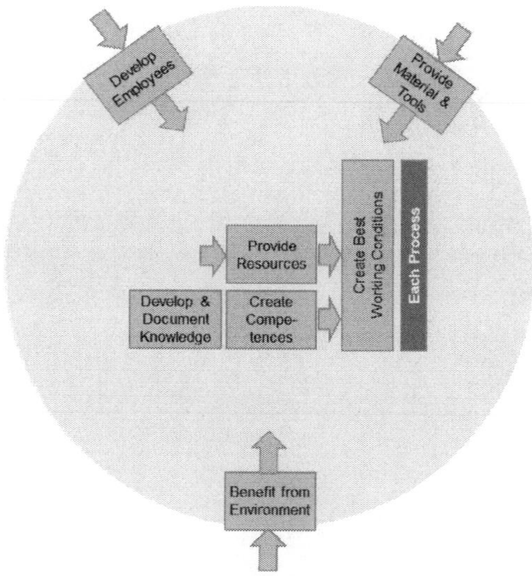

Figure 7-10: Develop Resources. The system has to bridge the gap between the resources that are available outside and the resources that are needed inside. Three functions develop these available resources such that provide resources can allocate them.

The processes form a network[431] which starts from non-self-created-objects and contains all transformation processes that create factors that can be used to create best working conditions for a process.

---

[430] The development of resources is the translation of metabolic pathways. Metabolic pathways start with freely available molecules like carbo-dioxide and transform them to amino-acids or nucleotides. See section 5.3 for more details.

[431] In biology the input molecules for the amino-acid synthesis pathway or the nucleotide synthesis pathways are connected. In business this is not the case, because people development processes and material/tool processes work on strictly separated objects.

7.5.5 Protect Resources

The purpose of protection is the reduction of losses. If resources are not getting lost or destroyed, then there is no need to create them again. This is how protection contributes to an increased net-growth (Figure 7-11).

The first protection mechanism is the outer boundary, which keeps everything together in high concentration. In organizations this is realized by the doorman who protects the company site, or by cybersecurity activities which protect information, or lawyers and insurances that protect the rights of the organization. This protection mechanisms are related to the discussed conceptual space.

Other protection mechanisms try to avoid losses inside the organization. Examples are health and safety activities which help to protect people, or the data backup that reduces the risk of losing important data.

Protection mechanisms are always related to the type of objects they want to protect, and related to the space they are located in. This makes it difficult especially for the protection of information. Information can be stolen in the physical world (by stealing a server), or by hacking a server in the cloud, or by social engineering.

If an expert has exclusive knowledge about processes of the organization, this is also a risk. Protection mechanisms have to ensure that the know-how is documented or handed-over to a second person, even if this is not creating any value yet.

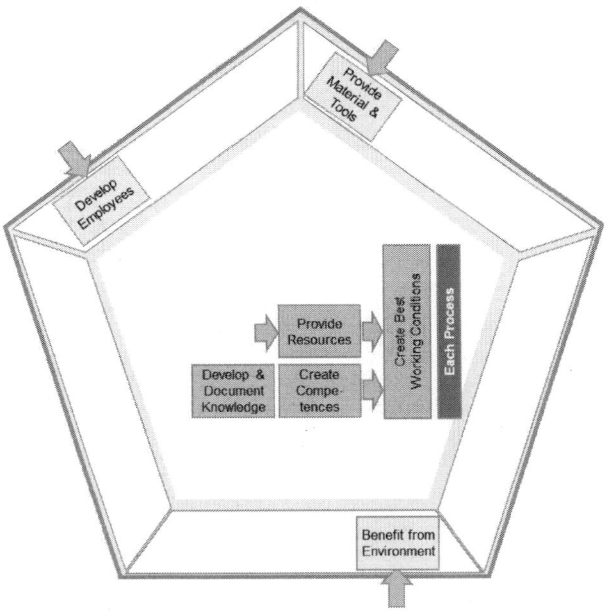

Figure 7-11: Protect and allocate resources. The outer boundary shows that everything inside is part of the organization, has to be protected and be available when needed.

7.5.6 Balance the System

Until now, we have discussed all functions that are needed to create best working conditions for each process of the organization. This forms a directed graph of processes starting from the objects coming from outside and ending with the best working conditions for each process. This complex graph can convert the inputs as fast as its weakest process. The bottleneck can be a limited resource outside or an inefficient or ineffective process inside.

The purpose of adaptation is to identify this bottleneck and strengthen it by assigning more resources or by increasing the knowledge (continuous improvement of the process).

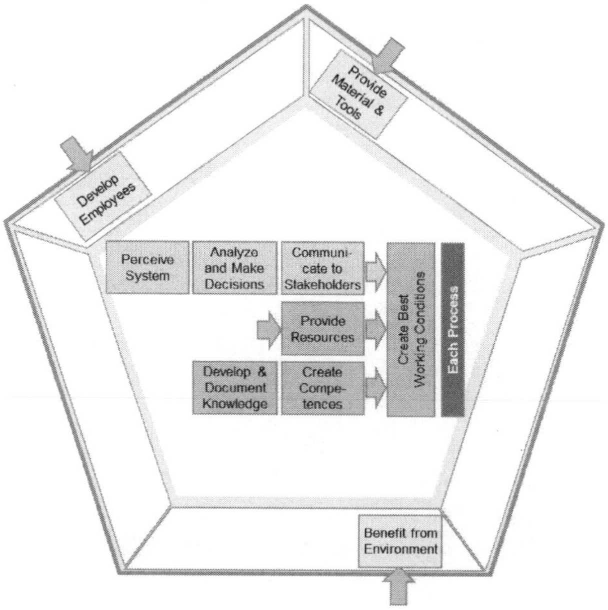

Figure 7-12: Perceive system, analyze and make decisions. Perception has to collect all available data. Analysis and decision making is about the identification and elimination of bottlenecks.

Thus, it is important that the organization is able to perceive its own resources and processes (Figure 7-12).[432]

1. The first step of adaptation is the perception of the own processes and the available resources. Perception has to deliver information. It can be very formalized like with KPIs or very informal.
2. Then this information has to be analyzed. Does the information indicate that the organization has a bottleneck? Which function is it? What is the current state of the function and why does it lead to the critical observation? If there are several critical functions: What is the most important one?
3. If the functions, that causes the bottlenecks, have been identified, then the organization can decide, to temporarily assign more resources, and/or to strive for a future state that promises better results and contributes to better KPIs.[433]

These steps are analogue to the control mechanisms of a cell. [434]

---

[432] See section 5.6 for the translation of adaptation mechanisms

[433] See (Goldratt, 1990) for the Theory of Constraints. Goldratt has explained the same theory in his famous business novel *The Goal.* (Goldratt, 1992)

[434] The main difference are the states. The cell has no understanding about the structure of its catalysts and is not able to change it. This kind of adaptation is done by evolution and not by the signaling mechanisms of the cell. Evolution has no defined future state that explain how molecules have to look in the future.

All these functions have to be done regularly. They are not limited to a yearly ritual; instead they happen simultaneously on several levels.

7.5.7 Create Value for Stakeholders

Stakeholders are important for us. They can give us the resources we need for our own growth or for providing value to other stakeholders. In return stakeholders want to get something from us. This give and take is a mutually beneficial symbiosis.

Stakeholders can be organized in stakeholder groups. Viable Business uses the five standard stakeholder groups *investors, employees, suppliers, customers and environment* and thus follows the recommendation of ISO 9004. The functional overview diagram contains two abstract functions that have to be defined for every stakeholder group.

Create value for stakeholders is about the creation and delivery of value for a stakeholder group. The most important point is the focus on the creation of value rather than on delivery of an object. At the end of the day the delivery needs to have benefits for the stakeholders. The most prominent example is the function *create value for customers,* which requires a large part of the organization's resources.

Get resources from stakeholders is about the import of resources from stakeholders. This function is always merged with the development of resources. This way, the functions contain all processes starting from the import of resources until the resource can be allocated as a part of best working conditions.

7.5.8 Acquire Stakeholders

The give and take with stakeholders can be realized as contracts or informal agreements. Such contracts or informal agreements have to be formed. The organization needs to have the abstract function *acquire stakeholders* for every stakeholder group. This function is responsible to form mutually beneficial contracts and thus creates responsibilities and claims.[435]

The organization has the responsibility to give something to the stakeholder, and it has the claim to receive something from the stakeholder (Figure 7-13).

---

[435] An excellent description of innovation in product management is provided by Fuchs and Golenhofen (Fuchs, et al., 2019). They describe the identification of markets, the elaboration of jobs-to-be-done and circumstances of customers, the definition of market segments and (prioritized) product functions that have to be implemented, the design of modular platforms and the implementation with agile teams. Fuchs and Golenhofen connect the dots and present tools to implement the entire chain from marketing to development. Absolutely worth reading! In the Viable Business model all these processes and methods, except the development activities, are part of function "Acquire Customers". Their purpose is to get more contracts.

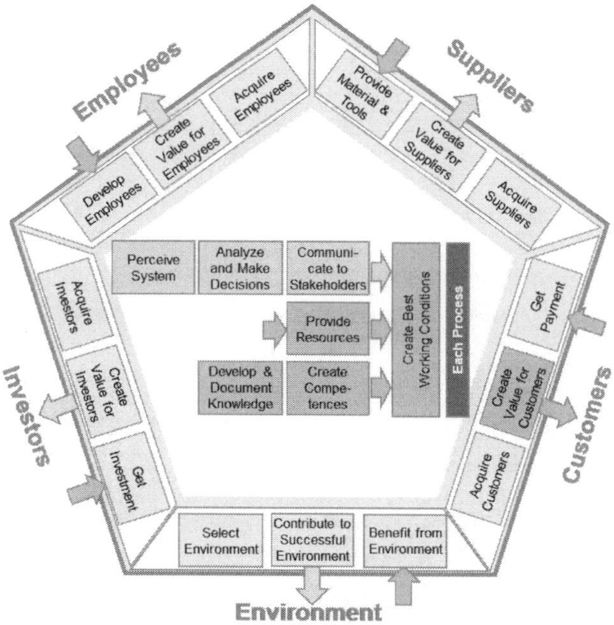

Figure 7-13: Functional Overview Diagram. The diagram shows all functions that have to be implemented to provide best working conditions for each process.

The relationship with stakeholders can be improved if we understand that the stakeholder is a Viable System too. The stakeholder needs something from us, because it is part of the best working condition for one of its processes. The more we learn about the stakeholder, the more we understand the bottleneck in the other viable system. This allows us to have a precise value proposition, that creates high value for the stakeholder, but can be realized with low efforts. This results in high revenue with low direct cost of sales, or in other words a high gross profit.

On the other hand, it is good if we understand our own needs and are able to communicate it precisely to all the stakeholders. This allows the stakeholders to provide exactly what we need – also with low efforts.

Both directions have the same target: Reduce losses at the relationship to stakeholders that results from missing information. It helps to reduce transactional cost.

7.5.9 Communicate to Stakeholders

The specific communication with stakeholders is already part of function *acquire stakeholders*. This specific communication has the target to form a symbiosis with a stakeholder.

This specific communication is based on mutual trust. Each stakeholder wants to have a high confidence that the collaboration will become successful.

If the organization communicates inconsistently with the different stakeholder groups, this reduces the trust. It is not possible to tell the story of sustained growth to one stakeholder group while telling the opposite to other groups.

The organization has to understand itself and communicate consistently to all stakeholder groups. This communication is independent from specific stakeholder groups and thus a central function.

7.6 Responsibilities, Authorities and Relationships

7.6.1 Viable Role

Responsibilities and Authorities

Viable Business can be implemented by a single person who is responsible for an organization and has the necessary authorities (Figure 7-14). If the organization is the entire company, then this person is the CEO. If Viable Business shall be introduced in a department (e.g. engineering department or R&D department), then the person is the department head. It is also possible to implement it without having an own department. In each case the person can see itself as the CEO for a given scope and decide to fulfill the role in a viable way.

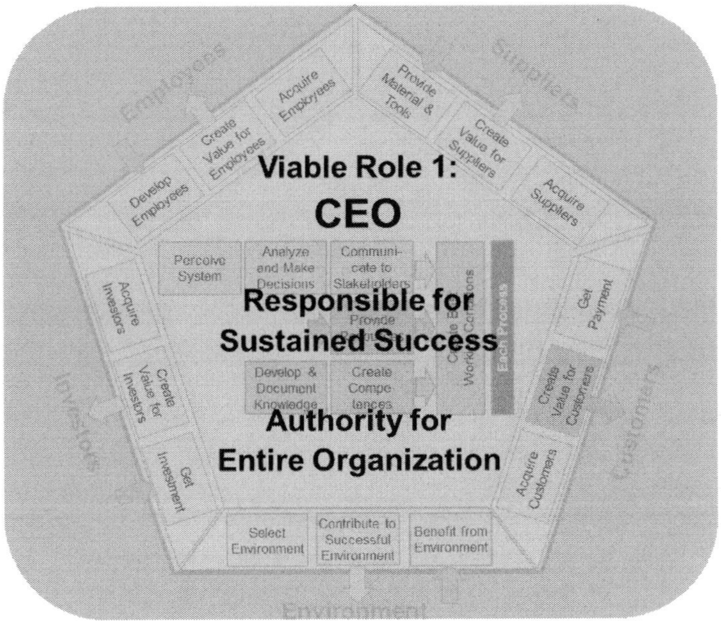

Figure 7-14: Viable Role. The CEO of an Organization is responsible for the implementation of all functions.

Typically, the organization, department, or a simple role gets responsibilities and authorities, when it is created.

Authorities include resources like the time of the person that fulfills the role, the time of others, tools and materials, the empowerment to decide something, access to information and other resources. Some of the resources are directly available, others are rights to get something from other parties.

Authorities are the resources needed to influence something. [436] The limits of authority are the boundary of the system. Like a cell, the Viable Role should try to get the highest concentration of

---

[436] This fits to the first habit of "7 Habits of Highly Effective People" which recommends working from the center of the own influence. (Covey, 1989)

the right authorities, to be able to create best working conditions, especially to fulfill the responsibilities.

Responsibilities describe the value that has to be given to someone else. Thus, they are about the creation of value for stakeholders.

How can all the responsibilities be fulfilled with the help of the given authorities? Previous chapters showed that the Viable Role has to fulfill all viable functions to create best working conditions, especially for the fulfillment of responsibilities. This requires sufficient time of the CEO and the right competences (Table 32).

	Competences of a Viable Role	Function
	Competences for self-organization	
1	Ability to perform excellent work preparation and to use best practices (Prepare before Do)	Create Best Working Conditions
2	Ability to acquire new competences (by observing processes, by communicating with others or by reading documented know-how)	Create Competences
3	Ability to document knowledge (for later use or for sharing it with others)	Develop and Document Knowledge
4	Ability to manage resources (own time management)	Provide Resources
5	Ability to develop basic competences (systematically identify and develop the basic competences that)	Develop Employees
6	Ability to create a perfect tool environment for the work based on available standard tools	Provide Material & Tools
	Competences for adaptation	
7	Ability to create a clear picture of own processes and resources	Perceive System
8	Ability to identify current and future bottlenecks	Analyze and Make Decisions
9	Ability to eliminate bottlenecks by deciding about objectives and usage of resources	Analyze and Make Decisions
	Competences for stakeholder relations	
10	Ability to identify all stakeholders	Perceive System
11	Ability to understand stakeholders needs, express own demands and to create mutually beneficial relationships.	Acquire Stakeholder
12	Ability to create high value for stakeholder as promised (including the ability to say no when additional responsibilities are added without additional resources)	Create Value for Stakeholder
13	Ability to get the needed resources from stakeholders	Get Resource from Stakeholder
14	Ability to create a consistent communication with all stakeholders	Communicate with stakeholders

Table 32: List of needed competences for a Viable Role

The competences provide the potential to do the things right, but they are not sufficient. Additionally, it needs some stable structures to apply them.

Structures to Realize Viable Functions

The CEO needs to realize many functions and provide best working conditions for all of them. All the functions have to be performed permanently. Since the CEO has one brain only, he has to do one thing after the other. Thus, all the functions have to be reflected in his calendar (Figure 7-15).

It is a challenge to ensure that no function can be forgotten and that each function uses the known best practice and is done with the right focus. A good solution is to use defined timeslots for all activities which are not about the creation of stakeholder value. The calendar gives timeslots for the realization of functions.[437]

Time	Mo	Tu	We	Th	Fr
08:00	Daily Planning & Work Preparation				
09:00					Weekly Review
10:00	Create Value for Others				Weekly Planning
11:00					Renew Energy
12:00					
13:00	Work Preparation for Afternoon				
14:00					Learn Work Preparation
15:00	Create Value for Others				Learn
16:00					

Figure 7-15: Timetable for the Viable Role. Example for the usage of stable substructures in the calendar

The most important timeslot is the one for the daily planning and work preparation. It contains the daily routines of the functions:

1. perceive system
2. analyze and make decisions
3. communicate with stakeholders
4. provide resources
5. create best working conditions.

Methods, like Getting Things Done, show how these tasks can be done practically and provide some useful tools.[438]

Excellent learning and teaching need their own protected environment. Therefore, a second dedicated timeslot is about creation of new competences. It contains the functions:

1. create competences
2. create documented knowledge

Inside the standard timeslots there should be a fixed agenda which ensures that each function is done with the best working conditions.

---

[437] See section 5.8 for the discussion of substructures. It explains the translation of cell-organelles to the calendar of the manager.

[438] See (Allen, 2013)

All other functions can be done in the remaining time.

7.6.2 *Viable Team*

In larger organizations the CEO is not able to do everything alone. Thus, the next step is to invite colleagues to contribute to the objective of sustained success. This leads to the first Viable Team (Figure 7-16).

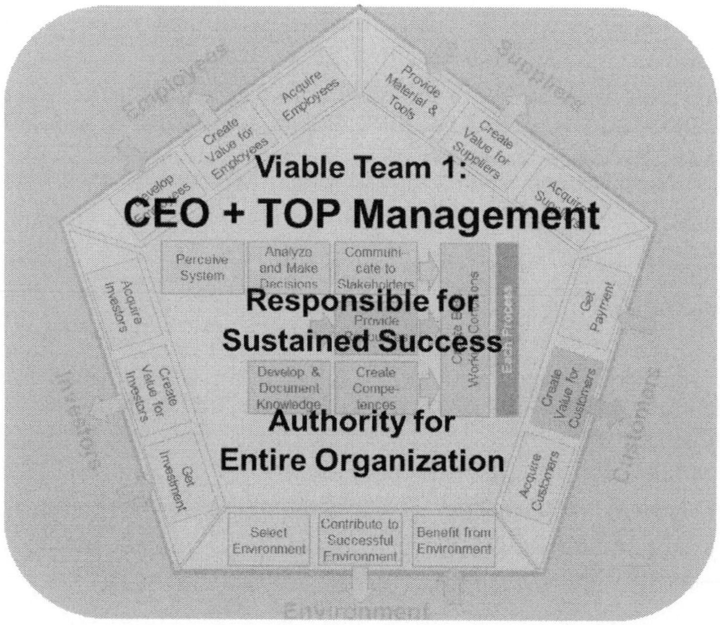

Figure 7-16: Top management team as the first Viable Team. Top management has to implement all viable functions.

The Viable Team as a whole is now carrier of responsibilities and authorities. The team can assign them to the team members or to sub-teams. This leads to a demand of communication between team members.

This communication is not part of the payload of the team. Instead, it is a part of the self-organization and adaption functions, which are already implemented by Viable Roles, that has to be done together.

Lean and agile methods can help to implement the functions. The sprint meeting contains methods to assign responsibilities and authorities to team members. The daily standup meeting contains methods to synchronize the team members. The frequency of both meetings depends on the team.

Sprint Meeting

The sprint meeting consists of review and retrospective of the old sprint and the planning for the new sprint. This planning part includes the assignment of tasks to team members and thus the assignment of responsibilities and authorities.

Standup

At a standup every team member answers the three questions:

1. What did I do yesterday?
2. What will I do today?
3. What are my impediments?

Obviously, it needs preparation work prior to the daily standup to answer these questions. For a software developer who works on a complex task, this is easier than for a manager who has many requests that come in daily. A person who doesn't overlook the own calendar or the mail inbox is not able to speak about the plan of the day. It is also not possible to understand the impediments.

For Viable Roles this is no problem at all: In the daily self-organization hour, the day has already been planned and all activities are well prepared. Everything that avoids or reduces best working conditions is an impediment.

If the self-organization meeting is directly prior to the daily standup, then the status, the plan for today and the found impediment can be reported.

Support

The reporting of impediments alone doesn't help. Someone has to solve it. In SCRUM this could be given to a SCRUM Master or other team members.

Viable Business is designed such that there are clear responsibilities for each part of best working conditions. All functions are designed by the root cause diagram. If the team has assigned all team-level functions to team members, then it is clear whose job it is to solve an impediment and it is possible to address the impediment during the meeting.

Sharing Information

Team members exchange information during their meetings. The verbal communication between team members has some disadvantages. At the end of a discussion a lot of information is forgotten, other information is understood by team members in different ways.

A good practice is to write down information and visualize it. This creates a stable representation of information and makes it easily available for all team members.

Practically, all tasks of a team are written on cards and placed on a Kanban board. All relevant information like KPIs can also be visualized. This leads to a set of boards containing the most relevant information, which has to be shared by all team members.

Sharing Resources

Not all resources are used exclusively by team members. Some tools, material and information need to be shared. One example are shared documents.

Until now it was no problem to ignore naming conventions or file structures. The more team members enter the team, the more it becomes necessary to agree on rules. Software developers not only have naming conventions for files. They also have detailed coding conventions. Other professions, like accountants, have conventions as well.

The team has to establish some standards that help the team members to collaborate efficiently.

7.6.3 *Large Viable Organization*

After the agile working mode in the Viable Team is established, the team can integrate more and more people into the team to handle a growing workload. This leads to a growth of the Viable Team, which becomes too large and needs to be split (Figure 7-17).

Replication of Teams

The creation of new teams by splitting existing teams has several benefits. The members of existing teams already have experience with all the lean and agile methods, and the collaboration between them has been trained before. Inside the existing team they already share the same targets.

Initially, one team was responsible for all functions. If it splits, it needs to delegate some functions to the descendant teams. This delegation requires the empowerment of the descendant team and the provision of sufficient resources This delegation gives the descendant team the necessary authorities to work successfully.

If this split is done several times, then hundreds of teams can be created which fulfill functions and are responsible for the corresponding stakeholders and KPIs.

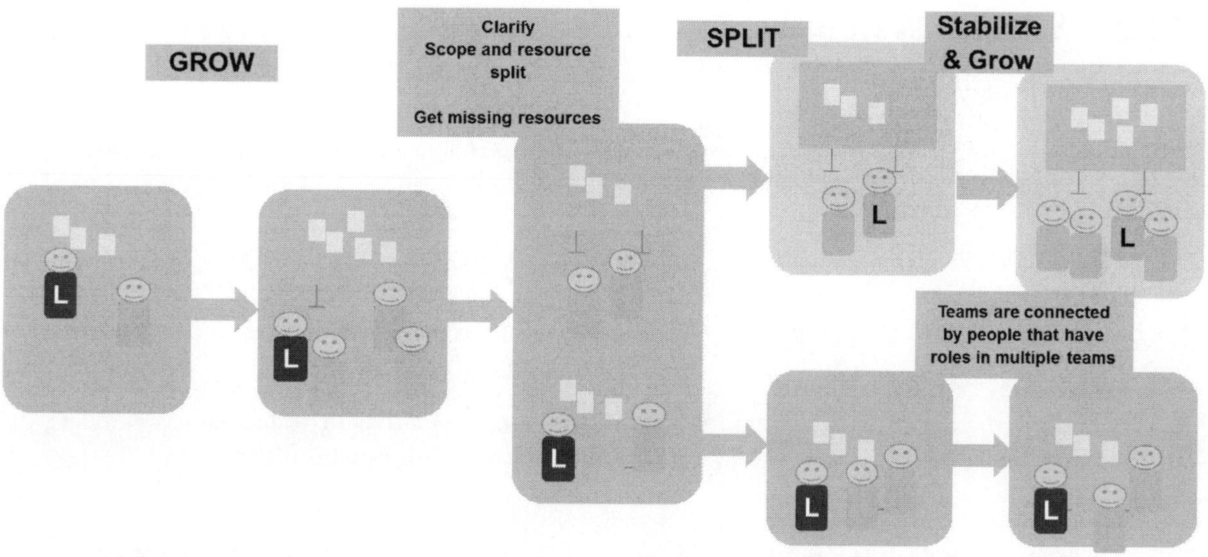

Figure 7-17: Replication of teams. Split the team requires to duplicate all resources (people, Kanban board)

Specialization of Teams

Every newly created team has a focus on specific functions. This leads to the functional specialization of teams. It is not needed that every employee has all skills. Employees can specialize and become more productive.

Team Types

If the organization grows and creates many teams, it turns out that some of the teams are very similar. Project teams, for example, use the same set of project management methods, even if they have unique project targets. The same is true for leadership teams. They also use a set of management methods.

This similar team design allows to speak about team types. The usage of team types allows us to design a Large Viable Organization based on a library of some predefined team types, which are already understood.

In Chapter 8 we will discuss a specialization that has been tailored for project business and describes a set of pre-defined team types.

7.7 Identification and Elimination of Bottlenecks

If there are no limitations in resources and processes, the organization had sustained growth with exponential growth rate. In reality, this growth is impeded by internal and external bottlenecks.

Thus, management has the task to identify and eliminate all the bottlenecks step-by-step. This identification and elimination results in policies and objectives for all the functions, especially for the provision of resources.

Let us see how this works in detail:

7.7.1 Perceive System

Purpose of Perception

The adaptation of Viable Business starts with perception of the inside and outside world. The purpose is to get sufficient information to identify bottlenecks with high reliability.

Perception of the inside world is about processes and structures. The organization can measure quality of processes, resource consumption or duration of processes. It also includes availability of resources like available know-how.

Perception of the outside world refers to the needed resources from outside, the stakeholder groups and the markets they are in. Important information can be the current satisfaction of stakeholders, which can be measured by surveys. Other information is about changes in the markets or activities of competitors.

Which information helps to identify the bottlenecks? We can start with the overall objective of sustained growth.

Growth

Growth is about self-creation. The question is: What is the growth rate per year? It can be measured in percent per time unit like 10% per year.

An alternative definition could be: How long does it take for an organization to duplicate its size? This would be measured in time units like “7 years”, which means a growth rate of 100% per 7 years. A negative duration, like -7 years, indicates that the organization had the double size seven years ago.

Both approaches do not speak about money, but they can be easily converted in money by multiplying with capital. The ROCE (return on capital employed) is measured in percent, but it is normally reported yearly and thus means percent per year. Every investor can multiply it with the value of his shares and gets the gain of value in the period of one year.

The system dynamics revealed that all the functions are contributing to the overall growth. Each function is needed to provide a resource for the best working condition of processes. If the entire system wants to duplicate itself, the throughput of each function also has to be duplicated.

The functions create customer value and get payment, for example, provide revenue. This revenue can increase every year, under the condition, that there are enough new orders from function acquire customers and enough people with the right qualification to do the work. If the organization is not fast enough to hire and develop new colleagues, then it has a bottleneck. It has new orders, but it is unable to execute them. Having many employees but no offers is even worse.

The question for the organization is now, whether it should use the time of the experienced people or to get new projects or to train employees. This question about the best resource allocation needs information about the throughput of each function.

The best answer is to allocate the resources in a balanced way, such that there are no large bottlenecks in one function while other functions have a surplus of resources.

Throughput of Functions

For some function the organization already uses performance indicators. Order intake is always measured, but other functions like the development of people need to be observed as well. The output of this function are qualified people that can be staffed. It is good to understand the demand, the current stock of qualifications, and the throughput of development for each role. How many project managers can be allocated today in one year?

Project Business typically has the problem to have not enough employees with high qualifications like project manager or system architect. If there is a bottleneck, then the entire organization is not able to grow.

Stocks

One of the problems is to understand the time behavior. Some functions produce stocks. Function *acquire customers* adds new contracts to the order backlog. Function *create competences* adds something to the already existing knowledge and skills of people.

If an organization stops all training activities, then there are still all competences available that have been developed in the past; but without new training, the average competence goes down. Experienced colleagues retire or leave, and new colleagues perform work without explicit training.

Stocks decouple demand from supply. They allow a function to deliver, even if the function does not produce fast enough. If the demand is permanently higher than the supply, the stock shrinks, and the organization has a severe bottleneck.

Thus, the organization should understand the size of stocks, and their losses. This information helps to understand, whether a stock is big enough to temporarily reduce the efforts for its creation, or not.

Duration of Functions and Concurrency

A similar question is: How long does it take to develop a new employee to a project manager? The duration of a process gives valuable information of potential bottlenecks. If it takes ten years to develop a project manager for large projects, and the demand for the next years is known, then it is possible to calculate the number of people that should be on this career path, in order to fulfill the future demand.

This connects to growth rate of the organization via the throughput of functions with the duration of a function. A long duration requires to do many things in parallel. If this parallelism is done by only one person, then this person is less focused, compared with doing one after the other. He has to switch focus between many activities. Even if this parallelism is distributed to several people, this requires more management attention.

A long duration has an additional disadvantage. If it takes 10 years to develop project managers, then the organization has to start to develop them now for the demand in 10 years. If the organization is stable or has a small growth rate, this is a challenge but possible. The higher the growth rate the more people have to be trained now.

Thus, it makes sense to reduce duration of all functions as much as possible. This contributes to a high throughput of a function by minimizing parallel work and it reduces the work in progress for each function. As a consequence, the organization should measure the throughput and the duration of the critical functions.

Typical KPIs for the measurement of duration are the project duration and the duration of acquisition.

Scarce Resources

Economy is about the usage of scarce resources. In our system the scarce resources have to be balanced to avoid bottlenecks. Thus, the organization needs to understand the available resources and the current resource consumption of each function. It also needs to understand how the throughput of functions depends on which resource.

Most functions do not need money. Instead, they need competent people with the right tools. The organization should have a list of all available human, physical and financial resources and understand their usage. This picture helps to see current or future bottlenecks.

Since money is always scarce, the cost of functions is of interest. If cost in projects can be reduced, then financial resources can be invested in other functions. Thus, project cost and bid cost are typical KPIs.

Value for Stakeholders

Value for Stakeholders can also be measured, if the value is well-understood. The value propositions for the stakeholder relationships can be used as a starting point.

One example is the engagement survey, which asks employees about their satisfaction with the employer. The questionnaire contains several questions, where employees can rate how good an aspect is fulfilled. They can also rate the importance for each aspect.

The happiness weighted with the relative importance create a performance indicator, which can used by the organization to improve the relationship to employees.

Viable Business Dashboard

The Viable Business Management System can be used to locate all this information in one picture and to relate it to the organizational functions (Figure 7-18).

The structure of the dashboard is different to the Balanced Scorecard, because it is not about the implementation of a strategy. Viable Business uses the dashboard to identify and eliminate the bottlenecks of the business-as-usual. Here performance indicators are related to functions, and functions are organized by stakeholder groups.

Thus, the structure of the dashboard is identical with the functional overview diagram.

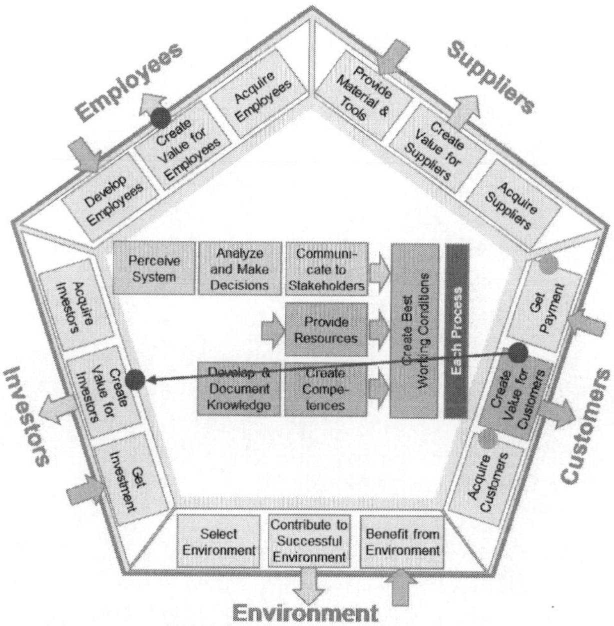

Figure 7-18: Perceive system. Circles represent performance indicators that are related to functions. Red circles show KPIs with poor values. Here the EBIT for the investor is too low, which is the consequence of the high NCC in customer projects. Employee satisfaction is also rated low.

If the organization has problems with some of the KPIs, like EBIT, non-conformance cost or project duration, these KPIs can be red-lighted in the functional overview diagram. This shows where the symptoms are.

The KPI system tells the same story as the functional overview diagram. It explains with numbers, data, facts how the value of stakeholders is linked to the available resources which can be used to reduce duration and increase throughput at the bottleneck of the system. All this structuring work can be seen as the preparation for the perception process.

Perception is about collecting all the numbers, data and facts. This information is the starting point for the discussion, what the reasons for poor performance values are.

7.7.2 Analyze and Make Decisions

Identification of Bottlenecks

In a second step the team can analyze the causes for poor KPIs (Figure 7-19). In many cases there is more than one reason for the high non-conformance cost. The reason can be the discipline of employees or their motivation. The reason can also be a wrong supplier selection or the wrong product.

The management team can use the Viable Business blueprint to discuss, which functions should be in the focus. When the critical functions are identified, it must be understood, how the functions are currently realized. What is their current state? Which processes and structures are regularly used to produce the results? This understanding links the way of working to the local effect that the function has an insufficient output and a remote effect in a different function.

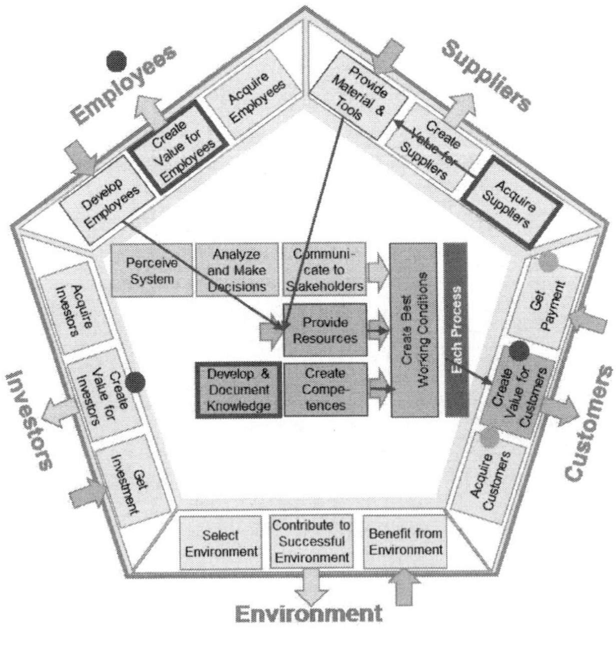

Figure 7-19: Analysis of bottlenecks. The diagram shows several reasons for the NCCs. The supplier selection process might lead to poor material quality, a poor culture leads to low motivation and discipline, and the developed product is not state of the art. All factors contribute to poor working conditions in projects and thus to NCCs. The diagram highlights the functions, which have to be detailed.

One example is wrong product portfolio management processes, which lead to the definition of the wrong product requirements and at the end to a product that cannot be easily sold. The remote effects are customer projects that are already under pressure when they start. This leads to high NCCs and low EBIT as remote effect.

In many cases the causal chain between wrong product portfolio management and EBIT is very long. Thus, it makes sense to document this chain in a so-called benefit map and make it visible for discussions.[439]

Viable Business contributes to the discussions by providing a shared picture of the organizational functions which has a clear relationship to the overall goal to have sustained success.[440]

When the problematic functions are identified, and their current states are understood, then the future states can be defined (Figure 7-20).

Elimination of Bottlenecks

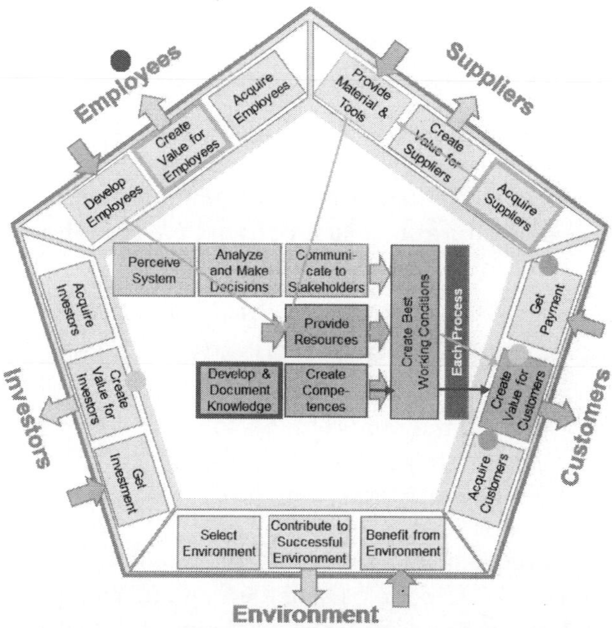

Figure 7-20: Definition of targets and future states. Management has selected the supplier and culture problem first and defined a future state for the functions. Now it needs resources for the processes which defines the new processes in detail and provides best working conditions for the changed functions. If the functions are changed, the NCCs should go down, and management can focus on the next bottleneck.

How should the organization work in one month, in one year or in three years? The future state gives a target for the function. The future state of a function should clearly refer to performance indicators of the function. This makes the future state a hypothesis that can be checked. The organization jointly believes that it would have better performance indicators if it changes the process.

---

439 Benefits maps are explained in section 8.5.10. A detail description can be found at (Axelos, 2011a).

440 Here the functional overview diagram realizes the 5th discipline. It is a shared understanding how all the activities are related and how they contribute to the shared objective. The 5th discipline is described in the bestseller of Peter Senge (Senge, 2006)

Future states can also be about stocks. What should be available in one month, in one year or in three years?

The picture of a better future is based on a clear understanding of the current state and the available numbers, data and facts.

This picture of the better future can also be seen as the vision or objective of the organization. From the perspective of Viable Business this vision is only a intermediate objective which eliminates bottlenecks and contributes this way to the real objective to live long and prosper.

How Management Achieves the Future State

In simple cases, the organization can create a new policy that explains how to work and realigns the resources to the bottleneck. Typically, these resources are not available. This makes it necessary to free resources at other functions.

In other cases, the future state of an organization can only be established, if some additional capabilities are made available. Here the organization has to invest resources to create the capabilities to be ready for a change of state.

Chapter 8.5 explains the mechanisms of improvement very detailed.

For the moment it is sufficient to understand that management has processes to identify the bottleneck of growth, to define a future state that does not have this bottleneck anymore and to establish policies to achieve the future state.

7.8 Viable Business Management System

In section 7.1.2 we already discussed the requirements for management systems. The ISO 9000 standard defines a management system as a "set of interrelated or interacting elements of an organization to establish policies and objectives, and processes to achieve those objectives"[441].

Sections 7.3, 7.4, 7.5, and 7.6 detailed this "set of interrelated or interacting elements of an organization". These elements are the people using tools and following best practices to fulfill Viable Roles. These Viable Roles realize all necessary functions to create best working conditions for each process.

Section 7.7 explained how the functions *perceive system* and *analyze and make decisions* establish policies and objectives in order to achieve sustained growth.

In other words: Viable Business fulfills the definition of a management system, even if it is abstract and has to be detailed for more specific situations.

---

[441] See (ISO 9000, 2015 p. 36)

7.9 Summary

Let us recapture the main concepts

1. Viable Business has the main objective to “live long and prosper”. It is the base for sustained success.
2. Viable Business utilizes the system dynamics of living organisms for its overall functional structure.
3. The functional overview diagram is the starting point for a functional decomposition of the system
4. The core of the system is the ability to create best working conditions for each process. Viable Business is process-oriented.
5. All relations to stakeholder have the same importance. They are symbiotic relationships.
6. Management is about the identification and elimination of bottlenecks in the system dynamics.
7. Viable Business can be implemented by a single person. It can grow organically to a Viable Team and to a large multi-team organization.

8 Viable Project Business

8.1 Introduction

8.1.1 People Perspective

The CEO and the top management team start the first Viable Team. Their goal is to make the entire organization sustainably successful. For this purpose, they use the Functional Overview Diagram to discuss how to create best working conditions for each process. They identify bottlenecks, define needed countermeasures to eliminate them, assign tasks to team members. If they want to gain speed, they can integrate additional team members. This leads to a growth of the team and consequently the team can do more work. Unfortunately, the effort for communication grows quadratically with the number of team members and reduces efficiency of the team. Normally, such a large team tends to build smaller subgroups inside (Figure 8-1).

After some time, maybe with more than 10 team members, the team size has become too big for successful teamwork. The team decides to split itself and creates additional Viable Teams based on the existing subgroups. These newly created teams stay connected by one or more people which participate in multiple teams. This allows to transfer information between teams.

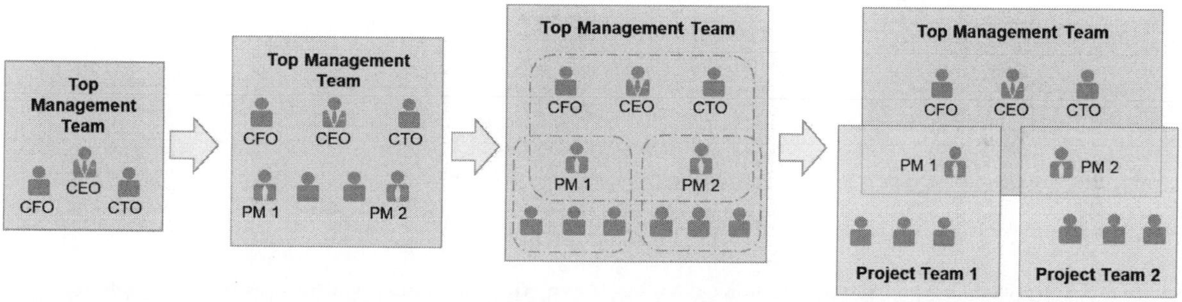

Figure 8-1: Growth and replication of large teams. When the teams become too large, these subgroups are the nucleus for new teams. Before the split, it has to be ensured that every subgroup has the ability to run an independent team.

For example: A startup, which executes two customer projects, starts to grow. All the employees work together in one large team without internal subgroups. After a while, there are three internal subgroups. Some employees are more interested in project one, others in project two. The owners of the startup want to discuss other topics like finances, recruiting or acquisition of the next project and are not so much interested in project details. The three subgroups become more and more independent and the entire team can split into three descendent teams.

What are the preconditions for a successful split? The subgroup or the single person that intends to form the new team, must have all abilities to run a Viable Team. If all members of a new team had a lack of self-organization then the new team has a high risk to fail. The people remaining in the original team must also cover all needed abilities; otherwise they would fail, when all the self-organized people are leaving.

C. Dachs, *Viable Project Business*, Contributions to Management Science, https://doi.org/10.1007/978-3-030-62904-5_8

Thus, the first step of replication is to ensure that the needed abilities have been duplicated. This requires training and coaching and takes its time. It is much easier to introduce new persons to the teams, than to train them. The speed of competence development is the limiting factor of growth.

8.1.2 Functional Perspective

Viability on Multiple Levels

The organization is now a Large Viable Organization, which consists of Viable Teams (Figure 8-2). The entire organization needs to implement all functions, and also each team has to implement them. Why is it needed to have all functions on both levels?[442]

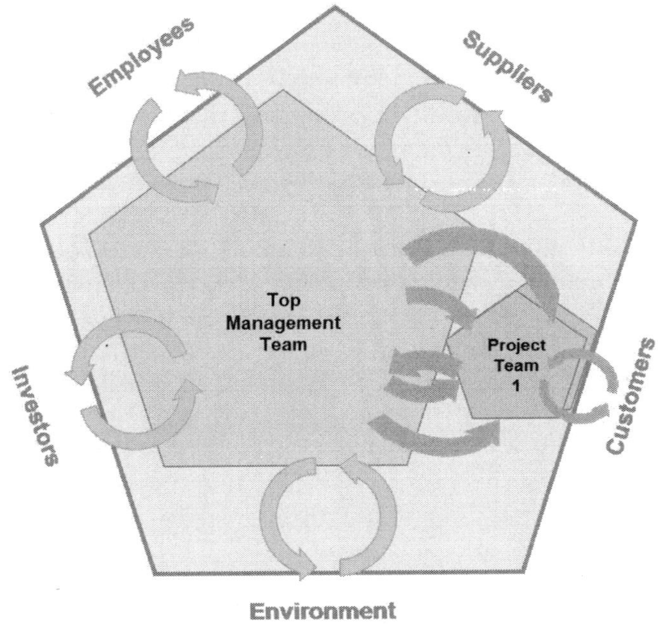

Figure 8-2: Large Viable Organization consisting of Viable Teams. Top management team provides best working conditions (man, machine, material, money, method and Environment) to the project teams. Project teams provide gross profit to the top management team.

For the Large Viable Organization nothing has changed. It needs all functions to connect the best working conditions with the outside world. Additionally, every Viable Team has to implement all the functions. This is necessary because every team has processes which need best working conditions that require resources from outside or from other Viable Teams. Consequently, every team can use the Functional Overview Diagram as a functional map for itself.

Internal Stakeholders, Exchange of Resources

Each team is now specialized on a group of functions, it can create more resources of a certain type and is thus able to provide these resources to others. This leads to internal supplier / customer relationships between the teams (Figure 8-3).

[442] This is similar to multicellular organisms. The entire organism is alive and each of the cells is also alive. There are n+1 living organisms.

What is the benefit for each team? A team can get several resources from the same stakeholder group. Project teams, in our example, get all the needed financial, human or physical resources from the top management team. Since the relationships to neighbor teams are stable, it is not needed to acquire new partners. It is much easier to maintain existing relationships. This reduces transactional cost and allows the project teams to focus[443] on creation of customer value. This allows to simplify the functional overview diagrams by reducing the number of stakeholders and by removing the acquisition functions.

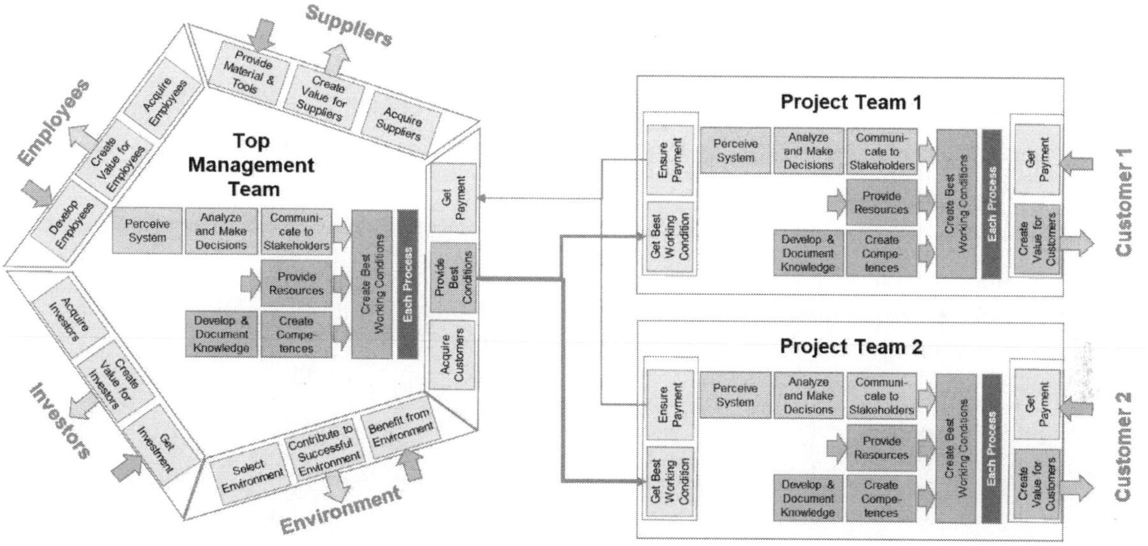

Figure 8-3: Collaboration of top management team with two project teams. The top management has to provide best working conditions for both projects. The projects deliver gross profit by creating and delivering customer value. All teams need to be Viable Teams, they have to implement the same functions. The difference is in the stakeholders they have.

This simplified view emphasizes that specialization can reduce transactional cost, but it is not quite correct. People work is simplified by the top management team. The project manager does not have to take care about work contracts, unions or basic qualifications. All these things are done by the top management team. This allows the project manager to fully focus on the project and to spend more time on customer value creation. Even if the project team gets the high-qualified, motivated people from the top management team, the project manager has to ensure that there is an excellent culture inside the team. He is also responsible that people learn during project work. Thus, it is better not to combine the stakeholder groups and omit the acquisition function. Instead, every team should use the functional overview diagram and identify all stakeholders and their needs individually.

Purpose of Teams

The collaboration between teams is about exchanging resources that are needed for best working conditions and about exchanging information. Like for the entire organization, it is possible to describe the stakeholders and the give and takes for each team. The project teams, for example,

---

[443] Increase concentration.

have the top management team as an internal stakeholder. They take resources from the top management team and give gross profit.

Similar to the entire organization, each team has a mission or purpose statement that describes what the team gives for the main stakeholders. Especially the teams that are connected to external stakeholders have an external and an internal purpose. The project teams, for example, create value for customers and provide gross profit for the own organization.

This give and takes of a team and the team purpose can be used as the base for the KPI system. The key performance indictors allow to measure, whether a team is able to fulfill its purpose. They can also measure the resource consumptions of a team.

8.1.3 Organizational Perspective

Teams

A more simplified version shows teams in a black box view. Every team is represented by a box. The collaboration between the teams is indicated with arrows (Figure 8-4).

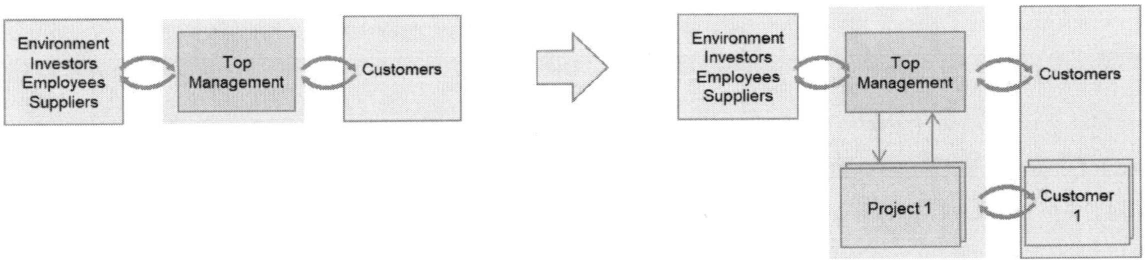

Figure 8-4: Assignment of stakeholder relationships from top management to project teams. This Large Viable Organization consists of a top management team and two project teams. The symbiosis between teams is shown with arrows. Every unidirectional arrow represents the function to give something. Bidirectional arrows represent a give and take.

Team Types

Some of the created teams are very similar, because they have the same purpose. This leads to a set of team types (Figure 8-5).[444] Team types can be used to collect all characteristics and methods that are shared between many teams. They allow to document standards for similar teams, share knowledge and to systematically train all team members. [445]

---

[444] This model is similar to biology, where every cell of a multi-cellular organism is a living organism on its own, but it contributes to the overall organism by realizing one function, while it gets needed resources from other cells.

[445] This is the concept of object-oriented design: The type – instance relation is similar to class – object relation in UML. Base types are base classes.

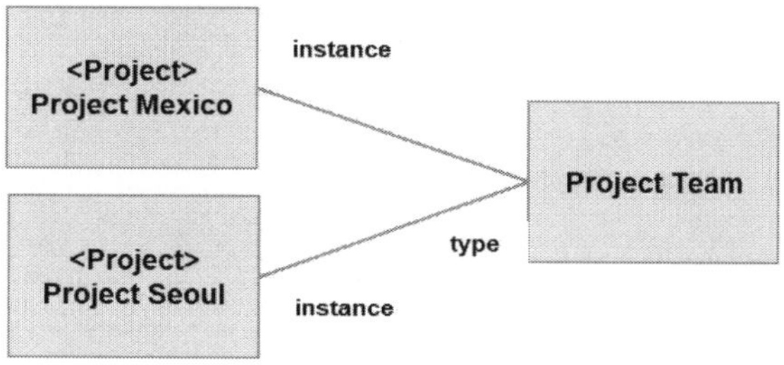

Figure 8-5: Team Type with instances. Projects Mexico and project Seoul have both the same team type.

Viable Project Business uses a set of pre-defined team types to design large collaborations.

Specialization

Some knowhow, like the elementary self-organization, is important for all team types and can be shared in a base team type. The basic team type is the Viable Team, which contains all the basic mechanisms to run a sustainably successful team, independent from the specific purpose. Every team has to be able to create best working conditions for itself, it has to be able to create competences, manage resources, and perceive itself and its stakeholders, analyze, decide and communicate. The basic principles of stakeholder relationships are also the same. Every team member of every team needs to be trained in these basic mechanisms (Figure 8-6).

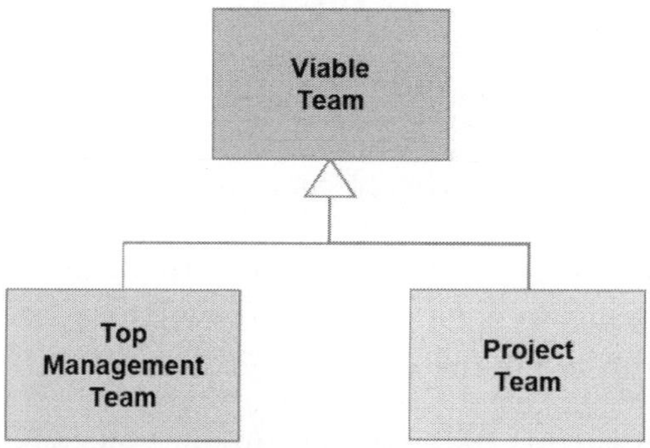

Figure 8-6: Specialization. Team Types can be specializations from others (Project Teams are specialized Viable Teams)

Specialized team types have additional or modified methods. In our example we have used the team types "Top Management Team" and "Project Team". Both are derived from "Viable Team" and add additional methods. Project teams need to implement project management techniques like planning, scheduling, quality management, documents management or change & claim management. Top management needs other methods like the market research or SWOT analysis.

The team types are different in several aspects:

1. Each team type has a specific purpose.
   Bid teams, for example, have the focus to acquire new customers. Team members need to have a much higher skill-level to connect with persons from other companies, than team members of other teams.

2. Each team type has other internal stakeholder relations.
   Instances of the same team type have the same typical collaborations with other teams of the same or different team type.
3. The team types have different lifecycles.
   Projects are temporary endeavors and the project teams exist temporarily for the duration of the project. The top management team is a permanent team.

Specialization is only a relationship between the team types, not between the teams. It is helpful to model common aspects, but it does not explain any collaborations. The collaborations of teams can be modelled by aggregation and associations.

Aggregation

The second relation between the team types is the aggregation relation. Aggregation means that one team type is the parent of another one (Figure 8-7). This relation can, in opposite to the specialization, be seen in the collaborations of teams.

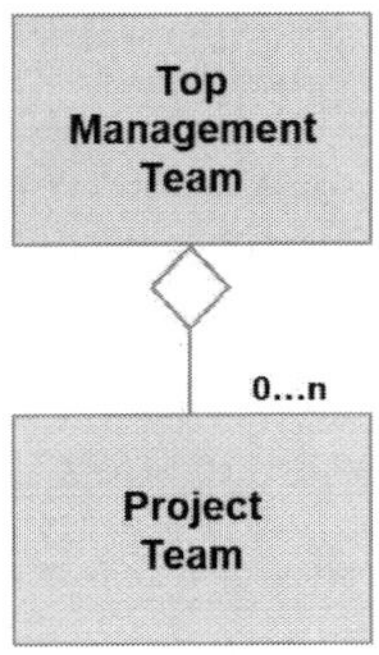

Figure 8-7: Aggregation. A top management team can collaborate with zero to many project teams.

The top management team, for example, has the overall responsibility for the organization. It can create dedicated project teams or stop them. Even if the project team is responsible for a single project, this does not reduce the accountability of top management.

Practically, the parent / child relationships show the path of responsibilities. Each team, except the top management team, has a parent team, which delegates authority and responsibilities to the child team.

Association

The third relation type is the association, which shows a typical collaboration between teams (Figure 8-8).

Figure 8-8: Association. The bid team has a collaboration with the project team.

One example for this relationship is the handover from the bid to the project team, after the contract has been awarded. Bid Team London has then a collaboration with Project Team London.

Object-Orientation

Viable Project Business uses the object-oriented design techniques to model teams and their collaborations. Section 8.6.1 contains a list of all pre-defined team types, with the used base types and the parent / child relationships.

Section 8.5 focuses on the team type "Continuous Improvement Team" and explains the main methods that have to be implemented there. This shows, how these team types can be refined.

Viable Teams

In Chapter 1 we have already discussed, that the members of a Viable Team have to communicate regularly in a team meeting. This led to a synchronization of the calendars for all team members. Regular meetings and fast escalation allow a fast communication.

Since each team of Viable Project Business is derived from a Viable Team, we can assume that this team-internal communication works quite well. Now we need to focus on the communication between all these teams.

8.1.4 Time and Location Perspective

If there is only one team, then all the team members need to have a regular meeting to exchange information. The team members can agree on the best time and location. The meeting is the best environment to exchange information and to make joint decisions, if all team members attend, all information is available, and thus the duration of know-how-transfer is short.

If there are many teams, then all these meetings have to be synchronized. Synchronization reduced the duration from one meeting to the other meeting and thus makes decisions and information flow faster.

In our example, the project manager is the link between the top management team and the project team. He has to attend both meetings to be able to transfer information between the teams. To allow fast communication between the teams, the regular meetings have to be synchronized. Project Teams have their meetings shortly before the top management meeting.

In small organizations the regular meetings can be done at different times at the same location. This allows short escalation times and the possibility to participate in every meeting (Table 33).

Team	Lead	Time	Duration	Frequency	Location
Project Mexico	John	Tu 10:00	1h	Weekly	Room 1
Project Seoul	Susan	Tu 11:00	1h	Weekly	Room 1
Top Management	Joe	Tu 13:00	1h	Weekly	Room 1

Table 33: Time and location of sequential project meetings

Unfortunately, the serialization of meetings extends the escalation time with every new project. A solution is the parallelization of all project meetings in different rooms. All projects meet in parallel and the project managers can meet with top management one hour later (Table 34). The escalation time is short, but members of the top management cannot participate in every project meeting anymore. Practically, the top management visits one team per week on a rotation basis.

Team	Lead	Time	Duration	Frequency	Location
Project Mexico	John	Tu 11:00	1h	Weekly	Room 1
Project Seoul	Susan	Tu 11:00	1h	Weekly	Room 2
Project Paris	Adam	Tu 11:00	1h	Weekly	Room 3
Top Management	Joe	Tu 13:00	1h	Weekly	Room 1

Table 34: Time and location of parallel project meetings.

The Tuesday 13:00h slot for the top management team has been used in some of the case studies. It allows to do the meetings of directly connected teams on the Tuesday morning, whereas the indirectly connected teams all meet on Monday. This allows having an escalation time of 29 hours with 40 teams using 6 team rooms.

In practice, it is a big challenge to defragment the individual schedules of teams. Since all calendars are already full, it requires to move many meetings to different times, which often results in a location problem as well. Teams can shift their meetings, if they use other meeting rooms.

In large organizations this defragmentation of time and meeting rooms needs a central planning and can take several months. Such a planning describes the future state of meeting slots and shows in which intermediate steps this future state can be reached.

8.1.5 Team Collaboration Diagram

The functional and organizational perspective showed how several teams can collaborate to create best working conditions for each process. They show the "Why", "What" and "Who". The time and location perspective gave answers to the "When" and "Where" questions.

The team collaboration diagram presents all the answers und can be used to communicate the internal structure of a Large Viable Organization. It shows all Viable Teams and their main relationships (Figure 8-9).

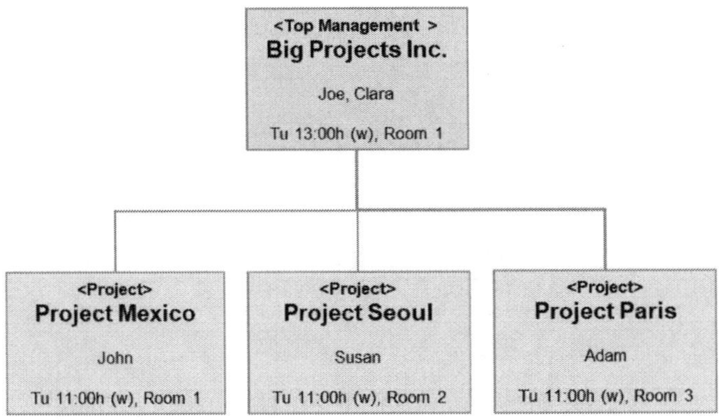

Figure 8-9: Team collaboration diagram. The diagram shows the team type, team name, team leader (or contact person) and the time, frequency and location of the regular meeting

Since the relationships between teams is the same for all teams of the same type, all the arrows can be omitted.

8.1.6 Greenfield or Brownfield?

The previous sections showed the first steps of a greenfield approach, where the company starts from scratch and establishes new teams when needed. The mechanisms of Viable Business showed how to create best working conditions in the first team. By splitting this team, creating junctions and mutually exchanging the missing resources, we can create the best working conditions for each team.

The greenfield approach is very untypical, since all the projects and departments already exist in large organizations. Here we need a brownfield approach. The brownfield approach also starts with the top management team. Additionally, it has to consider how to integrate all existing teams in the system.

Let us see how this works.

8.2 Ontogenesis 1: Create Suborganizations

Most large organizations already use the team approach for their projects. They have project teams (and maybe subproject teams) to execute the project work.

Viable Project Business focuses on the entire organization instead of single projects. It implements the system dynamics of biological organisms and founds the needed teams. If the teams already exist, like the project teams, then Viable Project Business adds the missing aspects to these teams and integrates them to the overall collaboration.

One focus of the collaboration is on the improvement machine, which allows to solve known problems and implement process changes quickly. The improvement machine helps to transfer knowledge from project to project and thus to become more productive and competitive.

Practically, it is not possible to start or change all teams in parallel. Viable Project Business follows a specific sequence that allows both to create the needed system dynamics and to integrate the existing teams. Everything starts with the top management team.

8.2.1 Team 1: Top Management Team

The top management team is responsible for leading the entire organization. This responsibility includes the relationship to all stakeholder groups. The CEO has the full authority, is also legally responsible for the organization, is the face to the public environment (Figure 8-10).

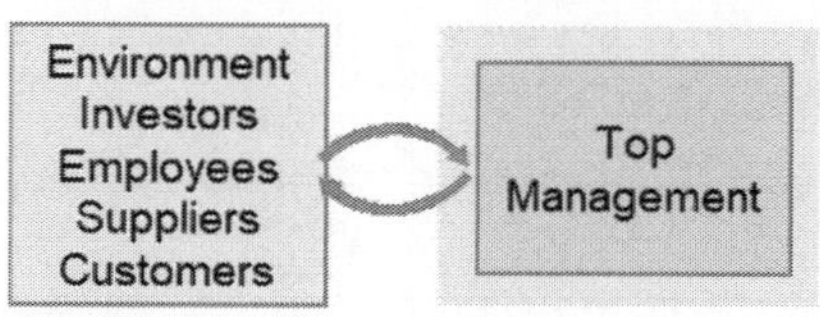

Figure 8-10: Top Management with its Stakeholder. Top management team is fully responsible for the entire business

The top management team typically consists of the CEO and the direct reports. At least at the beginning the team should meet weekly to perceive and analyze the state of the organization and to make decisions.

Base for this perception and analysis is the functional overview diagram which can be used as a map for all observations, measures and responsibilities. Section 7.7 shows how the functional overview diagram can be used to identify and eliminate bottlenecks in the organization.

8.2.2 Team 2: Leadership Team

Situation/Challenge

Ideally, the top management team is the anchor point of the organization. The team should meet weekly with high reliability.

This ideal situation is sometimes not possible. Typical reasons are:

1. The CEO is frequently travelling to customers and cannot participate on a weekly base.
2. Some of the top management members (e.g. CFO, Head of Strategy, Head of Communication) are not involved in the daily business.
3. Some people (e.g. Chief Engineer) should attend the weekly meeting, even if they are not part of the top management team.
4. The top management team consists of the owners of the organization only (typical for startups). It needs additional people for the weekly operational meeting.

New Team: Leadership Team

For this reason, it might be reasonable to start a leadership team additionally to the top management team. The top management still has the overall responsibility for the company, but the daily business is managed by the leadership team (Figure 8-11).

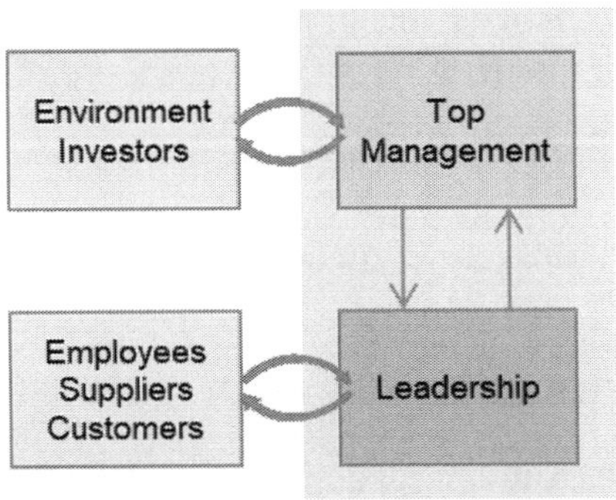

Figure 8-11: Top Management and Leadership Team. Leadership team is responsible for the daily business

The leadership team can be jointly led by the heads of project management and engineering. It is now the anchor for the daily business. It covers:

1. Bids
2. Customer projects
3. Services for customers
4. Resource management
5. Knowledge management
6. Internal development activities
7. Relationship to employees, suppliers and customers

The top management gets the authority and investment from the investor and has the responsibility to deliver a return on investment. For the daily business it gives a part of the financial resources and the authority to the leadership team.

The leadership team uses the transferred authority and financial resources for the daily business and ensures that the projects make profit. This profit is given to top management, which has to pay taxes and interests.

The collaboration between both teams is about authority, responsibility and finances. Both teams together realize the system dynamics of the company.

Comment

This team design is now very different to the classical organizational chart. There is no department which can be mapped to the leadership team. It seems to be a strong disadvantage to create a shadow organization additional to the classical department structures.

Practically, this is no problem: The head of engineering and the head of project management meet regularly in most organizations to solve joint issues. They already have the joint responsibility to make bids and project successful. The team collaboration chart of Viable Project Business emphasizes this need for collaboration and makes the authority of the team visible.

8.2.3 Team 3: Portfolio of Bids and Projects

Situation/Challenge

With this new responsibility the leadership team has to take care of all bids and projects. If there are not too many of them, the leadership team can directly take over the responsibility for multi-project management. The project managers are the link between their project teams and the leadership team. This team design has already been discussed in section 8.1. If there are too many bids and projects, then it makes sense to organize that indirectly via a portfolio team.

New Team: Portfolio of Bids & projects

The portfolio team "Bids & Projects" can take over the responsibility for all bid and project activities. The team is typically lead by the head of project management. Typical members of the team are the head of sales, the bid managers and project managers.

The benefit of this portfolio team is to reduce complexity for the leadership team. The leadership team does not have to be involved in every small bid or project. Now it can be more focused on the relationships to employees & suppliers, the development of people, improvement of processes and work preparation.

The portfolio team "Bids and Projects" can be focused on bids and projects and thus be very productive (Figure 8-12).

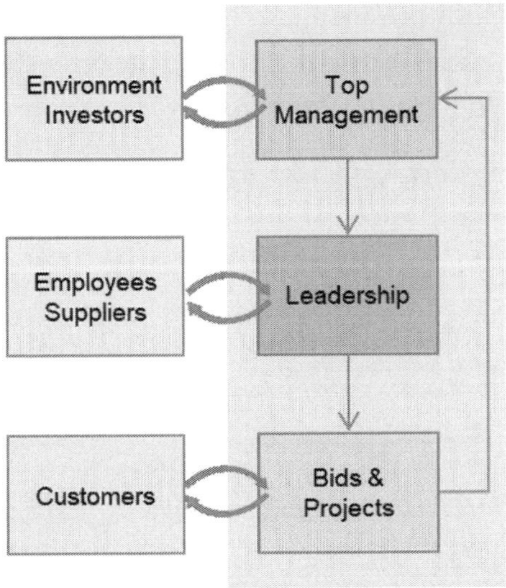

Figure 8-12: Portfolio of bids and projects. Bids and projects have direct contact to customers. They get all other needed resources from the leadership team and deliver profit to top management.

New Collaborations

The collaboration is about best working conditions. The leadership team has to provide all success factors to make projects successful. In the terms of the Ishikawa diagram: They are responsible for man, machine, material and method. Their team purpose can be described as: "Provide best working conditions for bids and projects". This makes the projects successful and they can earn money.

8.2.4 Team 4: Continuous Improvement

Situation/Challenge

The provision of best working conditions includes two main aspects: Provision of resources (man, machine, material) and provision of know-how. Especially the development of know-how and the trainings of employees is both time-consuming and very trade-specific. It needs trade-specific process experts to do that. In large organizations it is a full-time job.

If this overall responsibility for processes requires too many resources and leads to a too large leadership team, it makes sense to split the responsibilities:

New Team: Continuous Improvement Team

Managers continue in the leadership team. They remain responsible for employees, suppliers and resource management. Experts continue in the continuous improvement team. Their job is to develop and provide all know-how to bids and projects. Both teams have to be linked by the leader(s) of the continuous improvement team, which are members of the leadership team.

The continuous improvement team is responsible to make the process transparent and to organize know-how. A starting point is to detail the functions "Acquire customers" and "Create customer value" by a value stream analysis to have a more detailed map for this part of the organization (Figure 8-13).

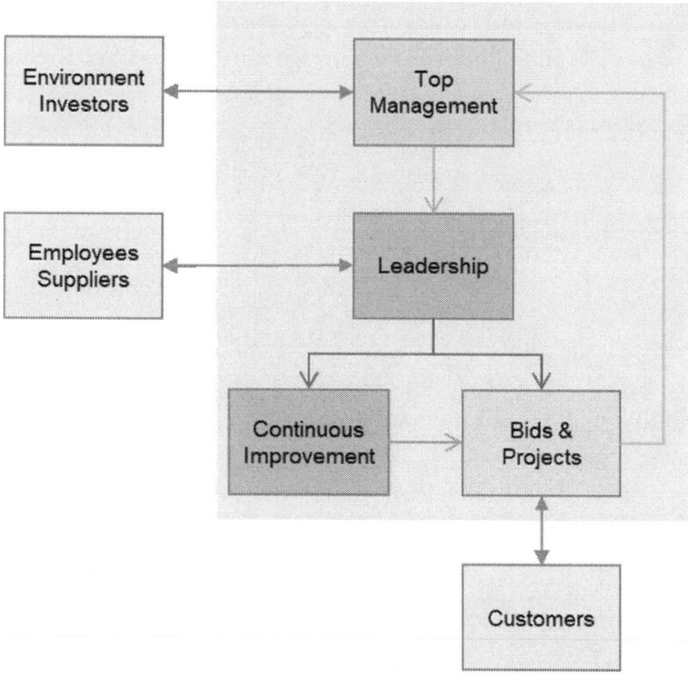

Figure 8-13: Continuous Improvement Team. The Continuous improvement Team provides best working conditions for bids and projects

Like the leadership team and the bids & project team the continuous improvement team has an overall-responsibility. Both together have to provide best working conditions for each process in bids and projects.

Comment

This split of responsibilities between the leadership team and the continuous improvement team has several consequences. One of the case studies shows that the split allows to create a fulltime job for continuous improvement. This head is responsible to create the overall transparency by making a value stream analysis, mapping KPIs, defining future states, orchestrating all improvements. The explicit definition of the team and the selection of the right team allows to make fast progress. As a disadvantage this setup creates two parallel tree-structures later, which makes collaboration more difficult.

An alternative approach is the integrated approach. The leadership team stays responsible for continuous improvement. When the leadership team grows, then it adds all department heads and group leads and experts. It forms a tree-structure, where the teams are connected by managers. The structure is then similar to the official org-chart.

The difference will be discussed in the case studies.

8.2.5 Team 5: Business Excellence

Situation/Challenge

Bids & projects get best working conditions from the joint work of the leadership and the continuous improvement team. This is only possible if these two teams have best working conditions too. If this is not given, then continuous improvement is too expensive and has not sufficient impact. Who does the job of continuous improvement for the other teams?

New Team: Business Excellence

It is the job of the business excellence team, to develop and provide the know-how that is needed by the continuous improvement teams (Figure 8-14).

They provide:

1. Trainings, coaching in Viable Business, lean and agile methods
2. Infrastructure (e.g. setup of obeyas and kanban boards, digital boards)
3. Templates and tools (e.g. for value stream analysis or A3)
4. ... and much more

All this helps to make the continuous improvement teams highly effective and efficient and, indirectly, help to make bids and projects successful.

The business excellence team needs a lot of know-how, to do its work. It is helpful, to start this team early enough and train each team member in Viable Business, lean and agile methods.

Like bids & projects also the leadership and top management teams need to have best practices. Business excellence can help them directly by defining management processes or indirectly by helping to establish specialized continuous improvement teams which are focused on the management activities.

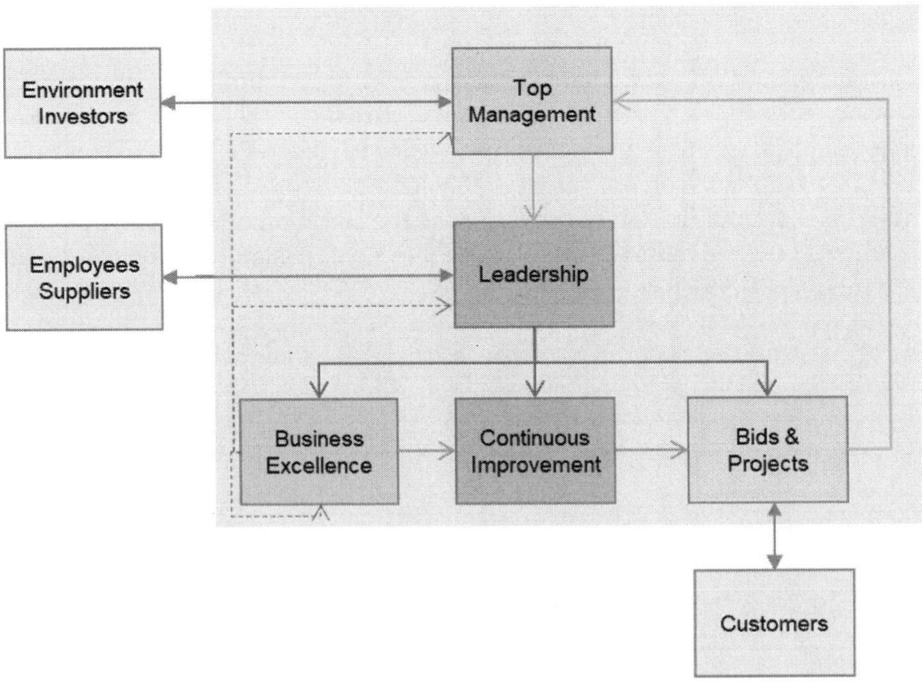

Figure 8-14: Business Excellence Team Business excellence provides best working conditions for continuous improvement. They can also provide best practices for the leadership and top management team (e.g. methods of self-organization).

8.2.6 System Dynamics, Teams and Collaborations

If we compare the five teams with the functional overview diagram, we can see that their collaborations are directly related to the system dynamics. Each team is highly specialized to provide one resource to other teams.

As explained in Chapter 1 the entire system grows with the speed of the bottleneck of the system dynamics. The top management has the responsibility to identify and eliminate it. A part of their consideration is to balance between success in the current year and success in the following years. This is very often the reason for a typical problem:

A very typical bottleneck in project business is a lack of methodic know-how and a lack of resources in continuous improvement. The symptoms have already been discussed in the introduction: projects have recurring problems and require more resources than originally planned. This leads to a permanent cost pressure.

Figure 8-15 show that the leadership team has to reduce resources in bids & projects to assign them to continuous improvement. This immediately weakens the working conditions for projects, whereas the benefits of improvement come with some delay. Even if it is clear to top management that many problems in projects are the result of insufficient training, processes or tools, it is difficult to change it. Management would need to decide not to start a new project or to invest less resources in bids. This is always difficult to explain to the investors.

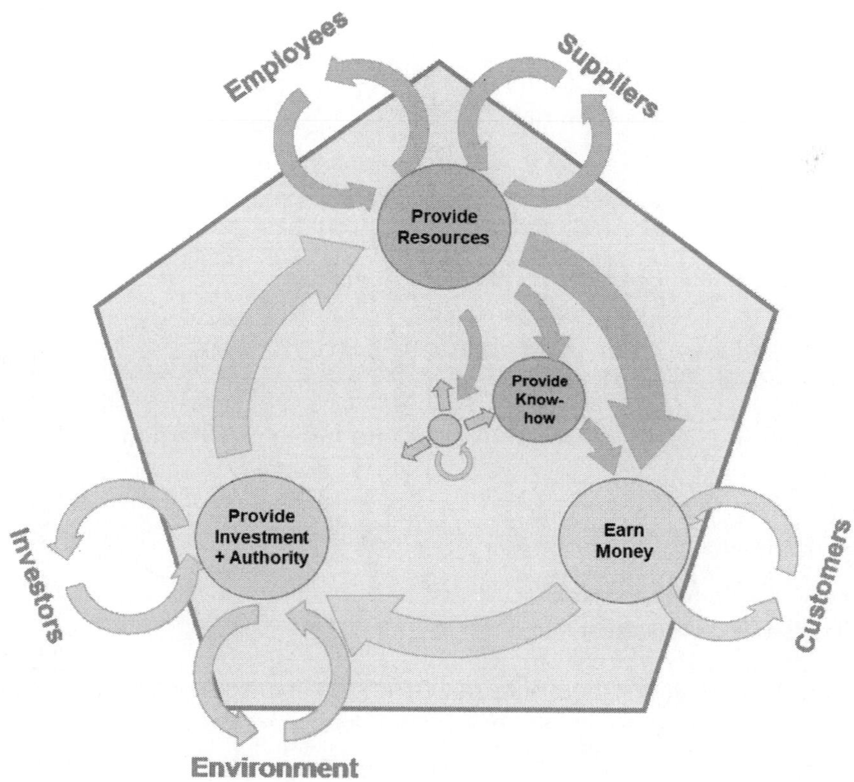

Figure 8-15: System Dynamics with colors of Team Types. The five teams are responsible for a part of the system dynamics.

Thus, continuous improvement has to provide good business cases which show that it is worth to invest the resources. It is the job of business excellence to provide the necessary know-how for the continuous improvement team to create such business cases.

The collaboration of the five teams ensure that each team gets all needed resources to create best working conditions. The needed resources are either coming from external stakeholders or from one of the five teams. [446]

The team can use the Ishikawa-diagram (fishbone diagram) and determine, which resource is problematic or missing. If the team uses the Functional Overview Diagram for its own team, it can find the function that does not work properly, or it can determine that one of the other teams has not delivered the resource.

The following table shows the exchange of resources in detail (Table 35).

Who provides which resource for the five teams	**Top Management**	**Leadership**	**Business Excellence**	**Continuous Improvement**	**Bids & Projects**
External stakeholders					
Environment	Authority				
Investors	Authority Money				
Employees	Man	Man			
Suppliers	Machine, Material	Machine, Material			
Customers					Money
Neighbor teams					
Top Management		Authority Money	Authority	Authority	Authority
Leadership			Man, Machine Material Milieu	Man, Machine Material Milieu	Man, Machine Material Milieu
Business Excellence	Method	Method	Method (!)	Method	
Improvement					Method
Bids & Projects	Money				

Table 35: Exchange of resources between team types

The first five teams have clear collaborations now. The design of the organization explains how each of the teams gets all the needed resources to create best working conditions. Practically, this

[446] This is analogue to multicellular organisms. Each cell needs to get all resources, either directly from outside the organism, or from another cell. Multicellular organisms like the human body manage that 100 billion cells get all resources. The resource management task of the leadership team can be compared with the cardiovascular system of vertebrates.

design is only a good starting point. The main work begins here: The top management and the leadership team have to systematically analyze the organization and to identify and eliminate the bottlenecks in the teamplay by changing structures or policies or by balancing resources. They are responsible to keep the entire system balanced.

8.2.7 Origin of the First Five Teams on a Timeline

The design of organizations by starting with the top management team, growing and splitting the team several times, is similar to the ontogenesis of a living organisms (Figure 8-16). The original top management team is the stem cell for the entire organism.

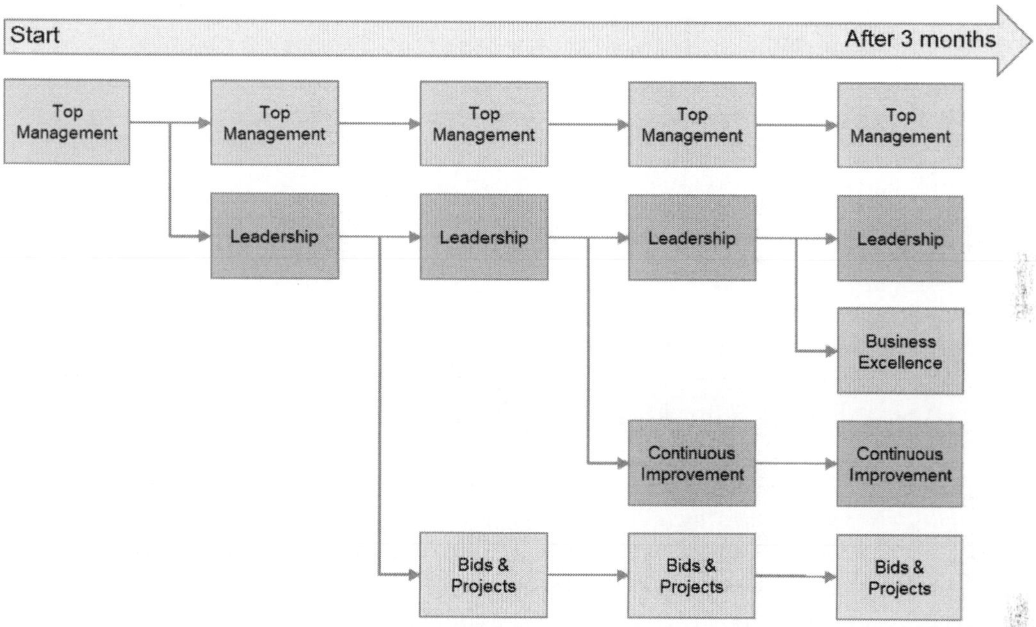

Figure 8-16: Timeline of ontogenesis 1. It takes several months to grow the top management team, split it and create the first five teams, which share the responsibilities. The phase is finished when all five teams are stable and have stable collaborations.

Each of the five teams delivers important factors to the other teams. Thus, it is important that each and every team is a Viable Team. If teams have internal problems to manage resources, knowhow or to collaborate with internal or external stakeholders, then the entire organization has a problem. How is it possible to ensure, that all of them are Viable Teams?

First of all, the top management team has to be trained carefully. Every team member needs to understand the functional overview diagram in detail. The top management team needs to have all abilities to cover all functions.

Secondly, when a team grows, each new team member has to be trained explicitly. Otherwise, the know-how concentration will be reduced with every new team member.

Thirdly, it has to be ensured that new teams are never started from scratch. They have to be started by splitting an existing team. At least the leader of the new team and his deputy should have worked in the parent team. This allows them to see the bigger picture and the contributions of their new team, before they start.

The experience from the case studies shows, that it takes around three months to setup these first five teams. Since all teams (except the top management team) meet weekly, the team members from day 1 have 10-12 weeks to learn and teach the new teamplay.

After three months around 20-30 persons are integrated in the five teams. If the teamplay is strong enough, then the system can be extended.

The collaboration diagram already has the needed collaborations to ensure that each team gets all necessary resources, but the picture is still very unbalanced. Especially the bids & projects team has the majority of resources and responsibilities. It needs more than one team to get the job done.

Let us start with detailing the project node. Then we can see how the management teams and improvement teams can be detailed as well.

8.3 Ontogenesis 2: Detail Permanent Teams

8.3.1 Growth

Of cause, five teams are not sufficient to manage the entire organization. The top management and the business excellence team can remain at their size, but the leadership team, the portfolio team of bids and projects, and the continuous improvement team have to grow to do their job for the entire organization. When these three teams reach their maximum team size, they have to split and found sub-teams.

The main question here is about the split dimension and the split criteria. It makes no sense, for example, to split the continuous improvement team in the sub-teams "people", "bid processes" and "China", because they are overlapping and not exhaustive. The split should not only be MECE (mutually exclusive and collectively exhaustive). It should have a defined split dimension like "processes". This is necessary to give a clear, unique scope to each sub-team and to clearly connect all the sub-teams on a detail level. The split should also be balanced to create sub-teams of the same size.

The original teams 5 teams retain the overall responsibility for their functions. They coordinate their sub-teams, which do all the detail work, and focus on overarching tasks. If there are too many overarching tasks, they can create additional sub-teams, which have a different split dimension.

Splitting the teams does not change the overall system dynamics. The implementation of the system dynamics is only distributed over more teams.

8.3.2 Detail Portfolio Teams and Introduce Experience Exchange

The portfolio team for bids and projects wants to get best working conditions from the leadership team and the continuous improvement team. For the head of project management, it is not possible to define exactly what is needed. He needs to ask project managers and subproject managers to describe the needs. They need to become team members as well. If the team becomes too large, the team has to split.

Since it is a permanent activity to define the needs, and to prioritize them, it makes sense to start one experience exchange team per trade (Figure 8-17).

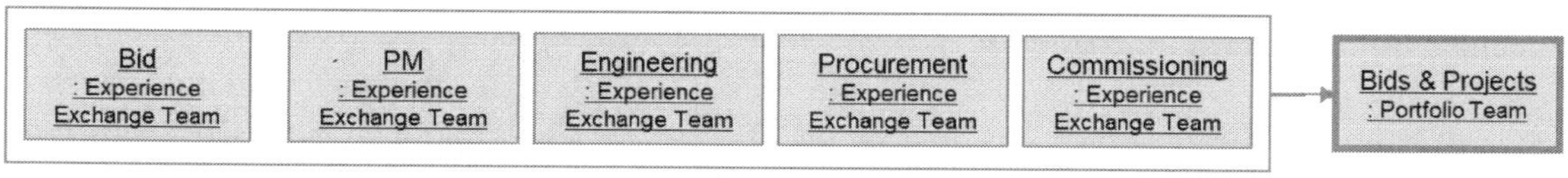

Figure 8-17: Teams for experience exchange. Portfolio of Bids and Projects creates one experience exchange team for each subproject.

Typical members of the experience exchange teams are the (sub-)project managers and experts of a trade.

8.3.3 Detail Continuous Improvement Teams

Each of the experience exchange team can define the demand for process improvement. Unfortunately, the team members have no time to improve the processes on their own. Their job is to make their projects successful in the shortest possible time. Thus, it needs additional people to do the work preparation and process improvement (Figure 8-18).

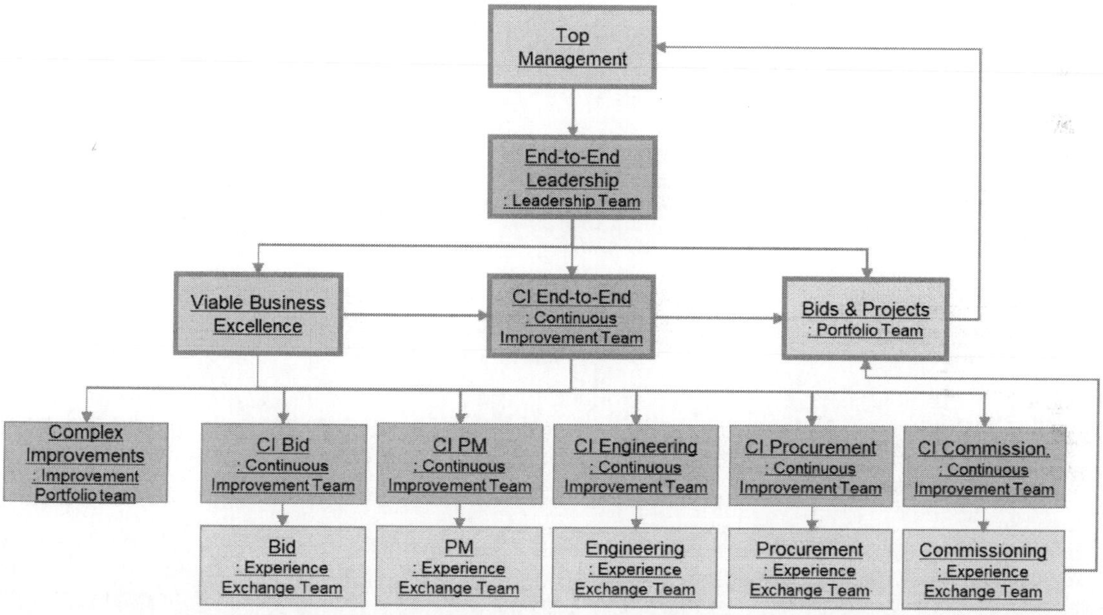

Figure 8-18: Connection of Continuous Improvement Teams with Experience Exchange Teams. Continuous Improvement teams are connected with the experience exchange teams. They fully cover the end-to-end process without overlap. The CI End-to-End team still has the overall responsibility. An additional portfolio team for complex improvements assumes responsibility for some of the overarching tasks.

There is already one team responsible for continuous improvement. Focus of the end-to-end continuous improvement team is to provide all know-how for all bids & projects. This responsibility requires a lot of resources, and the delivery of know-how must be very specific. It requires many experts to do that.

Sub-Teams

Thus, it makes sense to create dedicated continuous improvement teams that are assigned to the subprojects. Typical members of the trade-specific continuous improvement teams are the group lead and the best experts. Their job is to make the process transparent, define process standards, train and coach the subproject teams and continuously improve all the processes.

Continuous Improvement End-to-End

These new teams allow the CI end-to-end team to focus on overarching tasks like creating the overall transparency or driving large cross-functional improvements. This team should be led by the head of engineering. Additional members are some experts as full-time team members, and (maybe technical) representatives of the sub-teams.

Portfolio of Complex Improvements

If there are too many overarching tasks, the team "CI End-to-End" can create an additional sub-team "Complex Improvements", which manages the portfolio of large improvements. This sub-team has a different splitting dimension than all the other sub-teams. The internal customer of "Complex Improvements" are the other continuous improvement teams.

8.3.4 Detail Leadership Teams

The leadership activities include the relationships to all the employees. Recruiting, creation of value for employees, development of employees and their staffing are functions that have to be realized by leaders. This is a typical job of line management. Thus, the line managers on all levels have to be integrated in the leadership teams (Figure 8-19).

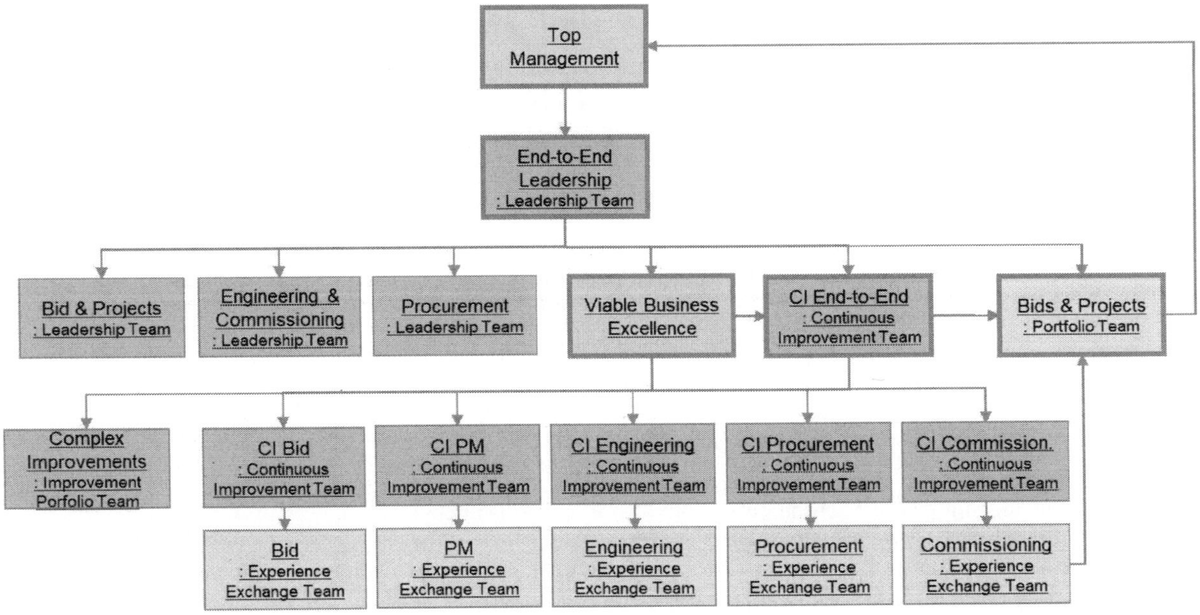

Figure 8-19: Growth of leadership teams. Leadership teams are connected to continuous improvement teams and experience exchange teams. They have the responsibility to provide the needed resources

A typical setup is to create a leadership team for every department. Members of the leadership teams are the department head, the first line managers (group leads), the commercial head of the department, maybe a resource manager and representative of the human resource department. If each group lead has 15-20 employees, and a department has 5 groups, this leads to one leadership team for every 100 employees. All the department heads are members of the end-to-end-leadership teams. Thus, the system should be able to manage several hundred employees with two levels.

The group leads connect the continuous improvement and the leadership teams. They represent the resource needs of CI on a management level. The technical heads of the CI-teams.

8.3.5 Further Detailing

In some cases, the continuous improvement teams need to be even more detailed. One of the case studies shows an organization which has 7 experience exchange teams and 40 continuous improvement teams. This makes it necessary to do improvement on three instead of two levels.

In such a scenario a continuous improvement team creates several sub-teams which work for the same subprojects. One example is CI Engineering, which can create several sub-teams for systems engineering, mechanical engineering, electrical engineering and software engineering, even if there is only one engineering subproject.

8.3.6 Growth Phase on a Timeline

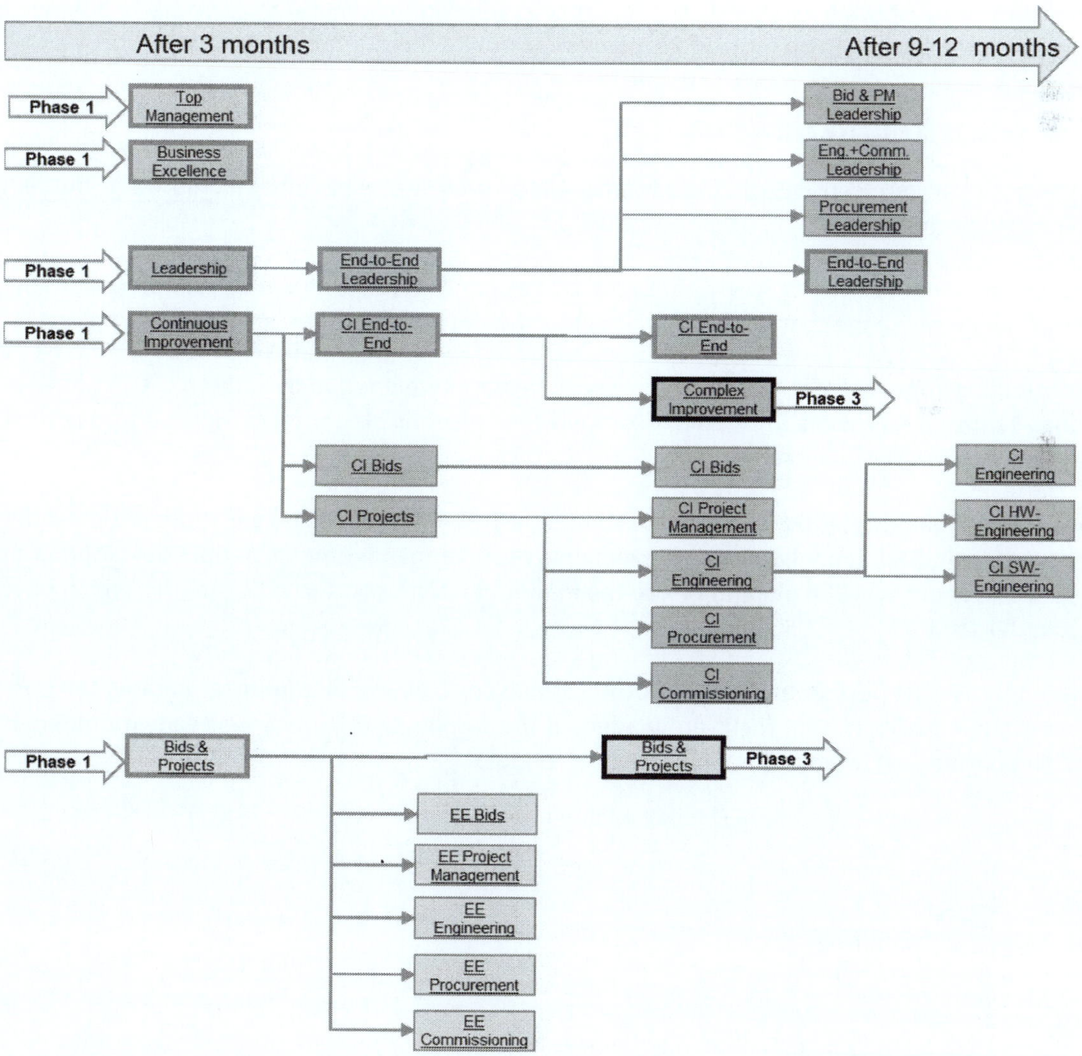

Figure 8-20: Timeline of ontogenesis 2.In the second phase three of the five teams split themselves and connect to each other.

Permanent Teams

The first five teams already introduced Viable Business for 20-30 persons. In a second step the organization details three of the teams: The portfolio of bids and project team has to create the experience exchange teams, which are able to represent the needs of bid managers, project and subproject managers. The continuous improvement team creates sub-teams, which are able to do work preparation and process improvement. The leadership team integrates all line managers in sub-teams (Figure 8-20). All these teams have to be carefully linked; not only on a team level, but also on a personal level. Persons connect teams.

One of the challenges in the growth phase is to keep the entire system balanced.[447] If all the teams are well-connected, then each Viable Team knows exactly where all the resources come from. Each team has team members, which are responsible to get a certain resource. These team members are the link to the internal supplier team, which has the explicit purpose to provide this resource.

Of course, this system needs control loops. Each team can report its existing bottlenecks to the parent teams (one of the five teams). The leadership and top management team try to balance all resources.

8.3.7 Growth and Balance

Now the entire system consists of 20-30 teams. All the new teams have been created by detailing the existing 5 teams.

The portfolio team has split its scope along the trades which typically form the subprojects. The new experience exchange teams already allow to share knowledge between different projects.

The continuous improvement team has created sub-teams which map directly to the experience exchange teams. It has also added a sub-team which is responsible for all complex improvements that should be managed centrally.

The leadership team has added a team for each department. This structure is more people-oriented than process oriented. It allows to make excellent resource management, people development and culture work. There is also a mapping from the leadership teams to the other teams, which have a resource demand.

All these teams exist permanently. They form a stable core, where all temporary teams can connect to. One of the benefits is, that the team leaders of the temporary teams are already members of one of the permanent teams. This simplifies the rollout.

---

[447] This is similar to growth of a living organism. The proportions of all organs have to be kept in a balance. A human embryo does not grow one organ after the other. All organs have to grow in parallel, and the organism has to be balanced at every point in time.

8.4 Ontogenesis 3: Add Temporary Teams

8.4.1 Add Existing Bid Teams, Project Teams and Subproject Teams

At this point in time, all bid managers, project managers and subproject managers participate in an experience exchange team. They already interact with the continuous improvement teams, which tries to solve existing problems. As the next step, the bid and project managers integrate their existing teams and use the methods of Viable Business (Figure 8-21).

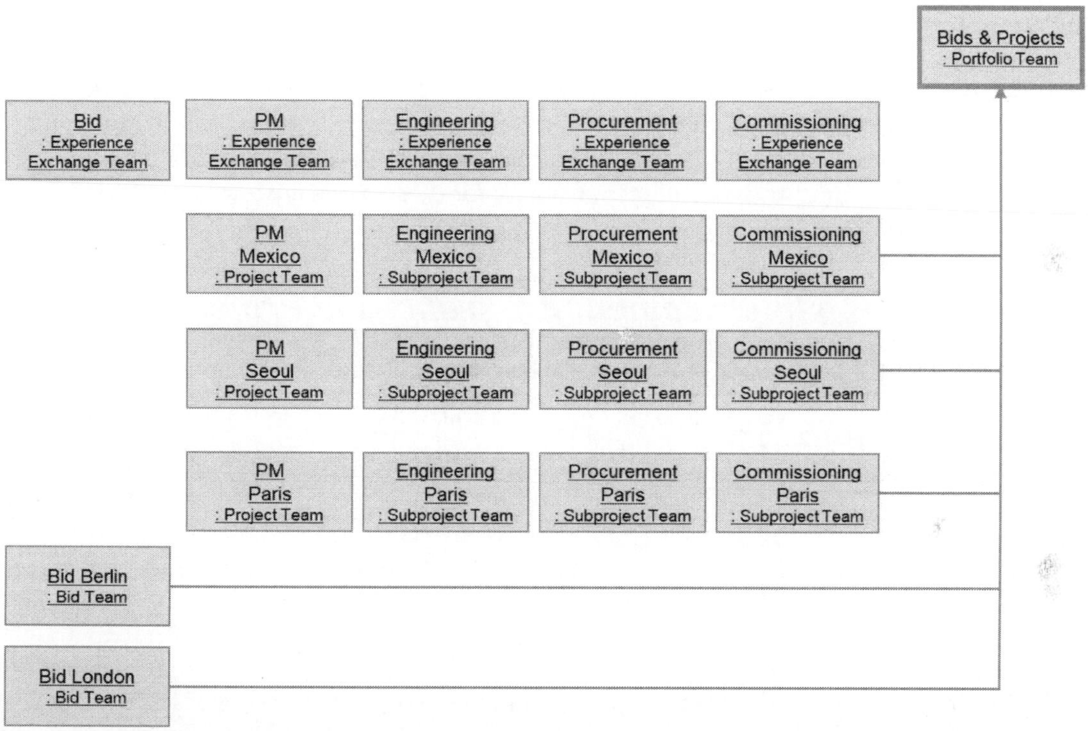

Figure 8-21: Portfolio team, bid teams and project teams with subprojects. Portfolio of Bids and Projects adds teams for each bid and each project. Later the teams for subprojects can be added.

Bid Teams

The bid teams are aiming to get the contract for a specific customer tender. The bid manager connects this team with the portfolio team and with the experience exchange team.

Project Teams

The projects are aiming to fulfill the contractual requirements and to create profit. The project manager can integrate the existing team to Viable Project Business. Typical team members are the project manager, the commercial project manager, subproject managers and the responsible experts for topics like scheduling and document management. The project manager links this team to the portfolio and the experience exchange team for PM. The subproject managers link it to the experience exchange teams for the subprojects.

Subproject Teams

When the bid teams and the project teams have been stabilized, the subproject teams can integrate their teams as well. They are slightly different to the project: They don't have to create profit directly. Their job is to contribute to the creation of customer value and to get an acceptance from the customer, whereas the project team also has to ensure payment.

Cascaded Portfolio

If the organization has more than 10 projects in parallel, or there are different types of projects, then there might be more than one portfolio team. The heads of the teams should be members of the operations team. If there are too many portfolio teams, they should be cascaded.

Temporary Teams Contribute to Specific Functions

The activities of the "green teams" can be located in the functional overview diagram. Bid teams contribute to the function "Acquire customers". Project teams contribute to the functions "Create customer value" and "Get payment". Subprojects contribute to "Create customer value". Since they all are temporary teams, the overall responsibility for these three customer related functions is in the portfolio team "Bids and Projects".

8.4.2 Add Existing Improvement Programs and Projects

The End-to-End team for continuous improvement has the responsibility to manage all overarching activities. This includes to create transparency for the end-to-end-process and to manage cross-functional improvements.

A3 Problem Solving

The basic problem of cross-functional improvements is, that one team has a problem that can only be solved by a different team. The details of improvement will be explained in section 8.4.4, here the focus is only on the team definition.

The CI End-to-End team creates a temporary cross-functional team with members of several CI sub-teams. This temporary team can analyze and solve this problem cross-functionally.

This setup can be used for many medium-size problems, where a small kaizen card is too small, and the program approach is too big. The A3 team is temporary and exists for a couple of weeks or months. Thus, it makes no sense to add all of these teams on the team collaboration chart.

Improvement Programs and Projects

Larger improvements which needs more resources and promise a high impact for several teams should be organized by a defined team. Typical examples for such complex improvements are programs for plant modularization or the introduction of a new tool chain.

A good approach is to use program management techniques. Some details of program management will be explained in section 8.4.4.

All existing large or complex improvements which require a dedicated team, should be added to the team collaboration diagram (Figure 8-22). This makes the ongoing activities visible for all other teams.

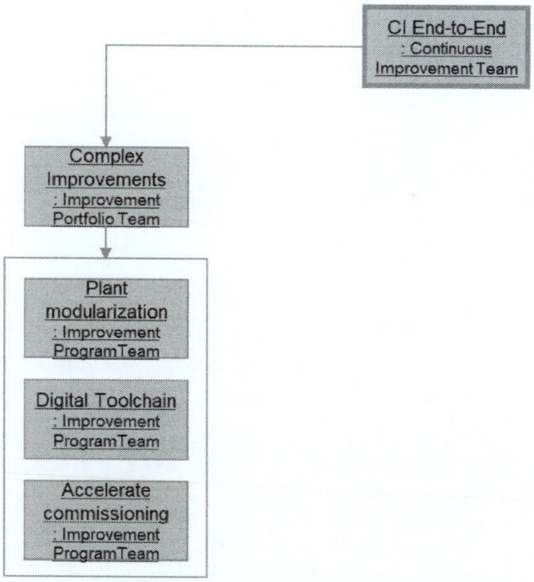

Figure 8-22: Improvement programs. The three complex improvements are added to the portfolio.

8.4.3 Team Collaboration Diagram

The team collaboration diagram shows all teams in one diagram (Figure 8-23). The color code emphasizes that the teams realize a certain role in the organization. Red teams provide resources, orange teams provide knowhow and green teams create customer value and earn money. The blue team has to balance the organization for the here and now and for the future.

High-Level Structures

These high-level structures have already been created during part 1 of the ontogenesis. Everything started with top management as the stem cell, the first five teams can be understood as the pluripotent stem cells which are the starting point for the organ systems.

The high-level structures determine the main collaborations, because they ensure that each team gets all needed resources.

Low-Level Structures

The low-level structures are needed to detail and to balance the system. These structures origin, when the first five teams start to create sub-teams in part 2 and 3 of the ontogenesis. Low-level structures are like organs and tissues, which contribute to an organ system.

Let us focus on the continuous improvement teams to understand how all the orange teams work together to jointly realize the organizational function.

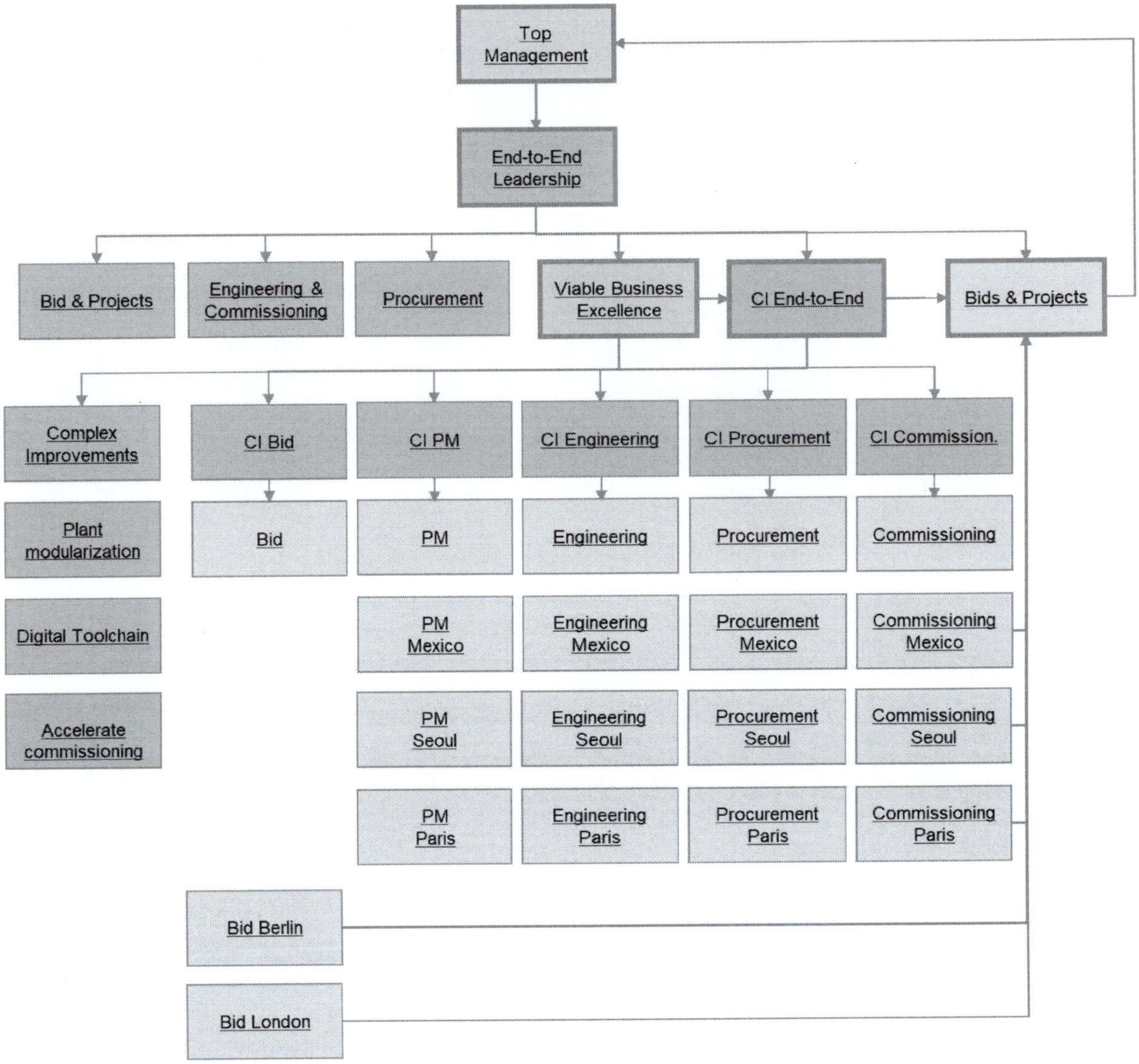

Figure 8-23: Team collaboration diagram with 36 teams. The organization has 36 teams including 2 bids, 3 projects with 3 subprojects each and 3 complex improvements. Normally, the diagram would also contain the name of the team leads and the time, frequency and location of the regular meetings.

The CI E2E team is the parent for all the specialized CI teams, which have to be designed MECE [448]. There is one responsible CI team for each process. This hundred percent coverage is important for two reasons: Firstly, CI Teams have to support the subprojects in each process.[449] Secondly, they need to have strong peer relations, if they want to improve the process.

If we compare this with biology, this group of teams has a similar structure to a mixture of epithelial tissue and muscle tissue. Epithelial tissue is used at all surfaces and creates a 100% coverage. Muscle tissue needs strong connections between cells to deal with strong contraction forces.

[448] Mutually exclusive and collectively exhaustive

[449] Work of continuous improvement teams will be explained in detail in section 8.5

These two features cannot be realized by the CI teams individually. It has to be realized by the entire group of teams. By using predefined team types, Viable Project Business, implements the necessary competences in each of the CI teams.

A different part of CI is the portfolio of complex improvements with all the programs.[450] These programs are not MECE at all. They can contain programs for "Culture", "China" and "Project Excellence", which are highly overlapping and not exhaustive at all. For programs this is not needed, because they do not have to cover a process. This group of teams is more like connective tissue, which is only loosely connected.

Both groups of teams together, realize the suborganization "Continuous Improvement", which has been created by the CI E2E team to provide the know-how for bids and projects.

8.4.4 Ontogenesis on a Timeline

Temporary Teams

Prepared with this strong foundation of permanent teams, the temporary teams can be added. This can be done in parallel (Figure 8-24).

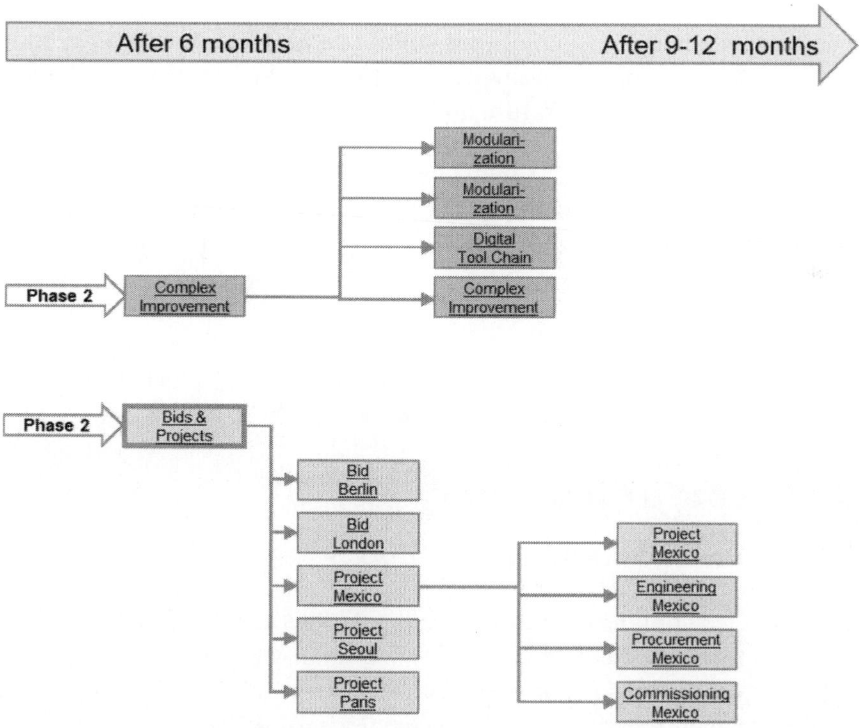

Figure 8-24: Timeline of ontogenesis 3. The organization adds all bids, project and subprojects to the bids and project portfolio. It also adds all complex internal improvements to the improvement portfolio.

Continuous Improvement can identify all ongoing internal activities and create program teams. The portfolio of bids and projects can start to add all bid teams and project teams. Later the subproject teams can be added as well.

---

[450] Programs will be discussed in section 8.5.10

Other Teams

The team collaboration chart is still not completed. Large organizations may add many more teams by detailing existing functions.

Examples are supplier integration teams, which connect the organization to suppliers. This can happen on a management level. In that case it is a sub-team of the leadership team, which is responsible for supplier relationships. It can additionally happen on a technical level. Here it makes sense that the CI End-to-End team creates such a sub-team.

Another example is a dedicated team for perception and analysis. Its job is to collect information about the organization and to provide a clear picture to management.

8.5 Improvement Machine

The previous sections described the overall anatomy and physiology of the organization and the ontogenesis. Now we have to detail the functions, teams and collaborations. Of cause, this work cannot describe all functions in every detail. We focus only on the needed processes and collaborations to continuously improve processes for bids and projects.

The detail mechanisms are realized with methods which are well-known in lean production or in program management. This approach has several benefits: It allows to use well-documented methods, which are already proven. Literature and trainings are already available.

The focus of Viable Business is to integrate the existing knowledge in predefined team types which contain all the needed methods and collaborations. These predefined team types can then be used by top management as the building blocks for the organizational design. The business excellence team can use the knowledge to systematically train, coach and support the continuous improvement teams.

Let us go through the various tasks of continuous improvement and collect the needed methods and collaborations, which have to be implemented by each of the CI teams.

8.5.1 Create Overall Transparency

Acquire Customers and Create Customer Value

A possible starting point for continuous improvement activities are two functions of the Functional Overview Diagram: The function "acquire customers" contains all marketing activities and the bid process. The function "create customer value" contains the entire processes from project start to project closure. Here we want to focus on "create customer value" only.

In Chapter 7 we have already seen, how the top management team has assigned the main key performance indicators for that function. Typical examples are:

1. Defects[451]
2. Planned duration

---

[451] ISO 9000 defines defect as "nonconformity related to an intended or specified use". See (ISO 9000, 2015 p. 40)

3. Unplanned duration (Delays)
4. Planned cost
5. Unplanned cost (Cost overrun)
6. Customer satisfactions[452]

Until now, these KPIs are assigned to a black box. It is the job of the CI End-to-End team to create transparency.

Value Stream Analysis

One of the first activities of the CI End-to-End-Team should be to detail the function "create customer value" (Figure 8-25).

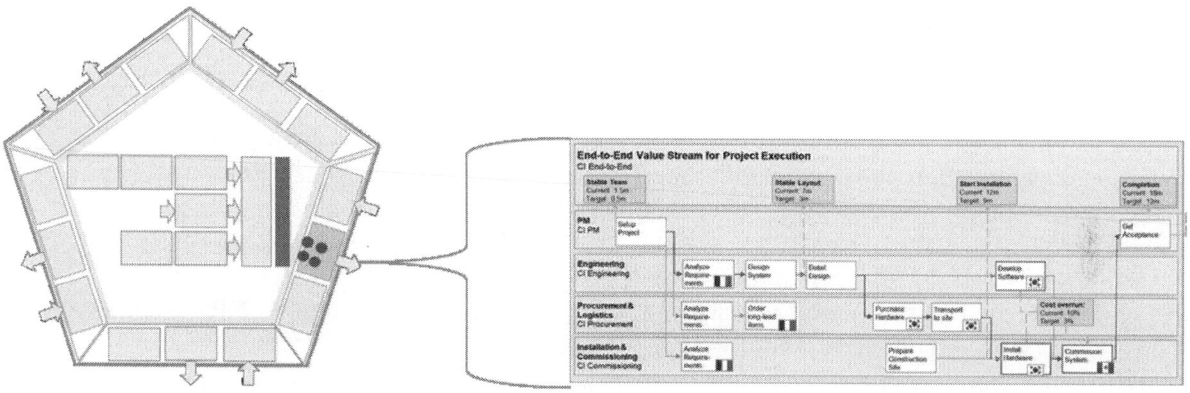

Figure 8-25: Value Stream as the detail view of the Functional Overview Diagram. The value stream analysis details the function "Create Customer Value"

The function contains the work of the project portfolio team and the work of the projects. Here we want to focus on projects only. Projects typically consist of three different types of activities.

1. Activities that have to be done one time and are then completed. These activities are typically described in a schedule.
2. Recurring activities like "Update schedule" or "Update risk register", which have to be repeated with a certain (e.g. monthly) frequency by the project manager.
3. Countermeasures that are created when recurring activities identify exceptions that have to be handled. They cannot be planned in advance.

The Value Stream Analysis can be used to describe the activities of the first type. For the recurring activities it is sufficient to make a register with all processes and their frequencies.

Let us start with the Value Stream Analysis (Figure 8-26):

[452] ISO 9000 defines customer satisfaction as the "perception of the degree wo which the customer's expectations have been fulfilled" (ISO 9000, 2015). See ISO 10004 for measuring and monitoring customer satisfaction (ISO 10004, 2018)

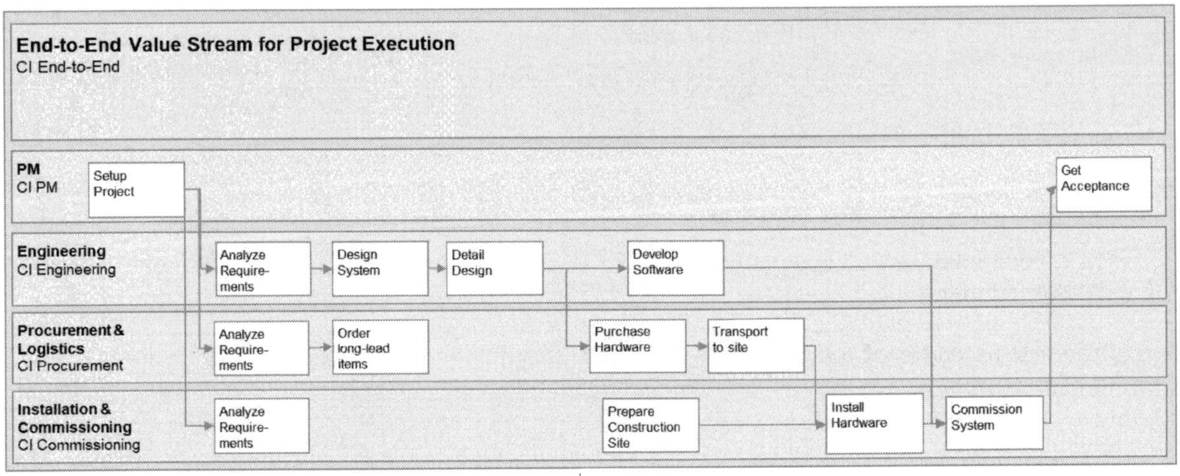

Figure 8-26: Value stream analysis. Value Stream Analysis has one swim-lane per team. Each swim-lane contains the processes that have to be performed.

In project business there are several difficulties to solve. Of cause, all projects are absolutely different and unique. The process that is described in the schedule of the projects depends on factors like project size, plant structure and the scope of supply of consortia partners and suppliers.

In such a case, the organization should start top-down to get a simple picture, which is shared by many projects, even if details are different. This overall description should contain the main processes of the project organized by the responsible subprojects. Since the CI-teams follow the same structure, each process can be drawn on a swim-lane, which has a responsible CI-team.

The value stream analysis should contain all major deliveries between the subprojects. This allows to detail the value stream for each swim-lane independently.

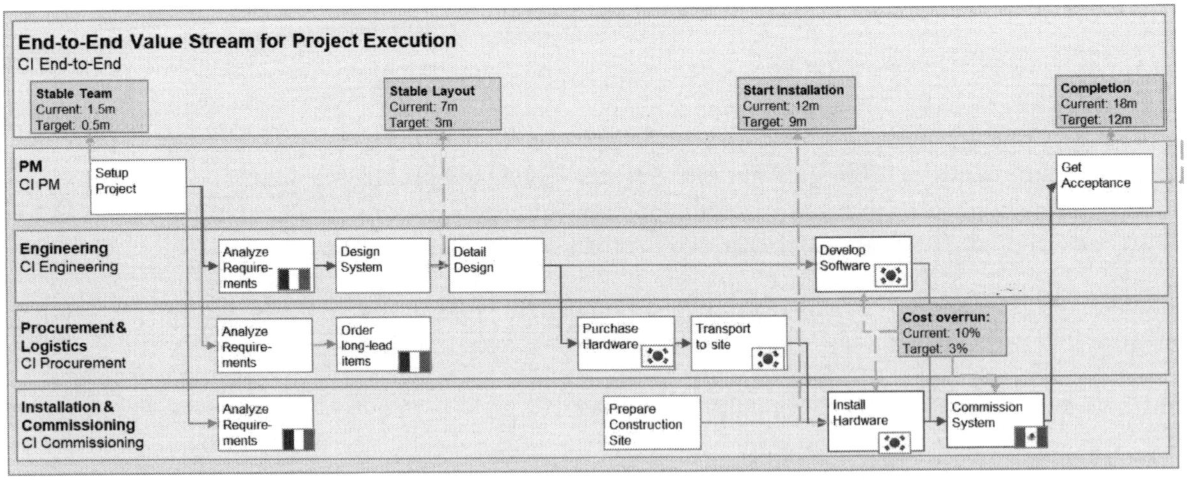

Figure 8-27: Value stream analysis with mapping to KPIs. It contains the critical path of the project and three processes which have a cost overrun. The blue boxes show the current and intended future timeline in months (m) and the target for cost overruns. The flags represent the current status of the projects.

Key Performance Indicators

In a second step, the CI End-to-End team should refine the main Key Performance Indicators, which have been assigned to the function "create customer value" by the top management team (Figure 8-27).

Defects create additional work and extend the duration. The value stream analysis allows to flag the processes that produce most defects. A good source for identifying recurrent defects are the quality management reports or customer complaints.

Planned and unplanned duration are KPIs can also be visualized in the value stream analysis. The overall duration is determined by the critical path or critical chain of the project. Since the critical path is different in every project, the value stream should show those paths which are critical in most projects. By adding some milestones, it is possible to show, which phases, and processes should be accelerated first.

Planned or unplanned cost belong to material or work. In both cases it is possible to identify the process, which has to be improved. Cost information is typically available in the standard cost breakdown structure. The data for cost overrun is also available. The challenge is to get a precise mapping to processes.

This first mapping of KPIs to the value stream analysis creates a link between the processes and the resulting indicators. This link helps each continuous improvement team to start with a detail analysis and to find process improvements that have an effect on the important KPIs.

Status of Projects

The value stream can also be used to visualize the status of the projects (if there are not too many of them). Flags can be used to show on which processes the subprojects are working. This visualization helps the continuous improvement team to understand what is urgently needed by which project. They can prepare the processes, which are coming next.[453]

Further Detailing

Such a value stream can be detailed by each CI team. At the end, the value stream will be described on several levels.

1. Top level is the function "create customer value" in the functional overview diagram as a black box which has some assigned KPIs.
2. Second level is the value stream analysis with the End-to-End perspective. It contains around 100 processes, 5-10 swim-lanes, all cross-functional dependencies.
3. If necessary, each CI-team can detail its swim-lane in a detail value stream analysis
4. On the lowest level, the value stream has processes which are linked to the work instructions.

---

[453] In the case studies this strategy has been called "Curling". The continuous improvement team cleans the way, directly in front of the subprojects.

8.5.2 Create Work Instructions

Best Working Conditions for a Single Process

Each of the processes on the value stream needs to have best working conditions. The links between the processes show, that the output of one process is used as input for the next one. It shows the flow of project specific material or information. According to the fishbone diagram, also the factors man, machine, method and milieu have to be considered (Figure 8-28).

This best working condition is directly related to the KPIs. If one of the factors is not available, then the process does not have best working conditions and is thus ineffective and creates too high cost or duration or it produces defects. In other words: If defects, cost and duration have to be optimized, the focus should be on the establishment of best working conditions.

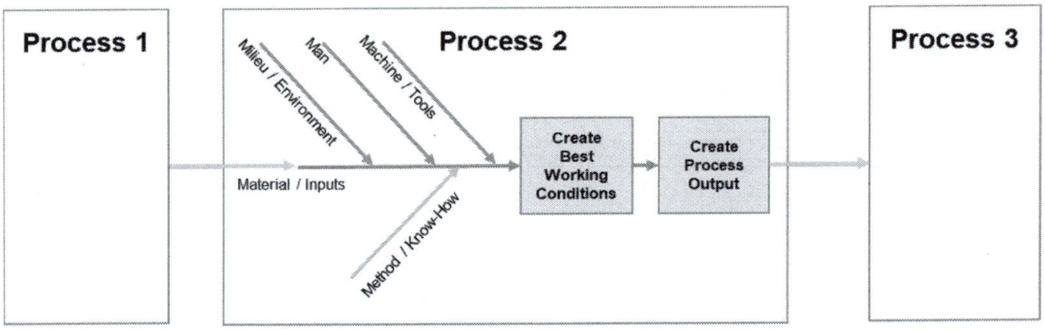

Figure 8-28: Process with fishbone diagram. Each process needs material and other factors to create best working conditions.

Mapping the Fishbone to Internal Supplier Teams

All the teams have been designed such that they map directly to the fishbone diagram. The leadership teams are responsible for resources, the CI teams provide the know-how and the project team organizes everything that is project specific. To have an exact mapping of factors to teams it is necessary to split some of the factors (Figure 8-29).

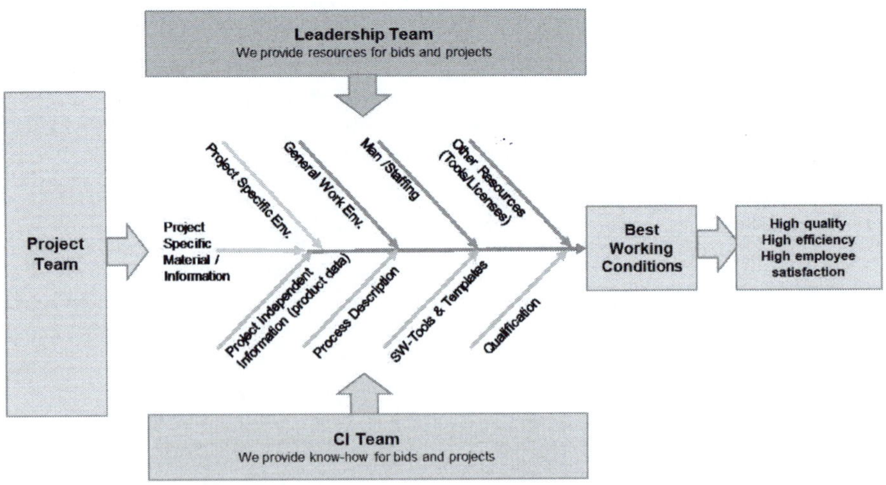

Figure 8-29: Fishbone diagram and supplying teams. The teams are clearly mapped to the fishbone diagram. For each process input it is clear, who has to deliver it.

Leadership decide about the usage of limited machines and tools like a fork lift, a notebook or a SW-license. This limitation is not given for self-created or open source software tools or self-created templates. The CI team can provide them, because there is no exclusive decision, which team gets the tools and templates first.

Leadership is also responsible for staffing. Each process needs to be staffed by persons with a sufficient basic qualification. The qualification for the specific process can be ensured by the CI-team by training, coaching and supporting the person. This distinction leads to a separation of staffing and process-specific qualification.

Leadership has to ensure the right work environment, which includes everything that is not covered by the other aspects. This includes to have an employee's badge, a workplace with a low noise level which fulfills health & safety standards.

But there is also some environment that has to be organized by projects. Project team members, for example, need to have contact to customers or they need to have a special badge to visit the construction site. Both cannot be classified as material or input, but it belongs to the project specific environment that has to be guaranteed by the project manager.

Information is also difficult to classify. Project internal documents, like circuit diagrams, have to be organized by the project. Additionally, there are many other documents which refer to standard products. It is the job of the CI team to make them available.

With this more detailed structure, there is a clear story: A process creates best results in the shortest time, if it has best working conditions. If the CI team collects everything that is needed and structures it with the fishbone diagram, it can map each factor to a team that is responsible to deliver it, before the process starts.[454]

This knowledge is the base for the description of work instructions.

Work Instructions[455]

A work instruction describes one process in detail. It is used to document and to communicate know-how about a process. It contains a short synopsis, a list of inputs, the process steps and a list of outputs (Figure 8-30).

The fishbone diagram helps to identify all inputs that are needed for the process and determine the internal supplier for the input. This knowledge can be documented in the work instruction. Later, during work preparation, it can be used to check, whether all inputs are available before start of the process.

The process steps follow the idea of Training Within Industry that has been refined by Toyota. The steps are like a cook recipe. Each step can have one or more key points, which describe, how to do the work correctly. These key points are the documented know-how of experts, who know how to

---

[454] For convenience reasons we only use the simplified Ishikawa diagram with four fishbones in the following sections. Practically, it is better to detail it and have clear responsibilities for each factor.
[455] Work instructions are standard documents that are explained in (ISO 10013, 2001)

do the job efficiently and without any defects. The reason for the key points is also documented. This helps the reader to understand why it is important to obey the key points.[456]

The output list contains the quality criteria of the output and shows the consumers. [457]

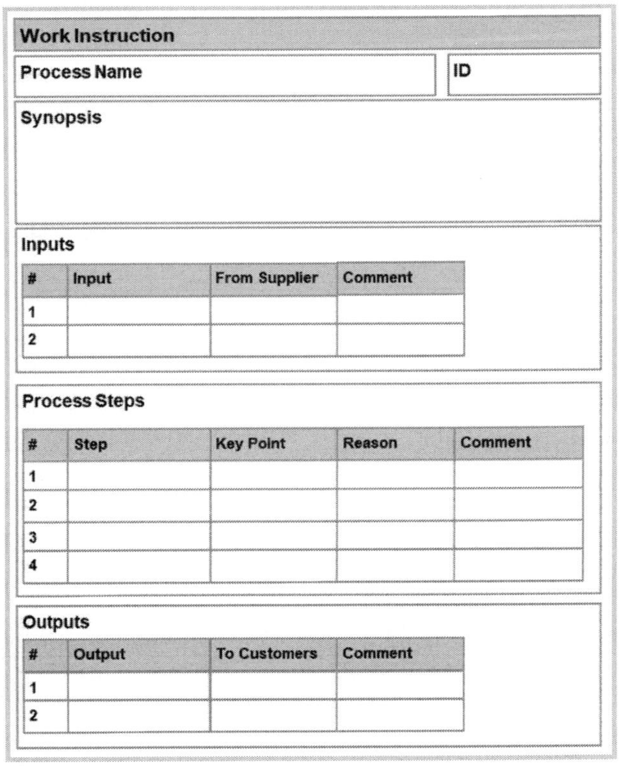

Figure 8-30: Work instruction. A work Instruction describes the necessary inputs, the process steps, and the outputs of a process.

The work instruction can easily be implemented as a Wiki-Page. This allows to link it to predecessor or successor pages or to the team-pages of internal suppliers and customers. In a Wiki it is also possible to add many more aspects like pictures, videos or links to special training material.

8.5.3 Provide Best Working Conditions

Collaboration with Project and Subproject Teams

All this preparation work can now be used, to fulfill the purpose of the CI team: Provide best working conditions for bids and projects. Each of the specialized continuous improvement teams, like the CI engineering team, uses the value stream analysis and work instructions to support the subproject teams.

This is only possible, if the subproject team is able to do an explicit work preparation itself. It must already follow the PDCA[458] logic to prepare a process, execute it, check the results, and learn from

[456] See (Training within Industry Program, 2009b) for the historical basics and (Liker, et al., 2008) for the more advanced model of Toyota
[457] See (Training within Industry Program, 2009b)
[458] PDCA stands for Plan-Do-Check-Act. It is also known as Deming Cycle or Shewhart Cycle.

experience.[459] If this is ensured, the process expert of the CI team can support the subproject in the following way (Figure 8-31):

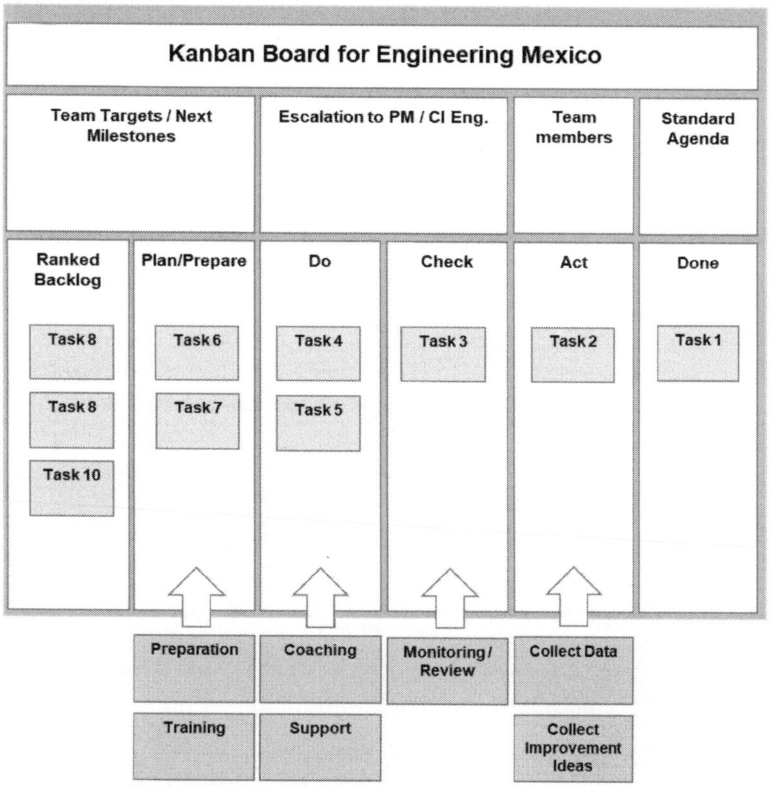

Figure 8-31: Kanban board. Work preparation processes of the continuous improvement team are linked to the phases of the Kanban board of engineering subproject Mexico.

Preparation

The subproject manager has a regular (biweekly or monthly) meeting with an expert of the corresponding CI-team. In this meeting, they discuss the ranked backlog of the subproject and select those tasks that have to be started in the next weeks. If these tasks belong to standard processes, the CI team already has prepared a best practice description, which contains the necessary preconditions for the process, organized by the fishbone diagram. This allows both, the expert and the subproject manager, to check, whether all preconditions are given, and, if not, who has to deliver them.

Training

One of the preconditions for a task is the qualification of the engineer who executes the task. If the task is new, then the CI team can provide a short training for the task, before it has to be started. This is especially important, if the task is on the critical path of the project.

TWI recommends a job instruction method, where the trainer demonstrates the correct execution of the task three times, before the trainee repeats it three times. The trainer has to check, whether the trainee executes the process correctly, observes all key points and knows all reasons.[460]

---

[459] See (Liker, et al., 2007) for a discussion of PDCA

[460] See (Liker, et al., 2008) and (Training within Industry Program, 2009a)

Unfortunately, this is not possible in project business, because the tasks are too big and sometimes there is only one repetition per project. Thus, the trainer can demonstrate it only in theory, maybe with examples of previous projects. In such a case, the training is just the best possible preparation, the main part of the know-how transfer is done during the real execution of the process, where the trainer coaches the trainee.

Coaching

The process expert can lead the engineer step by step through the work instruction. Sometimes it is sufficient to focus only on the difficult steps and those steps which have been changed recently. This coaching ensures that the engineer understands each step and knows, that he can ask, whenever something is unclear.

The written work instruction can be used to mark all steps, where the engineer should call the expert.[461]

Support

Even if the engineer does not need any coaching, there are many situations, where he needs support. This support cannot be planned in advance for each occurrence, but the expert needs reserved time and should be able to react quickly.

The fast escalation of problems is one of the key principles in lean production, where the employee pulls the Andon-cord to indicate a problem and get immediate support. Problems have to be escalated as fast as possible.[462]

Monitoring/Review

The engineer has to check the results of the process and compare it with the post-conditions which are described in the process. The process expert can support this process, because he knows the expectations of the downstream teams, which need the results as their input.

This monitoring is especially important, if the process or the process output has been changed by improvement work. Here it is not only important to check the process output. It also has to be checked, whether the process has been executed the new way.

Collect Data

At the end of the process, there is a chance to collect process-related data, which can be used for calculations. This data is also important to check, whether implemented improvements ideas work as expected.

[461] This approach is similar to running software code in the debugger step-by-step. The marks in the work instruction are the breakpoints.
[462] See (Ohno, 1988 p. 47)

Collect Improvement Ideas

Finally, the engineer and the expert can speak about usability of the work instruction description and discuss potential improvements. Additionally, the project can copy some of the results for future trainings. All this is an important input for the continuous improvement team.

8.5.4 Collect and Categorize Improvement Ideas

Sources for Improvement Ideas

The feedback of customer projects is an excellent source for improvement ideas. Projects are the users and expect best usability of work instructions and tools. They might have new ideas to increase quality or safe time and they can contribute with pictures or examples that can be used for training purposes.

Improvement ideas can also be created by the experts themselves, when they describe and discuss the processes, or by auditors or other parties.

In all cases the ideas for improvement have to be collected and registered. In a second step, it has to be considered, how formal the improvement has to be treated.

Categorization of Improvement Ideas

In the case studies we have used four categories

1. Ad-hoc improvement
2. Kaizen cards[463] for small improvements
3. A3 problem solving stories[464] for medium-size improvements
4. Improvement programs[465] for large and complex improvements

If an idea can be implemented directly by making minor changes in the work instruction, it can be done ad-hoc, without creating formal documents.

In other cases, the expert wants to discuss the ideas with colleagues, or needs more time to change some documents. If this takes more than one hour, it can be handled as a small improvement.

8.5.5 Improve with Kaizen Cards

Kaizen

Kaizen is the Japanese word for "change for better". It is about improvement. A simple way of improvement is the usage of kaizen cards (Figure 8-32). Each kaizen card represents a small improvement which can be handled inside a continuous improvement team. Kaizen cards can be used if the effort of improvement is less than some days and it needs no explicit funding.[466]

[463] See (Liker, et al., 2007) for a description of kaizen cards
[464] See (Sobek, et al., 2008) for a description of A3
[465] See (Axelos, 2011a) for a description of program management techniques
[466] See (Liker, et al., 2007)

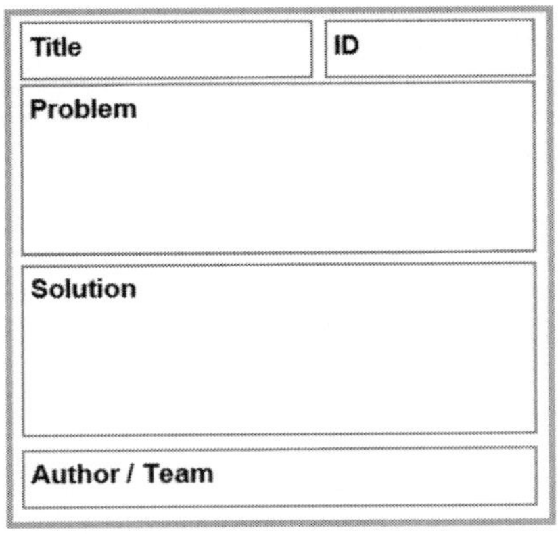

Figure 8-32: Kaizen card. A Kaizen card contains a description of the problem and a description of the potential improvement.

The kaizen card describes a problem in one of the processes. The problem description should refer to a process id and to one of the KPIs that shall be improved. This mapping allows to mark the process in the value stream to show the projects that this process has some known problems and needs improvement. The solution describes the idea that solves the problem.[467]

Even if the kaizen card does not document every detail, the improver should analyze the process using the fishbone diagram (Figure 8-33).

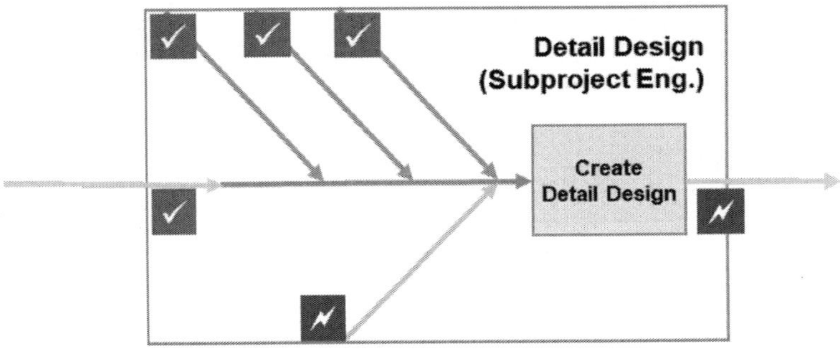

Figure 8-33: Diagnosis with fishbone diagram. The fishbone diagram helps to identify the factors that leads to poor process results or too high resource consumption.

Small improvements, which can be solved by a single CI-team, are often about problems with the method of the process, training material or self-created tools. The solution for the problem on the kaizen card is to create all documents and tools that are needed. These documents and tools can be used in the next project which executes the process.[468]

[467] See (Liker, et al., 2007)
[468] See (Liker, et al., 2007)

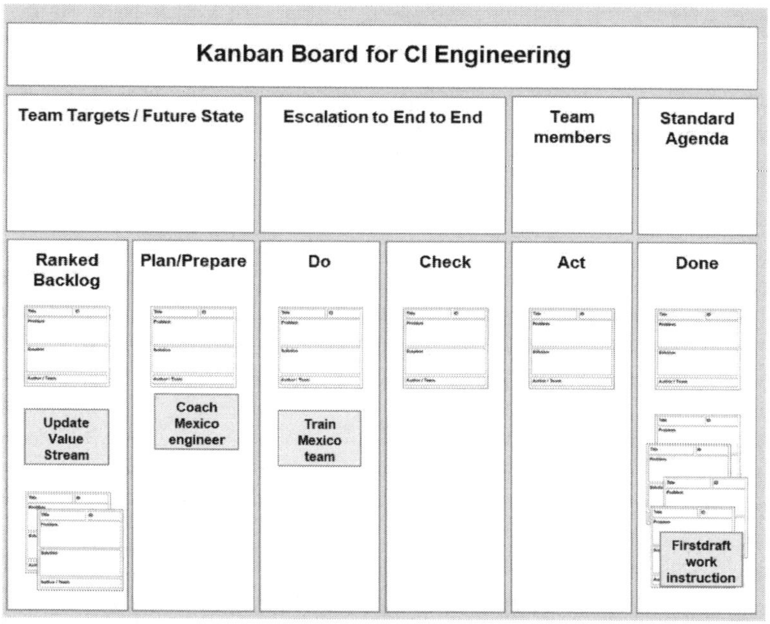

Figure 8-34: Kanban board of the CI team. It contains all improvements and planned work preparation activities

Like the subprojects, also the CI-Team has a Kanban board for its activities (Figure 8-34). This Kanban board contains all ongoing improvements, which are represented by the kaizen cards. It also contains other planned activities, which have to be done by the team. Examples are tasks to detail the value stream, to write work instructions or to train engineers in projects.

The phases of improvement work are linked to the subproject board. One of the starting points for kaizen cards is the feedback from customer projects, when their tasks are in the act phase. The CI-teams writes the card and adds it to the backlog of the CI-board, where its urgency and importance is discussed, and the card is ranked relatively to the other cards.

When the team pulls the kaizen card as the next improvement task, it moves it to plan / prepare. Preparation means to get all necessary resources. The most limiting resource is to get all needed team members together for a discussion, and the team has sufficient time to execute the task.

When everybody is ready and all information is available, the kaizen card can be moved to status do. During this phase, the work instruction can be updated, tools can be changed, and trainings material can be updated. Now everything is ready for piloting. The kaizen card can be moved to the check phase on the Kanban board.

In the check phase the kaizen responsible explains the process changes to the pilot project. This corresponds to the plan/prepare phase of a task on the subproject board. From the perspective of the subproject manager, he is about to prepare a task, whereas the CI-Team wants to check, whether an improvement is successful.[469]

The subproject and the CI team do the normal work preparation and follow the task. At the end they check, whether the project-task was successful, and they collect data and feedback.

---

[469] See Figure 8-31 for the collaboration of the CI team with the subproject team

The CI-team gets this as feedback and can now close the check phase of the kaizen card. During the act phase it can discuss, whether this improvement shall remain in the process, be changed or removed. After doing necessary changes or creating additional kaizen cards, the kaizen-lifecycle is finished, and the card can be moved to the status DONE.

8.5.6 Improve with A3 Problem Solving Story

Sometimes an improvement needs additional resources, or it needs cross-functional collaboration. In this case, the improvement idea has to be described more formally. The A3 problem solving story is a template for these improvements, which is well-known in lean production.[470]

Local Improvements

Local improvements are those, where a process is not efficient or produces defects, and the reason is in the process itself (Figure 8-35). Either the process description is wrong, or the templates and tools are not usable, or the training for the process is insufficient. In all of these cases, the CI-team for the process is responsible and can directly solve the problem. In most cases they use kaizen cards, but sometimes there is some additional investment needed to create new templates or tools. In such a case, they can use the A3 report to document the idea.

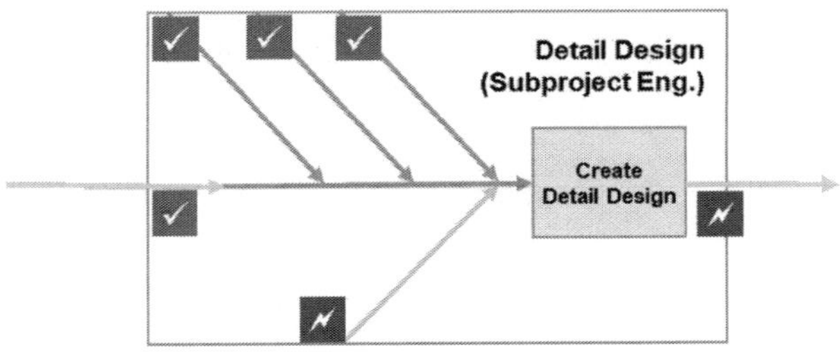

Figure 8-35: Local improvement. Cause and effect belong to the same process

Distributed Improvements

Sometimes, a process has a problem, which has its root cause in a different process (Figure 8-36). This other process often belongs to a different team. In other words: One team has to change a process and the other team has the benefits. Here we can refer to them as the change-team and the benefit-team.

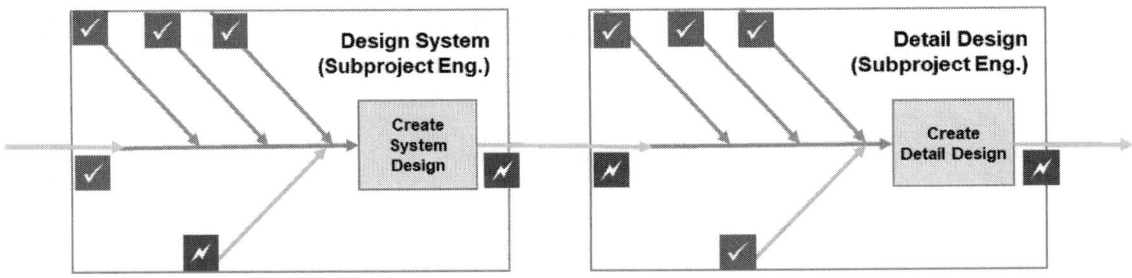

Figure 8-36: Distributed problem solving with local effect and remote effect. The cause is in a different process than the effect of the problem.

---

[470] See (Sobek, et al., 2008) and (Rother, 2010) for detail descriptions of A3 problem solving.

A3 Problem Solving Story

The A3 problem solving story is a template on A3 paper format and a related process helping to manage medium-size improvements (Figure 8-37). The process follows the PDCA logic. Sobek introduces different structures for problem solving and proposals. Problem solving reports can be used to analyze and solve problems in existing processes, whereas proposals can be used to introduce new ideas or even new processes Sobek proposes to start with A3 problem solving first, because it strengthens the ability to define and maintain a standard. If this ability is mastered, then the A3 proposal stories can be used to radically change these standards.[471]

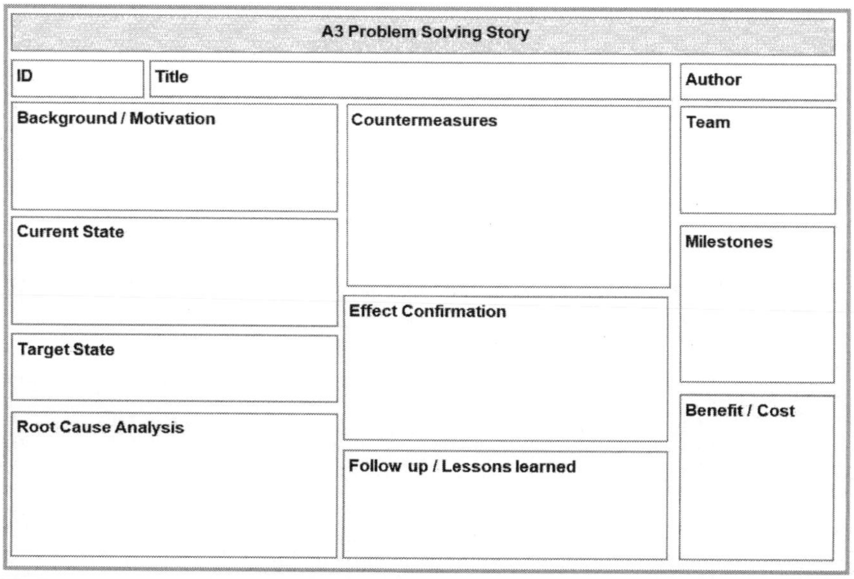

Figure 8-37: A3 problem solving story. The template is structured to explain, what has to be done, to change which process, to earn which benefit, with which overall relevance.

In the plan phase of an improvement the authors fill out the template to describe the problem the solution and give an estimate for cost and benefit. The A3 problem solving story will then be used to make a decision, whether the improvement shall be implemented.[472]

The template leads systematically though some questions and creates a strong storyline.

Background/Motivation

The background explains the reason for the A3. It can refer to the KPIs that have been published by the organization. With this reference the A3 shows that it contributes to the reduction of defects, cost, cost overruns, delays or duration of the overall process or to other aspects like customer or employee satisfaction.[473]

---

[471] See (Sobek, et al., 2008)
[472] See (Sobek, et al., 2008)
[473] See (Sobek, et al., 2008)

Current State

The current state shows one or several processes, where it is related to. For the reader it is easy to follow, if the section starts with processes from the end-to-end value stream and goes then into detail. Measures of the problematic process, like the defect rate or cost overruns, show the importance for the overall improvement targets.[474]

Target State

The target state explains how the process should work after the improvement and what the new measures (e.g. reduce defects by 90%) should be.

The difference of KPI-values of the current state and target state is the base for the first benefit estimates. It shows the gross benefits per project and can be multiplied with the number of affected projects in the next 3 years. This gives the overall gross benefit. [475]

Root Cause Analysis

The root cause analysis can use the fishbone diagram to identify all factors that contribute to the current problem. This "problem fish" is related to the "work preparation fish". Everything what can be detected in the root cause analysis influences the current process negatively. The process has is poor, because it currently has no best working conditions. Thus, every observation of the root cause analysis, leads to a necessary factor, that has to be provided for the process. [476]

Countermeasures

The countermeasures list is the plan for the do phase. It contains a list of all activities that need to be done. This includes to change other processes, or to create work instruction, tools and templates and training material for the process.[477]

Effect Confirmation

The effect confirmation is the plan for the check phase. It describes, where this A3 could be tested in reality, and which observations are expected. [478]

In project business this section has to describe which project will pilot the process change.

Milestones and Net Benefit

The estimates of the countermeasures and the selection of the pilot project give sufficient information to define the milestones for the do, check and act phase.

It is also possible to estimate the net benefit. The gross benefits are known from the comparison of target and current state. The cost of countermeasures is also estimated.

---

[474] See (Sobek, et al., 2008)
[475] See (Sobek, et al., 2008)
[476] See (Sobek, et al., 2008)
[477] See (Sobek, et al., 2008)
[478] See (Sobek, et al., 2008)

Management Decision

Prepared with that information, the leadership team can now rank the A3 problem solving story relatively to all the others. The ranking considers the net benefit, the duration of the improvement and the available resources. If all resources area available and the team is ready to do the countermeasures, the plan phase is finished, and the do phase starts.

8.5.7 *Manage a Stream of A3 Problem Solving Stories*

Like the kaizen cards, the A3 problem solving stories follow the PDCA process (Figure 8-38)[479].

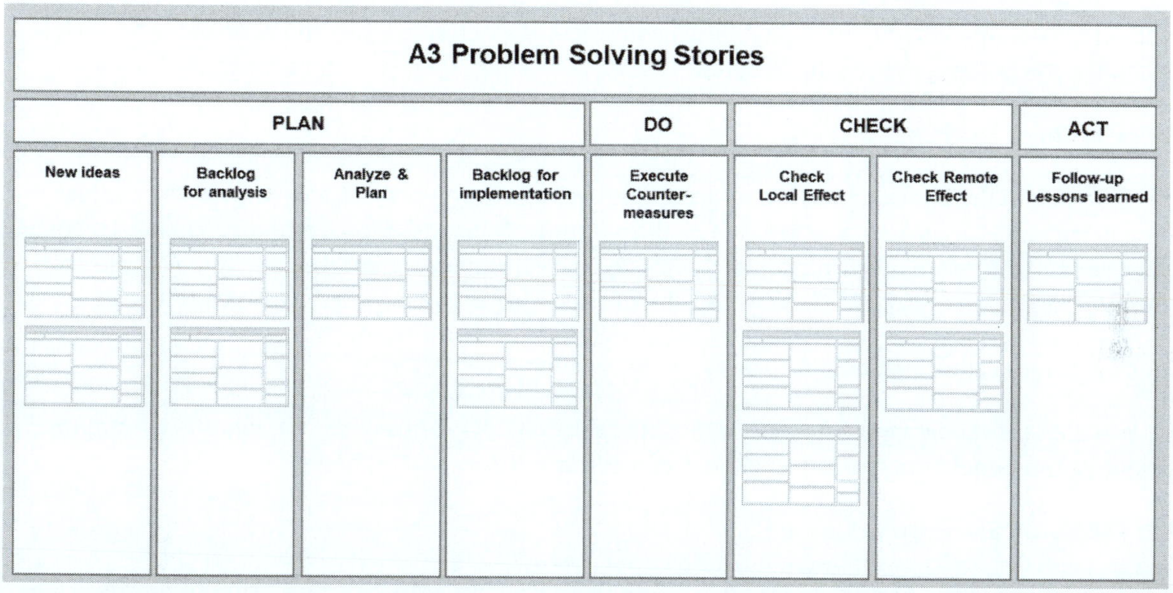

Figure 8-38: Workflow for A3s. The workflow can be visualized on a large board analog to the Kanban board of each CI-Team. This board needs more space due to the physical size of A3s. It also has detailed some of the phases to make bottlenecks visible.

Here is the full lifecycle of the A3 problem solving story as it is used by Viable Project Business in the case studies. The process details the classical PDCA process for the usage in project business,

Plan 1: New ideas

New ideas start in phase plan 1. In this phase the author describes the improvement idea and describes the background, current state and target state of the improvement. The author should also make a first draft benefit calculation.

Plan 2: Backlog for analysis

In the plan 2 phase the CI End-to-End team compares the new idea with other ideas and rank it relatively to those. This is important, to invest the analysis time for the most promising ideas.

Plan 3: Analyze and Plan

When all needed experts are available, they can start the plan 3 phase and discuss the new idea, analyze the deeper problems, define countermeasures, plan the pilot phase, estimate benefits, cost

479 See (Sobek, et al., 2008)

and duration of the improvement. The result is the fully described problem solving story, which can be presented to the leadership team.

Plan 4: Backlog for implementation

If the team has finished its work, the author presents the idea to the leadership team to get an approval. The leadership team can decide to accept or decline the idea. If it is accepted, the A3 will be ranked for execution. Now the leadership team has to ensure that all resources are provided.

Do: Execute Countermeasures

In the do phase the author manages that all countermeasures are done, and everything is available to change the process in the pilot project.

Check 1: Check local effect

The first step in the check 1 phase is to prepare the usage in the project and to train the project team members. This is part of normal work preparation. Then the improvement team has to ensure that the changed process is used correctly. This is done by coaching and support. Finally, the improvement team has to check, whether the process shows the expected positive effects.

Check 2: Check remote effect

If it was a distributed improvement, it needs a separate check 2 phase to see, whether the team expecting the benefits of the improvement can confirm the expected benefits.

Act: Follow-up and lessons learned

In the act phase the CI End-to-end team can use the data from effect confirmation to decide whether the changed process is the new standard or should be cancelled.

One of the problems in project business is, that the check 2 phase can be years after the check 1 phase. Examples are changes in the bid process or in engineering that have influences to commissioning work. In such a case the decision to rollout the improvement is often done after the check 1 phase. This not only means that a benefit is not confirmed, it also can introduce severe risks that a process change leads to significant disadvantages. Thus, it needs a risk assessment, to decide, whether the rollout shall be made after a successful check 1 phase or whether the results from check 2 phase should be awaited.

8.5.8 Improve the Improvement Process

KPIs on A3 Problem Solving

Of cause, everybody is interested to create the maximum improvement impact at the shortest time. The cumulative flow chart is used in agile project management to visualize progress.[480] It can also be used to show the numbers of A3 or the benefit of A3 grouped by their state at a certain point in time. This visualization helps to understand the flow of improvements and detect the bottlenecks.

---

[480] The cumulative flow chart is an alternative to the burndown chart

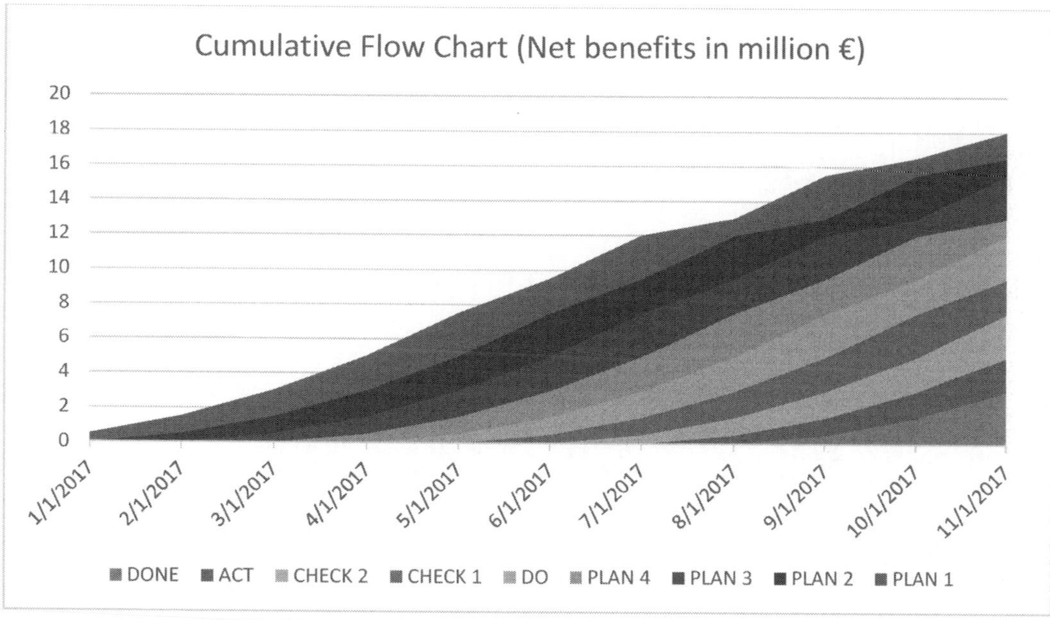

Especially in project business there may be a long waiting time between check 1 and check 2, because there is one year between changed process, like system design, and the benefit-process, like commissioning. The cumulative flow chart makes it visible, by showing that many improvements stay in the check 2 phase for a long time.

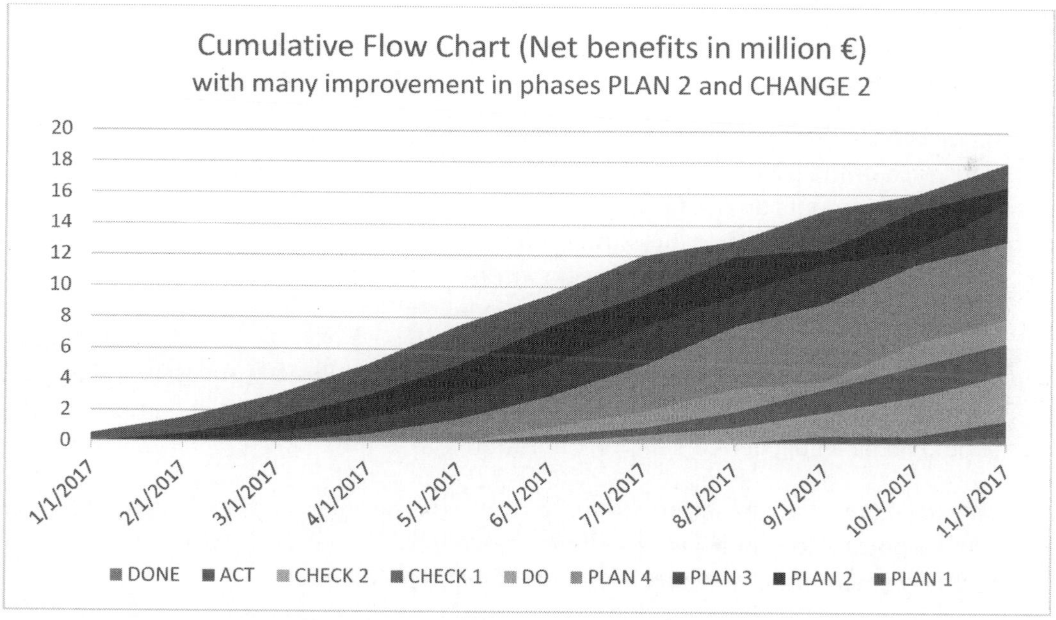

If there is a resource bottleneck in continuous improvement, this can be observed as well: Too many ideas remain in phases plan 2 or plan 4. This means, that there are ideas, but there are currently no resources to analyze them or to implement the solutions.

For a more detail analysis of the bottlenecks in the improvement process, the teams can use the same method, which they have used for the projects.

Improvement Value Stream

Business excellence can use the value stream analysis to improve the improvement process (Figure 8-39).

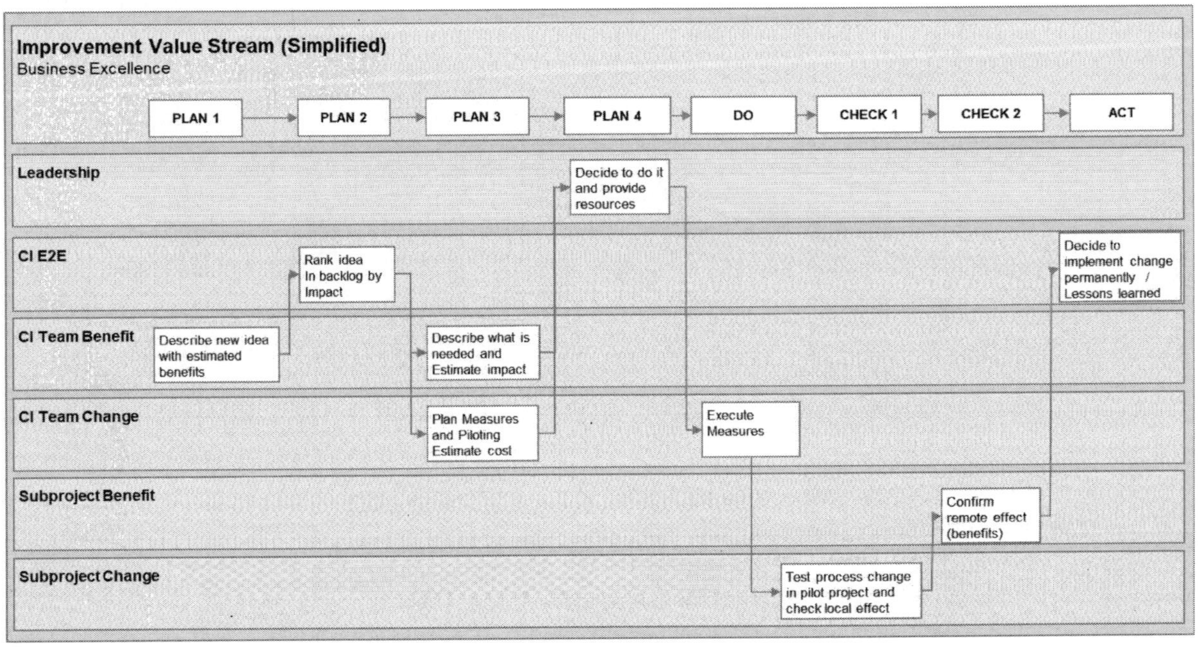

Figure 8-39: Improvement Value Stream. The improvement value stream visualizes all processes that are needed to realize an improvement.

The business excellence team can describe the entire improvement process and provide work instructions for each of the process steps. This helps to identify the bottlenecks of the improvement process. If the improvements do not bring the planned benefits: Which process is the reason for that? Is the execution of countermeasures the problem or has the root cause analysis been to narrow? Maybe the target state of the A3 was not properly defined or the current situation was wrong.

Other questions are about the cumulative value of improvements. Which processes influence the value? The idea generation process should focus on big benefits and the ranking and the decision processes should focus on big benefits in short duration.

With a detailed description of the improvement process, the business excellence team can analyze and eliminate the bottlenecks. Business excellence can improve their work instructions and training material for improvement processes and retrain all CI-team members.

Leaving the End-to-End Value Stream

If the business excellence team observes permanent problems in some processes of the end-to-end value stream, it can also go deeper and ask, why a CI team is not able to improve the processes. Obviously, the CI team has not the best working conditions. Business excellence can use the same methods as the CI team to find the deeper reasons (Figure 8-40).

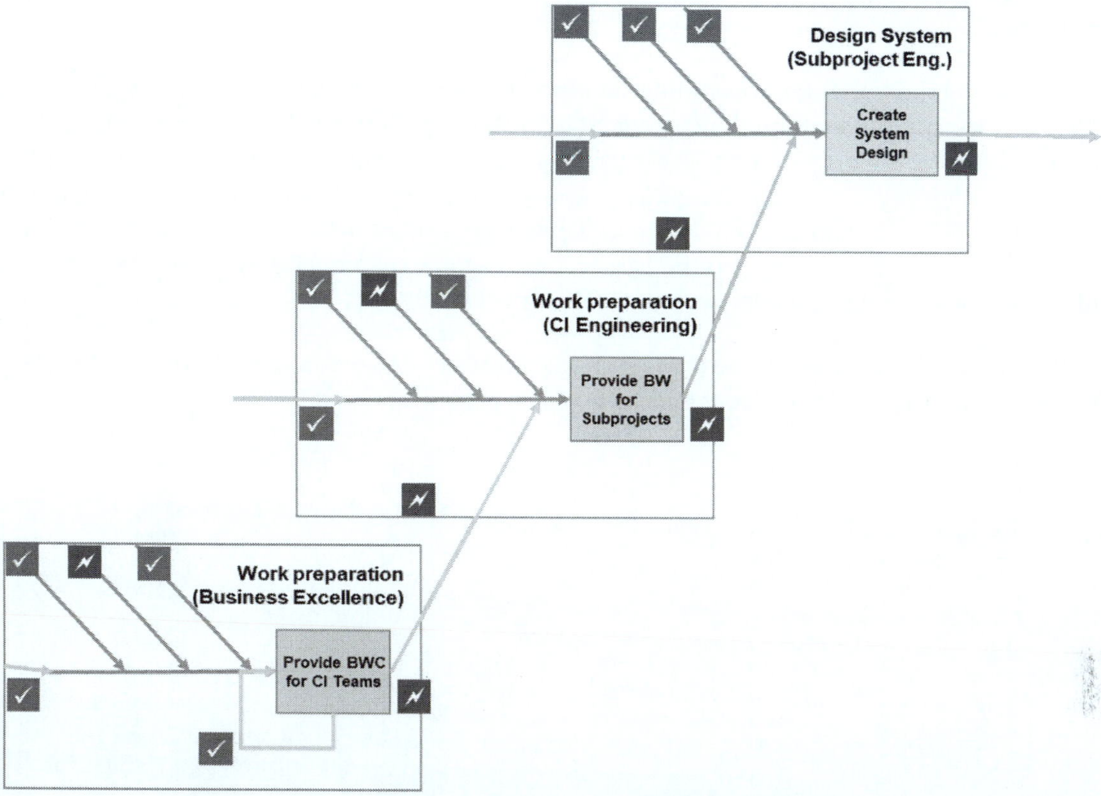

Figure 8-40: Chain of fishbones. Root cause of problems can also be the missing know-how. This requires improvements in the processes of the CI teams, which is the job of the business excellence team.

One typical reason are missing resources for continuous improvement. In this case, the business excellence should make this transparent for the leadership teams: Missing resources lead to a lack of work preparation and to problems in customer projects.

Another typical reason is missing know-how in the improvement methods. This would indicate, that the training or coaching was not sufficient.

Improvement of Improvement

All the described activities help to continuously increase the efficiency and efficacy of the CI teams. Thus, the CI teams create better working conditions for projects, and the projects can earn more money, which can be invested in the work of the CI teams – and, of course, the business excellence team.

8.5.9 Define the Future State

Of course, the organization wants to implement as many improvements as possible – everything in parallel. This introduces a new challenge: Even if all these improvements want to contribute to an improvement of KPIs, they may be contradictory. A simple example is that one team wants to exchange a tool chain and the other team repairs the old toolchain in parallel.

Future States

Rother and Shook show that improvements should not only by triggered by KPIs. Instead the organization needs to understand the current state of the organization and the relation to the current KPI values. If the organization wants to reach target KPI values, it has to design a future state that describes how processes should be in the future. This is typically done by making a value stream analysis (VSA), a value stream design (VSD) and a roadmap that describes how to reach the future state. Both, the target KPIs and the agreed future state, can now be used for the identification and alignment of improvements (Figure 8-41).[481]

This definition of future states can be done with various time horizons. Typical horizons are a 1-year short-term future state and a 3-5-year mid-term future state.

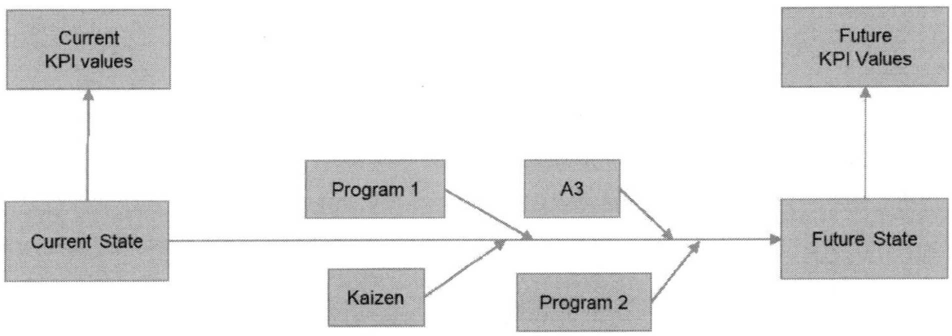

Figure 8-41: Future State. The agreed future state gives a clear direction for all improvements. The organization has already agreed that the future state helps significantly to realize the improved KPIs values.

Since the CI teams have the responsibility to improve processes, it is their job to find the best possible future states.

In large organizations this could lead to contradictory future states. Thus, it makes sense to coordinate all these activities.

Development of Organization-Wide Future States

The joint development of future states is described in the lean method "Hoshin Kanri", also known as lean policy deployment. Jackson describes the method of Hoshin Kanri very detailed. It starts with a definition of the seven nested PDCA cycles (long-term goals, midterm goals, annual goals, tactics, operations, kaikaku and kaizen). This chain connects every small kaizen improvement with the long-term goals of the organization.[482]

Jackson also explains the typical artefacts and processes that are used to implement Hoshin Kanri. The A3X templates link improvement measures to goals from the next higher levels. They are also linked to the responsible teams and to any KPIs that might be influenced by that improvement. [483]

---

[481] See (Rother, et al., 2009)
[482] See (Jackson, 2006)
[483] See (Jackson, 2006)

We will come back to this A3X in section 8.5.13, after we have discussed the implementation of large improvements, which contribute to the realization of the future state.

Let us focus on the joint development of a consistent future state first. Viable Project Business can use this concept the following way:

It is important to have a clear, consistent future state for the entire organization. The top management team is responsible for everything and thus has to identify and eliminate all bottlenecks. This includes the setting of future states, which help to improve functions that are already a bottleneck or will probably become one.

Since top management has delegated the End-to-End-process to the CI E2E team, it asks the CI E2E to detail the future state for all bid and project processes. The CI E2E team already has a value stream analysis and KPIs. Since the heads of all CI teams are members of the CI E2E team, they can define a future state for the entire End-to-End process (Figure 8-42).

All the CI-team can detail this picture.

Practically, this process can be implemented by a series of meetings, which starts with top management, via CI E2E to all the CI teams. After all future states are detailed, they need to be consolidated to one consistent big picture.

Jackson explains the catch-ball process, which describes how all the meetings have to be organized practically.[484]

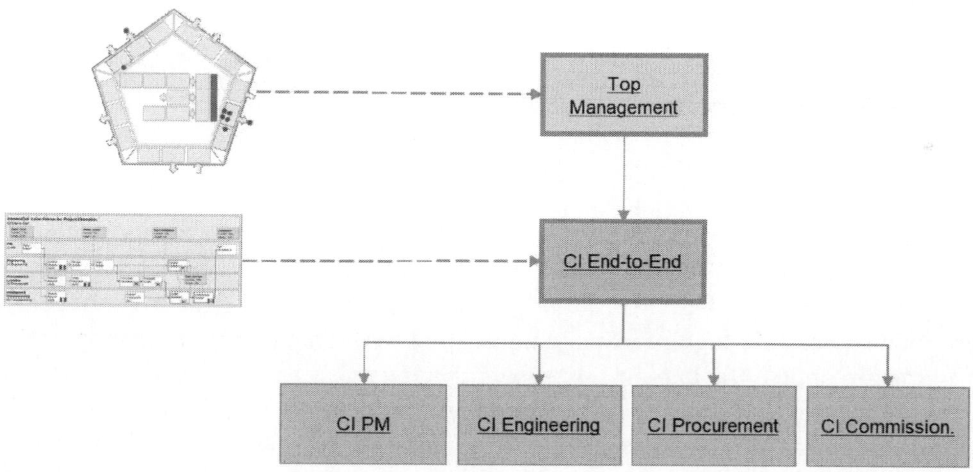

Figure 8-42: Future States on several levels. Top management is responsible for all processes, the CI E2E team is responsible for the processes in the end-to-end value stream, CI teams are responsible for their swim-lane.

At the end of this exercise, the organization has a clear consistent picture that can be used to align all improvements to the same direction.

---

[484] See (Jackson, 2006)

8.5.10 Manage Large Improvements with Programs

For some improvements the A3 problem solving story is not a sufficient description. This is particularly the case, if improvements need a high investment, introduce changes to more than one team or have a complex benefit structure.

These large and complex improvements can be professionally managed with program management techniques.

Program Management

Programs are temporary organizations like projects, but they have a different purpose:

OGC[485] defines a program as a "temporary, flexible organization created to coordinate, direct and oversee the implementation of a set of related projects in order to deliver outcomes and benefits related to the organization's strategic objectives."[486]

According to PMI a program is "a group of related projects managed in a coordinated way to obtain benefits and control not available from managing them individually. Programs may include elements of related work outside of the scope of the discrete projects in the program." [487]

In both cases the program coordinates a set of related projects to obtain benefits for an organization. This focus on benefits rather than on a deliverable makes the difference to a project. Programs have to deliver something, coordinate the transition to an organization and ensure that (organizationally relevant) benefits can be earned.[488]

Especially the management of benefits is important. In many cases the benefits have several dimensions (time reduction, cost reduction, customer satisfaction) that cannot be compared easily. Sometimes it is difficult to quantify benefits at all. OGC proposes "Management of value" as a set of easy-to-use standard techniques to describe value and benefits.[489]

Viable Project Business uses this program approach for all large and complex improvements. This allows to use well-documented best practices. The benefit of using Viable Project Business is the precise teamplay. The team architecture already contains all teams and structures that are a prerequisite for a successful program management: The Leadership teams are responsible for all resources a program needs. The CI teams represent the "business as usual" and are the customers of the programs. If a program starts, there are already clear KPIs, well-defined processes and clear targets. This gives an excellent environment for programs and helps to make them successful.

There are also some differences to mention. The OGC states that a program "is likely to have a life that spans several years."[490] For OGC the typical usage of programs is "to coordinate a set of related projects".[491] In these situations, program management is an excellent approach, but it can also be used for situations where the duration is less than a year and the deliverables can be

---

[485] Office of Government Commerce
[486] See (Axelos, 2011a p. 5)
[487] See (Project Management Institute, 2008 p. 318)
[488] See (Axelos, 2011a)
[489] See (Axelos, 2010b)
[490] See (Axelos, 2011a p. 5)
[491] See (Axelos, 2011a p. 5)

produced by a single project or by the program team itself. This simplifies the situation: A3s are medium improvements, everything that is larger or more complex is a program.

To understand the structure of programs it is helpful to start with one of the most important artefacts of a program:

Benefit Map

The entire delivery chain of a program can be described with a benefit map (Figure 8-43). It shows, which deliverables have to be created to provide a new capability, which is necessary to change the state of some processes. It also explains that the change of the process leads to some positive local effects and some positive indirect remote effects, which both contribute to the organizational objectives. The benefit map is one of the central artefacts of program management.[492]

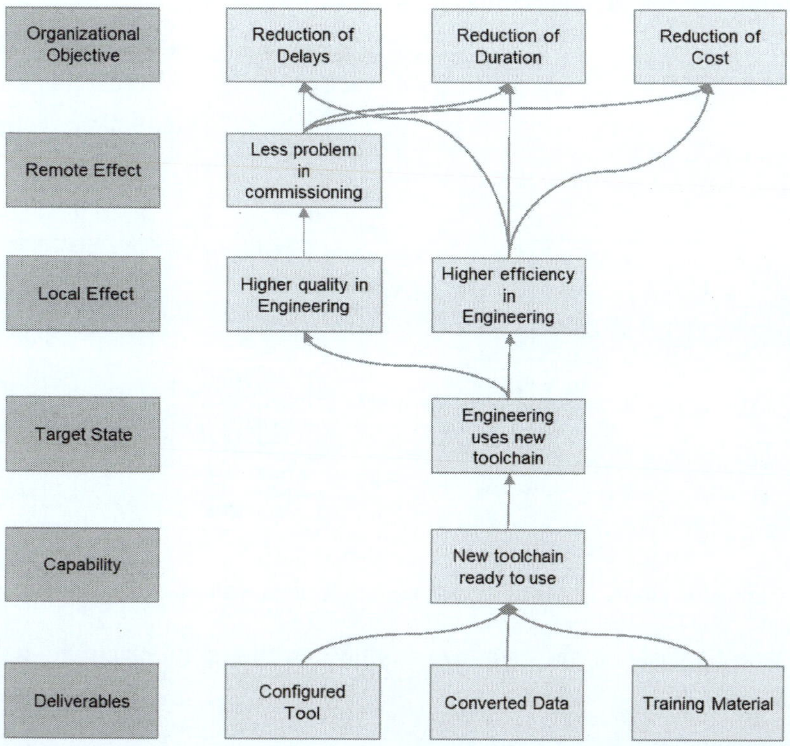

Figure 8-43: Benefit Map. The benefit maps shows how the creation of deliverables help to change a process and contribute to the strategic objectives

A program is responsible to manage the improvement as described in the benefit map. Since a program cannot create all deliverables itself and cannot ensure to establish the target state, it needs a collaboration with the other teams.

To explain this teamplay in more detail, we can follow the program through its lifecycle. This lifecycle is similar to the lifecycle of A3. Since the investment is much higher for programs, the plan phase takes more time and has higher cost. Thus, there is a phase that creates the best working conditions for a successful planning phase.

[492] See (Axelos, 2011a pp. 75-90) for benefit management

Program Identification Phase

The program identification phase starts with a new idea. If a sponsor is interested to implement the idea, the organization can spend some time to describe the idea in a program brief and to develop a program preparation plan. The program brief describes the idea and highlights potential benefits. The program preparation plan describes the needed resources to plan this program. Based on these documents, management can now decide whether to start the program definition phase or not.[493]

Program Definition Phase

If management is interested and all resources are available, a program team can be found, which has to detail the idea of the program brief. At the end, it has to show a clear benefit map, that explains what needs to be done to get the benefits. The OGC has a detail description for all the processes and artefacts of the definition phase.[494]

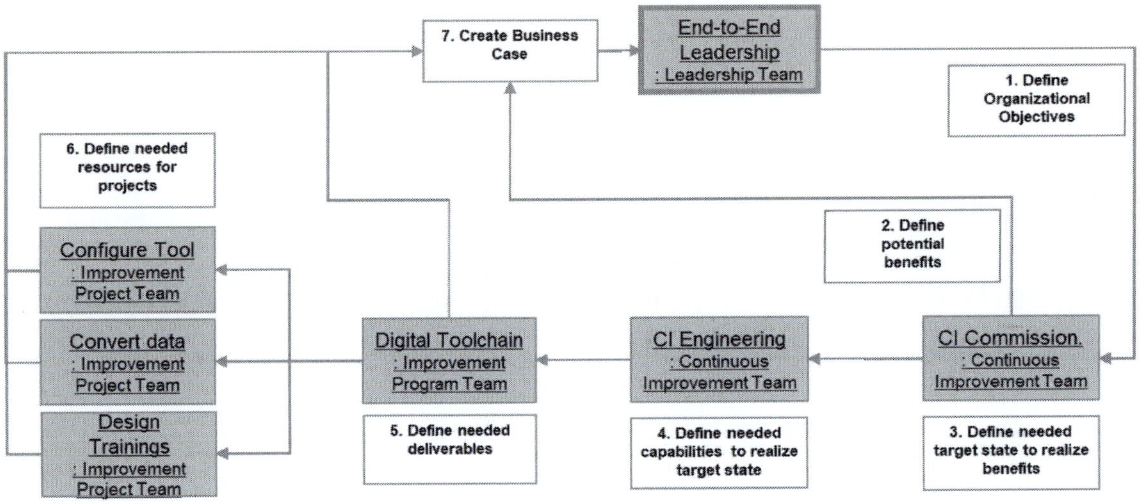

Figure 8-44: Program definition phase. Teamplay in the program definition phase

Here we want to take a special perspective that emphasizes the relations between the different roles (Figure 8-44).

1. The program starts by linking the initial program vision to the organizational objectives. These objectives are the improvement of KPIs or the achievement of an already agreed future state. This ensures that the program is strategically relevant for the organization.
2. The CI teams which are responsible for the benefits can now describe in a benefit register, how they can contribute to these objectives.
3. They also describe, which process changes are needed to realize the benefits. They give the commitment to realize benefits, if the target state is used.
4. The CI teams which are responsible for the change of the processes can now describe all missing capabilities, which have to be provided by the program to make the change possible.

---

[493] See (Axelos, 2011a p. 181)
[494] See (Axelos, 2011a pp. 189-197)

5. The program team structures these requirements and defines projects with clearly defined deliverables.
6. The projects estimate cost and duration for the creation of the deliverables.
7. The program uses benefits and cost to describe a business case, which can be decided by the leadership team.

This workflow creates four strong commitments, which have to be linked precisely.

1. The CI-teams having the benefits are committed to have benefits, if the process changes.
2. The CI-teams changing the processes have committed to change the process, if they get new capabilities.
3. The program team itself is committed to provide the capability for a change, if it gets budget and time to start the projects.
4. The project teams are committed to produce the deliverables in time and budget.

All statements together give the leadership team the opportunity to invest in the program to get a contribution for the organizational objectives. The leadership team can rank the defined program relatively to other programs and start it, when all resources are available.

Program Execution Phase

Execution starts when all resources are available. The execution phase delivers the contribution to the organizational objectives in the reverse order (Figure 8-45):

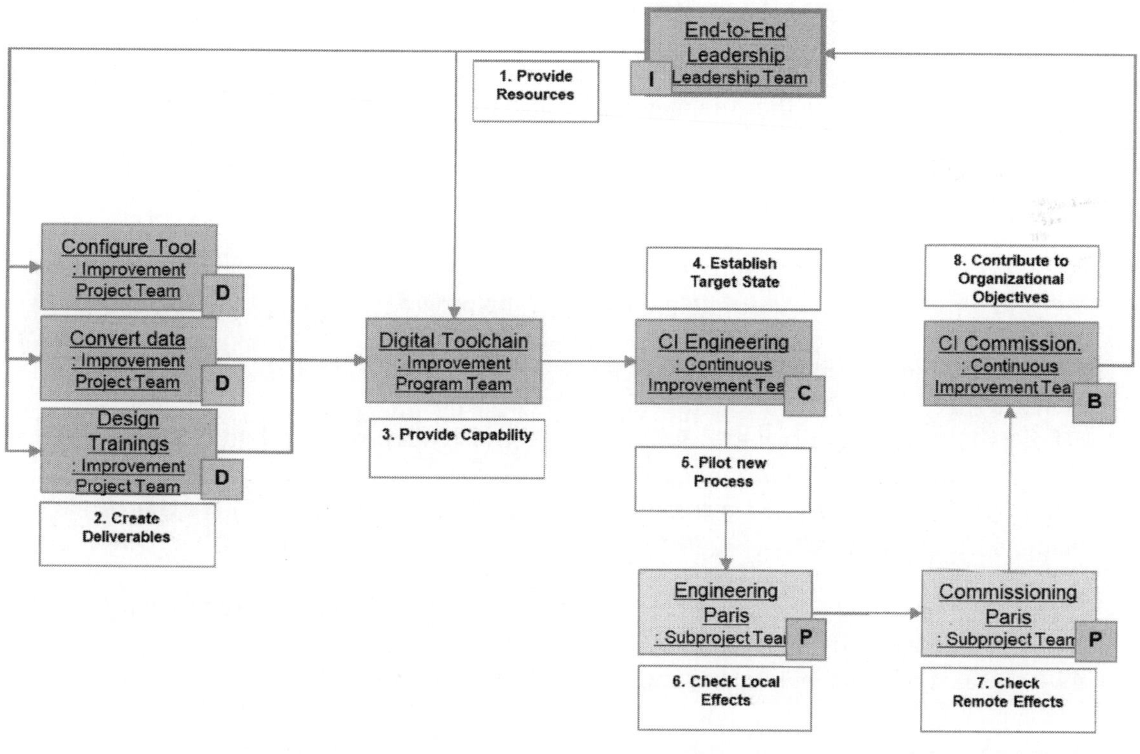

Figure 8-45: Program execution phase. Teamplay in the program execution phase with the different roles of the teams: I = invest, D = develop, C = change process, B = ensure benefit, P = pilot. Program management has to orchestrate these teams.

1. The leadership provides all necessary resources.
2. The projects create the needed deliverables.
3. The program team ensures to provide the needed capabilities to change processes.
4. The CI teams integrate the new capability to their work preparation and establish the new process.
5. Subprojects pilot the new process.
6. The CI-teams together with the subprojects can check the local effects of their process changes.
7. The CI-teams which have to committed to the benefits can check whether the positive effects occur.
8. If everything works as planned, all the teams have implemented the improvement and created a contribution to the organizational objectives.

In some cases, programs are too large for a single transition. In such a case, the program definition plans several tranches, which can be introduced separately one after the other.[495]

A3s and Programs

Viable Project Business uses the same basic processes for A3 and programs. They all follow the PDCA logic, but there are also some differences to mention:

The program identification has the same purpose as the plan 1 phase for the A3s with one big difference: It explicitly creates a plan for the definition phase that describes the needed resources for planning. Since A3s are smaller, they do not create an explicit plan for planning.[496]

The program definition phase has the same purpose and follows the same logic as the plan 3 phase for the A3s, with the difference that the planning results are much more detailed than a single A3 paper.[497]

Coordination of Programs

Typically, an organization wants to run several programs in parallel. Even if their organizational objectives and future states are absolutely consistent, the programs can create problems:

1. They can create bottlenecks in the CI teams, if they all want to change processes of the same trade.
2. They can double-count benefits. Two programs which have independent plans to reduce 50% of cost, do not achieve zero cost but 25%.
3. They can create bottlenecks in customer projects, if they want to pilot their changes in the same customer project.

Thus, all the programs need to be coordinated. If there are not too many programs running, this can be done by the CI E2E Team, which is responsible for the entire end-to-end process. If there are too many programs to oversee, it makes sense that CI E2E founds a dedicated portfolio team.[498]

---

[495] See (Axelos, 2011a)
[496] Compare with the A3 workflow in section 8.5.7
[497] Compare with the A3 workflow in section 8.5.7
[498] In large organizations this can be done in phase 2 of ontogenesis. The portfolio team has the first job to identify all large internal activities and understand their benefits and stakeholders.

8.5.11 Manage a Portfolio of Programs

Portfolio

The OGC defines a portfolio as "the totality of an organization's investment (or segment thereof) in the changes required to achieve its strategic objectives"[499] and portfolio management as "a coordinated collection of strategic processes and decisions that together enable a more effective balance of organizational change and business as usual." [500]

These definitions emphasize that its needs investment in change to achieve the strategic objectives, and that these investments have to be balanced. This balancing not only includes the avoidance of problems and bottlenecks that have been described in the recent chapter. It also includes to balance the investments to the set of strategic objectives, which requires a clear analysis of the system bottlenecks.

Practically, the portfolio team can be led by the heads of R&D or engineering. In many cases they directly represent the "develop", "change" and "benefit" roles. As members of the leadership team they are also responsible for the "invest". From a program perspective, the entire portfolio team is the sponsoring group.[501]

For each program one member of the portfolio team should be the senior responsible owner who is responsible for the success of the program. The senior responsible owner is then the direct contact to the program manager.[502]

The portfolio team can use a Kanban board similar to the one used for A3s to visualize the flow of programs.

Portfolio, Program and Project Office

One of the challenges is to consistently use the methods of Portfolio, Program and Project management on a high level. This is difficult especially in the program definition phase, because it takes a long time to train experienced program managers. One solution is to have experts in a dedicated team that help to structure all programs. This P3 Office can define and improve the standards and support the usage. This ensures that documentation is consistent on a high level [503, 504]. Such standards are also the base for continuous improvement of the portfolio and program management techniques.

8.5.12 Improve Other Functions

The previous sections explained how to improve the end-to-end process that creates the customer value. This end-to-end-process is the detail level for the function "Create Customer Value" in the functional overview diagram.

---

[499] See (Axelos, 2011b p. 131)
[500] See (Axelos, 2011b p. 131)
[501] See (Axelos, 2011a p. 37)
[502] See (Axelos, 2011a)
[503] see (Office of Government Commerce, 2008)
[504] See (Taylor, 2016)

All the other functions like "Acquire Customers" can be done in a similar way. Sometimes the value stream analysis cannot be used, because a function consists of many independent processes. This requires additional methods from the lean toolbox like the work structure analysis.[505]

Examples for improvements of other functions are:

1. Introduction of a standardized supplier selection process
2. Improvement of the resource allocation process
3. Introduction of a KPI system and dashboard
4. Introduction of a defined people development process to systematically develop project managers

All these improvement activities change the state of a certain function to improve a KPI. They have all been selected to remove bottlenecks of the organization.

For top management it is important to have a clear picture of all ongoing improvements, the related functions, KPIs and teams.

8.5.13 Make Improvements Transparent

The previous sections already showed, how programs are linked to the organization:

1. Each program belongs to a portfolio
2. Each program contributes to one or several future states of the organization
3. The future states are a description of a better future. They refer to a function and influence the KPIs of this function.
4. Functions and KPIs are mapped to responsible teams
5. Programs have teams as their stakeholders

The special approach of Viable Business is the mapping of functions and KPIs to the functional overview diagram. The functions and KPIs are those, which have been identified as systemic bottlenecks by the self-perception processes. The KPIs are not only about commercial numbers, they can also be about soft factors like employee satisfaction.

The future states describe, how the organization has to be at a defined time horizon to eliminate the bottlenecks. Detail descriptions can use methods like value stream design which show, why the future state improves the KPIs.

Each program contributes to one or many of these future states. Thus, each program helps to eliminate bottlenecks. Portfolios help to manager the ongoing programs.

The mapping of programs to teams shows, which team is consumer of the project results. In many cases, programs deliver to the CI teams, which use the results to improve working conditions for projects.

All this information can be documented as one big picture: The so-called A3X. It contains all KPIs, functions, future states, portfolios and programs, and teams in on picture (Figure 8-46). All the entities can be mapped.[506]

---

[505] See (Wiegand, et al., 2011) for analysis techniques and (Wiegand, et al., 2009) for design techniques
[506] See (Jackson, 2006) for a detail description of A3X

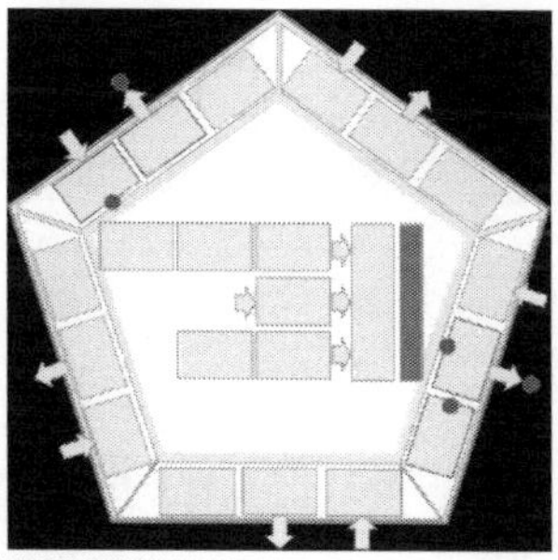

Future States 2020						Teams																
New leadership model	New career path for project managers	New bid process	New engineering process	Modular product		**Top Management**	**Leadership End-to-End**	Leadership Bid & PM	Leadership Engineering	Leadership Procurement	**Business Excellence**	**CI E2E**	CI Bid	CI Project Management	CI Engineering	CI Procurement	CI Commissioning	**Portfolio of Bids & Projects**	Paris	Berlin	London	NN
					Major Improvements																	
					Portfolio "Organizational Design"																	
x					Culture program	ICB	ICB	CB	CB	CB												
	x				Define New career path for project managers		ICB	CB										B				
					Portfolio "Improvements for Projects"																	
		x	x	x	Modularization program		I						CB		CB		B				P	P
			x		Digital Toolchain				I					CB	CB	CB	CB				P	P
		x	x		**Acquire Customers**					x		x	x					x				x
		x			Order Income					x		x	x					x				x
			x	x	**Create Customer Value**					x		x		x	x	x	x	x			x	
			x	x	Planned Duration					x		x		x	x	x	x	x			x	
			x	x	Delay					x		x		x	x	x	x	x	x	x	x	
			x	x	Planned Cost					x		x		x	x	x	x	x			x	
			x	x	Cost Overrun					x		x		x	x	x	x	x	x	x	x	
		x	x	x	Customer Satisfaction					x		x	x	x	x	x	x	x			x	x
x					**Create Employee Value**	x	x	x	x	x												
x					Employee satisfaction	x	x	x	x	x												
	x				**Develop People**	x	x	x	x	x								x				
	x				Number of project managers	x	x	x	x	x								x				
					Functions / KPIs																	

Figure 8-46: A3X. The A3X contains a mapping of functions, KPIs, future states, major improvements and teams. The letters in the mapping of programs to teams describe the role of the team (invest, change process, benefit, pilot change).

The A3X can be used to show the dependencies on one piece of paper, but it has some limitations. Commercial managers would like the see the cost and benefits of projects per fiscal year. It is also difficult to see how future states with different time horizons map.

In the digital age it makes more sense, to develop a multi-dimensional model based on the described entities in a relational database and create specific views and dashboards for all the users. This digital approach would allow to integrate all ongoing A3s to the same picture, what is not possible on an A0-poster.

8.5.14 Summary: Continuous Improvement as a Team Type

If we bring all methods and collaborations together, we have a description of the ideal continuous improvement team. This collection is independent from the specific trade and independent from the organization. It can be considered as a reference team type, which can be used as one of the building blocks to define large organizations.

The description of the team type has several aspects:

Team Purpose and Scope

The purpose of the CI team is to provide all necessary know-how to bids and projects. This contains everything that is not a resource (which comes from leadership) or project-specific.

Each CI team is responsible for a defined set of processes and related KPIs, which is described in the overall value stream analysis. The CI team has to create transparency in detail, provide know-how for (sub)-projects and continuously improve the processes towards the overall improvement targets.

Team Setup

The team consists of a group manager and some experts. The group manager is responsible for the resources and represents the team in the leadership team.

Lifecycle

The team should be started by detailing the CI End-to-End team. The team exists permanently. It can be merged with neighbor teams, if it becomes too small.

Methods and Collaborations

The previous sections showed several methods and collaborations of the continuous improvement teams. Some of the methods have to established before others. The ability of self-organization, for example, is prior to the ability to do work preparation for the customer projects and prior to all improvement work (Table 36).

Continuous Improvement Team			
	Area	Methods	Collaborations
1	Self-Organization	All functions of a Viable Team	Business Excellence Leadership Team (deliver resource)
2	Responsibility for Processes	KPIs, Value Stream Analysis, Work Instructions	CI E2E (define scope)
3	Work Preparation	Preparation, Training, Coaching, Support, Monitoring, Collect data & feedback	Subprojects (internal customers)
4	Improvement	Idea generation, Kaizen, A3, Root cause analysis	Neighbor CI Teams, A3 Teams, Subprojects, CI E2E
5	Future States	KPIs, Future States	CI E2E
6	Complex Improvements	Benefit maps	Program Teams

Table 36: Methods and collaborations of CI teams

The development of necessary competences should start with the basics first. On the other hand, it is not possible to achieve mastery in self-organization before the first improvement can be done.

Some of the methods, like A3, are also quite difficult to master. It needs some experience to write good A3 problem solving stories.

Additionally, there are many dependencies to consider. A good A3 improvement has a clear contribution to a KPI or a future state and refers clearly to a process. This is only possible if the targets are already clarified. Realistic target setting, on the other hand, is only possible, if the organizations has some experiences how to achieve it. Thus, it makes sense, to introduce some maturity levels for each area, which can be reached in small steps. [507]

It is the job of business excellence to train all the methods and to ensure the proper usage. They can define some maturity levels for each area. Level 1 for improvement could contain kaizen cards only, level 2 contains A3 problem solving in a simplified way. On level 3 the A3 should refer to a KPI.

8.5.15 How Does the Improvement Machine Help?

Let us go back to the initial problems from the introduction section and discuss the seven typical problems with improvement again (Figure 8-47).

Problem 1: Projects do not document problems accurately

Viable Project Business uses well-defined team types which have clear responsibilities. One of the responsibilities of subprojects is to follow the PDCA logic. They see missing preconditions in the plan/prepare phase of a task and to collect data and feedback in the check phase.
It is the job of the CI teams to train the subprojects in PDCA and to coach and support them if necessary. Whenever there are deviations, the subproject manager reports it to the right CI team

Problem 2: Teams do not exist or have insufficient resources / know-how

Since Viable Project Business has a 100% End-to-End coverage for continuous improvement, for each problem exists one responsible CI team (the default is the CI E2E team). This team is a Viable Team and thus able to produce results, if all resources and know-how are available. The resources are provided by the leadership teams, know-how for improvements is provided by the business excellence team.

[507] Such a capability maturity model can be designed similar to ISO 9004 (ISO 9004, 2015) or CMMI (SEI, 2018)

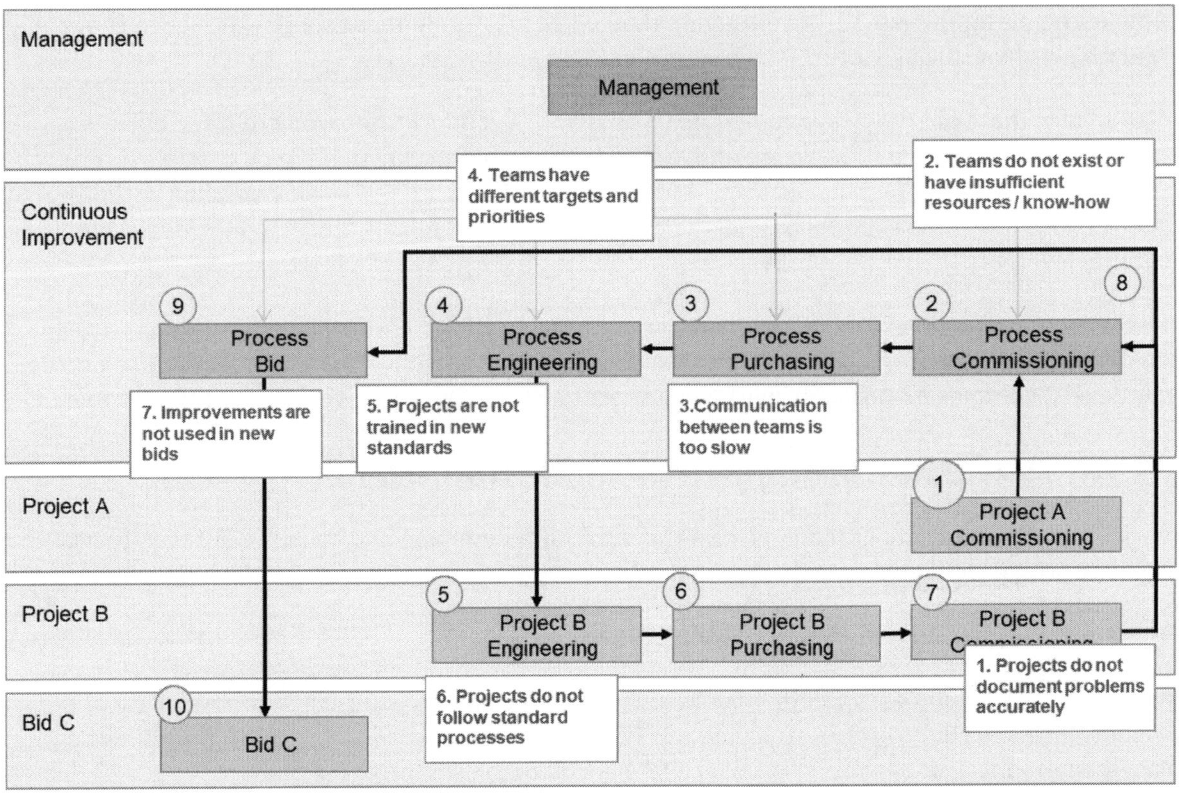

Figure 8-47: Typical problems with continuous improvement. Several collaboration problems lead to the inability to improve processes.

Problem 3: Communication between teams is too slow

All CI teams meet on a weekly base. The CI E2E Team with the heads of all CI teams meets after the specialized CI teams. This allows a fast communication and escalation.

Problem 4: Teams have different targets and priorities

The target setting starts from an organizational perspective. The end-to-end process of customer value creation has clear overall KPIs and a defined overall future state.

When the CI E2E team creates sub-teams for the different trades, it also aligns these sub-teams to the overall targets. For all CI teams it is important to make the best possible contribution to the overall success. New ideas can be given to the CI E2E team and globally ranked. If an improvement cannot be done immediately, the reason might be other, more important, improvements. The A3 Kanban board makes this transparent for everyone.

Problem 5: Projects are not trained in new standards

This is the main job of the CI teams: They improve processes and train and coach the projects.

Problem 6: Projects do not follow standard processes

The training, coaching and support of the CI teams help the subprojects to use the new standards. Additionally, the leadership, project and CI teams are aligned and fully agree, that subprojects have to follow the best practices form the CI team. If this is not feasible for some reasons, then they can agree on a project specific way. Sometimes the avoidance of standards can be a culture and discipline problem, which has then to be addressed by the leadership teams.

Problem 7: Improvements are not used in new bids

The CI teams use the data of their standard processes as input for new calculations. The CI Team for Bids ensures that this data is used in new bids.

Summary

The improvement machine addresses all of the problems mentioned in the introduction. It uses many well-known lean methods to organize successful improvements.

What is now the special contribution of Viable Project Business? It provides the fundament for improvement by establishing strong teams with a strong collaboration:

1. Each team is a Viable Team and thus trained to work efficiently if all resources are available.
2. For each team it is absolutely clear, who has to provide the resources.
3. The purposes of all teams are aligned, such that there are no open ends in resource delivery
4. All CI teams work to the same direction due to shared KPIs and a shared future state.
5. The collaboration between all teams are explicitly designed and integrated in the team type descriptions.
6. There is a clear responsibility to identify and eliminate bottlenecks in the improvement process.

All these features are a strong fundament for the continuous improvement work.

8.6 Practical Implementation

The implementation of Viable Project Business follows the ontogenesis that has already been described.

Practically, it makes sense to create a picture of the future organization, before starting with the team setup. This definition of the future organization is the job of the top management team.

Top management can create a team collaboration diagram of all the teams that shall be established. Three aspects have to be considered here:

1. Use predefined team types if possible.
2. The team collaboration diagram should fit to the specific needs of the organization.
3. The sequence of rollout needs to match the organizational needs.

8.6.1 Use Predefined Team Types as Building Blocks

Each of the teams belongs to a team type. Many of the characteristics of teams belong to the team type, rather than the individual team. Project teams, for example, are temporary teams, which deliver a defined output, whereas continuous improvement teams are permanent teams.

The collaborations between teams is also based on the team types. A3 improvements are temporary activities, which are started by a continuous improvement team. The CI team is the parent of the A3 team.

The identification of the team types allows to collect and share knowledge. The following table gives an overview of all team types (Table 37).

	Team Type	Base Type	Parent	Main Purpose	Customers	Perma-nent Team?	Typical team members	Creates Sub-teams
1	Top Management	Viable Team		Make Organization sustainably successful	Investors Environment 2	Yes	CEO and top management team	2
2	Leadership	Viable Team	1	Provide Resources for other teams	Employees, Suppliers 1,3,4,5	Yes	Department Lead + Group Lead	2 (if cascaded) 3,4,5
3	Portfolio of Projects and Bids	Portfolio Team	2	Create Customer Value, Earn Money		Yes	Head of Project Management	3 (if cascaded) 6,7
4	Continuous Improvement	Viable Team	2	Provide Best Working Conditions for Bids and Projects	3,6,7,8	Yes	Manager + Experts	4 (if cascaded)
5	Business Excellence	Viable Team	2	Provide Best Working Conditions for Continuous Improvement	4	Yes	Viable Business / Lean / Agile Experts	None
6	Bid	Project Team	3	Get new contract	3,7,8	No	Bid team	None
7	Project	Project Team	3	Create customer value and earn money	Customer 3	No	Project manager and subproject managers	8
8	Subproject	Project Team	7	Create customer value	Customer, 7	No	Subproject managers + team	None
9	Improvement Portfolio	Portfolio Team	4	Manage a portfolio of complex improvements	4	Yes	Head of Engineering, head of R&D	10
10	Improvement Program	Viable Team	9	Define and implement a complex improvement	4	No	Program Team	11
11	Improvement Project	Project Team	10	Create deliverables for a complex improvement	10	No	Project team	None
12	Experience exchange team	Viable Team	3	Exchange experience between projects and agree on standards	6,7,8	Yes	Project manager and subproject managers	None

Table 37: List of pre-defined team types

All of the teams are derived from Viable Team, Project Team or Portfolio Team. These base team types allow to collect project or portfolio management know-how and use it both for internal and external purposes.

8.6.2 Design the Team Collaboration Chart Based on Organizational Needs

The previous sections gave an indication how the organization can be designed, but it should not be understood as a fixed template. The organizational design has to be adapted to the company.

There are many aspects that have to be considered. Examples are:

1. Type of business
2. Size of the organization
3. Number of projects (influences whether project portfolio has to be cascaded)
4. Number of subprojects per project (influences how many CI-Teams are necessary)
5. Number of complex improvements (influence the number of portfolios)
6. Number of locations, maybe in different countries and even time-zones.
7. Authority over E2E-process. (Maybe some important functions are not under control of the organization e.g. procurement department is not part of the organization)
8. Impatience (The organization has time to systematically establish the continuous improvements teams or it is under severe pressure to report early results)
9. Available lean/agile knowhow

All this has influence on the organizational design and the sequence of implementation.

8.6.3 Find the Best Opening Strategy

Like in chess there are many possible opening moves available, which lead to same intended result. Here is a brief overview of some possible opening moves:

Standard Opening from Top Management

Section 8.4.4 showed a straight-forward opening, which starts with top management, creates five teams to cover the entire system dynamics, grows by splitting the permanent teams and adding the temporary ones.

This has the advantage that top management is involved and engaged directly from start. It allows to create a well-balanced system. This version is the preferred way to introduce Viable Project Business.

Opening with a Single Department

Instead of starting with the top management, it is possible to start with a single department. Experience shows that it might become difficult to get the end-to-end-coverage in a second step. This makes it difficult to make distributed project improvements.

One example is to start Viable Project Business with the commissioning team. They can structure their own work and provide best working conditions for each construction site. After a while, they will have a large backlog of problems which needs to be solved by the engineering department. Highlighting these problems may seem like to blame the neighbor department and cause negative reactions. It is very important, to involve the engineering colleagues as early as possible.

In the best case this single department can show their successful work and the method is used by the neighbor departments as well. The heads of the teams found the CI E2E team and organize their leadership tasks in a joint leadership team.

Opening with a Crisis Project

This approach allows to start with the highest top management attention. The strategy is to help a crisis project and to show that the problems are not project specific. One example is to help a crisis project during the system test phase and point out what the weaknesses of the test phases are. This allows to establish continuous improvement for the system test in general. This first CI team is the starting point to convince the neighbor departments also to start with CI.

This strategy leads to the same situation as described in the "opening with a single department". The main difference is the higher top management attention, which can be used to start with a CI E2E team earlier.

Opening with a Complex Improvement

Sometimes an organization wants to implement a complex improvement. Program management techniques help to structure this improvement and lead to the question, which teams ensure to change and maintain the process.

This leads directly to the question, who has to ensure that the standard is trained and coached, and who has committed the benefits. The answers to these questions are the CI teams.

Recommendation

It is possible to start with each of the four alternatives. The bottom-up variants have all the challenge to get an end-to-end-coverage. If it is necessary to start this way, it should be tried to use the first success stories to convince top management as early as possible to continue top-down.

8.7 Viable Business on Higher Levels

Large Organizations with Several Business Units

Large Organizations can also implement Viable Business with more than two levels. The large organization consists of business units and corporate units.

The business units have the purpose to earn money for the entire organization. Functionally, they have the same purpose as the "green teams" on a lower level. The corporate functions have the purpose to provide the necessary resources from outside the organization. This includes corporate human resources, corporate procurement and corporate finance. They are directly related to the stakeholder groups employees, suppliers and investors. Their activities ensure that the business units have all needed resources.

Other corporate activities are about knowledge. Corporate guidance creates know-how that can be shared by all business units. One example is the corporate guidance for project management. This group can define standards, train and certify project managers and provide support for the business units.

All corporate functions have a direct relationship to one of the Viable Teams in the business unit. In the case of corporate guidance for project management, this would be the CI-team for project management. The CI-team can use the guidance of the corporate governance to improve the PM processes. It is also responsible to ensure that obligatory standards are implemented. The same mechanism can be used for all other corporate-wide initiatives like engineering initiatives or manufacturing initiatives.

Internal Ecosystems

Large organizations can have several business units for project business, product business and service business. The product business develops and delivers the products that are used in project business to create a plant. After completion of the project, the service business provides operations and maintenance. All three businesses depend on each other.

Like on a lower level, it needs the ability to make distributed improvements, where one business has to do something to make the other business successful. Now this distribution spans several business units and the time delay between countermeasure and benefit may be several years.

A typical example is the "serviceability" of plants, which creates cost for the project business and simplifies life for the service business.

External Ecosystems

An even higher perspective would be the modelling of entire ecosystems, which include the 1st- and 2nd-tier suppliers in one picture. If companies are strongly connected, they have to understand the bottlenecks of the entire ecosystem. This ecosystem perspective can be independent from the corporate perspective.

8.8 Summary

The teamplay of Viable Project Business follows the anatomy of a multi-cellular organism.

1. Every team needs to work as a Viable Team to be successful and to be able to efficiently contribute to the overall system.
2. Every team gets all needed resources. This is like homeostasis in multicellular organisms, where every cell gets the perfect environment and thus can focus on specific tasks.
3. Every team has a purpose that creates value for a stakeholder or helps to provide the best environment for other teams.
4. And it has one more element: The top management and the business excellence team are responsible for continuously improving the teamplay by identifying and eliminating the bottlenecks.

After all the gray theory, let us see how this system works in practice.

9 Case Studies

9.1 Case Study as a Scientific Method

9.1.1 Scientific Methods in Social Sciences

Viable Project Business changes an organization as a social system. Yin describes five methods that are frequently used in social sciences[508]:

1. Experiment
2. Survey
3. Archival Analysis
4. History
5. Case Study.

An experiment requires control of the situation. In complex social situations this is typically not given. Survey and Archival Analysis are quantitative methods that depend on comparable data and statistical techniques. History focuses on the analysis of historical events and requires several sources, that can be analyzed. Case studies describe contemporary events qualitatively. Each of the methods has its own benefits and disbenefits.[509]

The case study method can be used in situations, where quantitative information is not available, and the situation is not under full experimental control. One example is the usage in a publication about standardization in large industrial plant manufacturing[510]. The situation in each company is unique and cannot be compared quantitatively with other companies. It is also not possible to control the modularization and standardization of plants like an experiment. However, it is highly important to learn from the approaches of organizations. In such a situation, the case study method can be used.

In the case of standardization, it would also be possible to make a survey. Surveys produce quantitative data. They can show how many organizations already have which degree of reuse on which level, but they cannot reveal deeper insights in the problems of standardization.

Introduction of Viable Project Business in an organization can be compared with modularization and standardization in large industrial plant manufacturing. It is an endeavor that takes several years. Viable Project Business influences even more people than a standardization program and is thus even less under experimental control.

The introduction of Viable Business happens in parallel to other measures. New managers with a new management style takeover responsibility, organizations are restructured, critical projects finish and release resources, some long planned new products are launched. This makes it difficult to isolate the introduction of Viable Project Business from its context.

---

[508] See (Yin, 2018 p. 9)
[509] See (Yin, 2018 pp. 5-13)
[510] Gepp describes several case studies about standardization (Gepp, 2014)

C. Dachs, *Viable Project Business*, Contributions to Management Science, https://doi.org/10.1007/978-3-030-62904-5_9

Implementing Viable Business is a journey of top management, a Viable Business expert and several hundred managers and engineers. It changes the way of working for several hundred people, who are already under high pressure. Thus, no implementation follows an ideal implementation plan and uses the current best knowledge. It is more a compromise – the best possible implementation under the given circumstances. Typical unplanned variations are the degree of involvement of top management and the available resources. On the other hand, the deviations from the racing line are a chance to understand, why some aspects are so important, and what would happen, if they are missing.

The missing experimental control and the inability to quantitatively compare the implementations in different organizations, lead to the case study as the method of choice.

9.1.2 Case Study Design

Yin recommends five components of the case study design:[511]

1. The research question that has to be answered by the case study
2. Proposition for a concept to answer the research questions
3. Selection of the cases
4. Design of data collection
5. Criteria for the interpretation of data

Research Question

The second research question in the introduction was: *Can the system dynamics of Life be used to design a management system that helps us to significantly improve project business?*

In sections 1 and 8 we've already integrated the translations from biology to business in one system and proved that it fulfills the criteria of ISO 9000 for being a management system. However, we still have to check whether it can be implemented in project business and whether it has significant benefits.

Propositions

According to Yin, the propositions help to connect the initial questions to the cases. They detail the question in a way that allows to systematically collect data to create evidence.[512]

Viable Project Business establishes a defined teamplay, which ensures that each team gets the needed resources from defined other teams. One important resource is the know-how of people. It is the focus of the Continuous Improvement Teams to develop and provide this know-how for all processes. If we want to check, whether this concept works, we have to detail the questions above and reflect the architecture

1. Did the organization establish continuous improvement teams with end-to-end coverage?
2. Which resources were available for the continuous improvement teams?
3. Which methods have been trained and coached?
4. How were the meetings organized to ensure fast organization?

---

[511] see (Yin, 2018 p. 27)
[512] See (Yin, 2018 p. 27)

5. Did the CI teams have aligned targets?
6. How did CI teams document knowledge and transfer them to projects?
7. How did CI teams ensure that projects follow the standards?
8. Which benefits could be realized?

The case study questions are more specific than the three sub-questions in the recent section.

Selection of Cases

According to Yin, the selection of cases requires defining and bounding the case. It has to explain what is included and excluded in the case studies. [513]

Possible Cases

Since Viable Project Business is a new model, which has been developed in this work, there is no organization that has already implemented it without the author. Thus, the case studies are a selection of organizations, where I, the author, significantly contributed to implement Viable Project Business.

This could lead to a conflict, since the scientific case studies could be confused with business success stories. To avoid this, the case studies mainly focus on the challenges and problems of implementation. These findings gave inspiration for improving the model. Commercial success stories are not included here.

Viable Business has been implemented in six organizations. One of the organizations was not project business but product business, another organization was too similar to case study three.

Selected Cases

Thus, four implementations of Viable Project Business have been selected for the case study:

1. Plant manufacturer for metal industries (Electric, Automation)
2. Plant manufacturer for metal industries (Mechanic)
3. Plant manufacturer for energy management
4. Plant manufacturer for logistics industries

In the selected organizations we want to study the eight case study questions.

Yin describes four types of case studies, which result from two decisions: Is it a single-case or multi-case study? Is it a holistic or embedded study? The holistic study focuses on the organization as a whole, whereas the embedded study focuses more on the subunits.[514]

In our case we have holistic multi-case study, which analyzes the introduction of Viable Project Business, the enhancement of the improvement capability, and the business results in four organizations.

Before we continue with the design of the data acquisition plan, some words about the timeline of theory development should be mentioned.

---

[513] See (Yin, 2018 pp. 27-34)
[514] See (Yin, 2018 p. 48)

About Parallelized Case Study and Theory Development

Yin recommends having a strong theory behind the case studies. The theory allows asking good questions and good interpretations. It is also important to generalize the findings in the case studies.[515]

In our case, the theory has been described very detailed in Chapters 4 - 8. This theory has been developed in parallel to the case studies.

Chronology

The idea of Viable Business evolved over seven years. Research, development and implementation has been done in parallel. Implementation raised research questions, and their answers influenced the development of new concepts. Thus, it is necessary to understand which biological concepts has been understood and translated to business at which time, in order to understand the knowledge level at the start of each pilot project (Figure 9-1).

Chronology of Research

The starting point in 2012 was the observation that many organizations in project business had problems learning fast enough. The transfer of knowledge from one project to the next was difficult, especially if many teams were affected. So, the basic idea was to focus on the learning organization rather than on the specific problems of the organization. Which structures and processes are necessary to get a fast learning process?

The focus on continuous improvement teams showed that there are many teams of the same type that have to work together – like cells in an organism. This observation can be seen as the birthday of Viable Business. It introduced a new perspective and nature as the ideal benchmarking partner.

In 2012-2013 the focus was on the understanding of tissues and organ systems in organisms and their translation to business. The model organism *C. elegans* with its neural network was used as bench marking partner. This phase was also inspired from the neuro-cybernetic works of Stafford Beer, who translated the nervous systems of humans to business. It turned out, that the neuro-cybernetic approach does not explain why an organism lives, and that the *C. elegans* is already too complex for a deep understanding.

In 2014 the focus moved to the cell level. It is impossible to understand the design of multi-cellular organism, without a deep understanding of biochemistry and molecular biology. Now the *Escherichia coli* bacteria and the *Saccharomyces cerevisiae* became the models for the translation. End of 2016 a set of 300 translations on cell level was available and ready for the integration in a functional model.

[515] See (Yin, 2018 pp. 34-42)

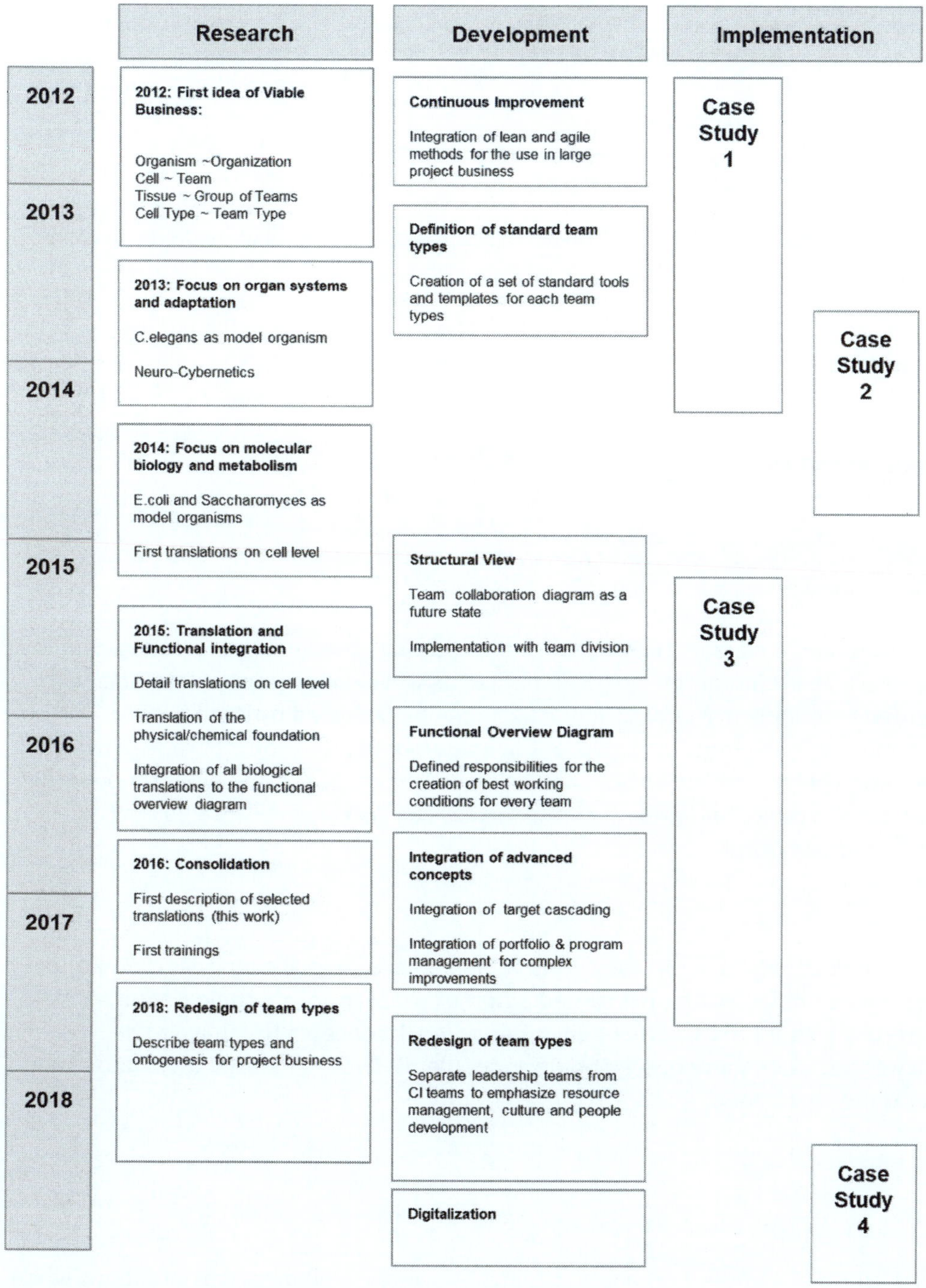

Figure 9-1: Timeline of case studies, development and research

This short history in Figure 9-1 shows, that the four case studies started on a very different base.

The first two case studies had no knowledge about cells or metabolism. The design of teams and team types was more driven by the need for continuous improvement and the activities of project business than by a deeper understanding of living systems. However, the understanding of cell junctions and tissues already allowed to create large organizations based on defined team types. During the third case study the translations and the functional design became much clearer and influenced the design of the organizations. The focus was much clearer on the provision of best working conditions for each process in each team.

The description of case studies started in 2018.

Collection of Data

How can data by collected to answer the questions? Yin proposes six sources of evidence in case studies:[516]

1. Documents
2. Archival records
3. Interviews
4. Direct observations
5. Participant-observations
6. Physical artifacts

In our case study we focus on documents, interviews with managers and coaches, and participant-observations of the author.

The degree of the end-to-end coverage, together with the meeting schedule, is documented on the team collaboration chart. The joint targets are documented as well. The content of lean and agile training is documented in training material. Other aspects can be collected by interviewing managers, coaches and the consultants. As a part of the implementation, I contribute with own perceptions. As described before, the interviews can be biased, because the implementation is the baby of all interviewees. On the other hand, the interviewees can also report those points that needs improvements in the concept.

Analysis of Data

Whereas, the collection of data tries to get facts from multiple observers, the analysis is about their interpretation. During the implementation it turned out, that some concepts could not be implemented. Reasons are either management decisions or weak concepts that have to be improved. The analysis discusses these implementation problems, tries to identify their causes and changes concepts for the next pilot.

Reporting

Anonymity

The ideal case study uses real names of organizations and informants. Yin describes situations when it can be necessary to protect the organization and the informants. One of the situations is, when the case study is on a controversial topic, another one is, "when the issuance of the final case report may affect the subsequent action of those who were studied."[517]

Anonymity of interview partners helps to get more information about that what does not work, even if a higher management has a different opinion. Anonymity of the organization helps to avoid

---

[516] See (Yin, 2018 p. 114)
[517] See (Yin, 2018 p. 239)

conflicts with the external communication of organizations, that want to describe the best picture only. Thus, the case studies are anonymized.

Structure of the Case Study

Each case study starts with a short description of the background of the business and its challenge. Then the implementation of Viable Business is described. The figures show a simplified version of the team collaboration diagram and the sequence of implementation. Next, the business results are presented. The observations describe the lessons learned. Finally, the conclusion shows the consequences of the insights. It describes new research questions and changes to the standard architecture, roadmap or toolset that are direct results of the case study.

9.1.3 Measures to Ensure the Quality of the Research Design

Literature describes four tests to analyze the quality of a research design. Yin has adapted it for the usage in case studies.[518]

Construct Validity

The question is, whether all the collected data is correct. Yin proposes to use multiple sources of evidence and to let informants review the draft case study report.[519] The following case studies have at least three sources: Documents, interview and participant-observations. The case studies have been reviewed with selected interviewees.

Internal Validity

According to Yin it is about establishing stable causal relationships. He recommend logic models and addressing rival explanations.[520] Our case studies are already based on a logical model, that has been detailed in sections 1 and 8. They contain explanations, that can be falsified in the case studies. Whenever feasible, rival explanations are used as well.

External Validity

Yin recommends using a theory to allow a generalization of the results.[521] The theory behind the case studies has been detailed in sections 1 and 8.

Reliability

Can the study be repeated? Yin recommends study protocols and a study database.[522] In our case it is critical to publish internal material like organizational charts, targets, existing problems or commercial numbers. Thus, researchers who work for the companies in the case studies, could get the data and repeat interviews with the same results. For others, the data will not be accessible.

---

[518] See (Yin, 2018 p. 43)
[519] See (Yin, 2018 p. 43)
[520] See (Yin, 2018 p. 43)
[521] See (Yin, 2018 p. 43)
[522] See (Yin, 2018 p. 43)

9.2 Case Study 1

Background

In 2012 a large plant manufacturer developed large plants for the metal industries. The project business was decentralized between two locations. Departments in one country were responsible for electrical engineering and automation of the plant. Departments in the other country were responsible for project management, mechanical engineering and installation. The introduction of Viable Business in electrical engineering and automation is covered in this case study, whereas case study 2 describes the implementation in the departments in the other country.

The organization had two very critical large projects at that time. In both cases the organization wanted to avoid delays and penalties. The organization also wanted to reduce project risks, cost and duration in all other projects.

The implementation project started 09/12 and ended 03/14.

Implementation

In summer 2012 we started with the focus on one of the two critical projects. At this time one of the two projects was in the middle of system test, which had to be passed before all automation equipment like cabinets and servers, could be sent to the construction site for installation and commissioning.

In this first case study we decided to start with a vertical breakthrough from the subproject to senior management, before we make a rollout to all processes (Figure 9-2).

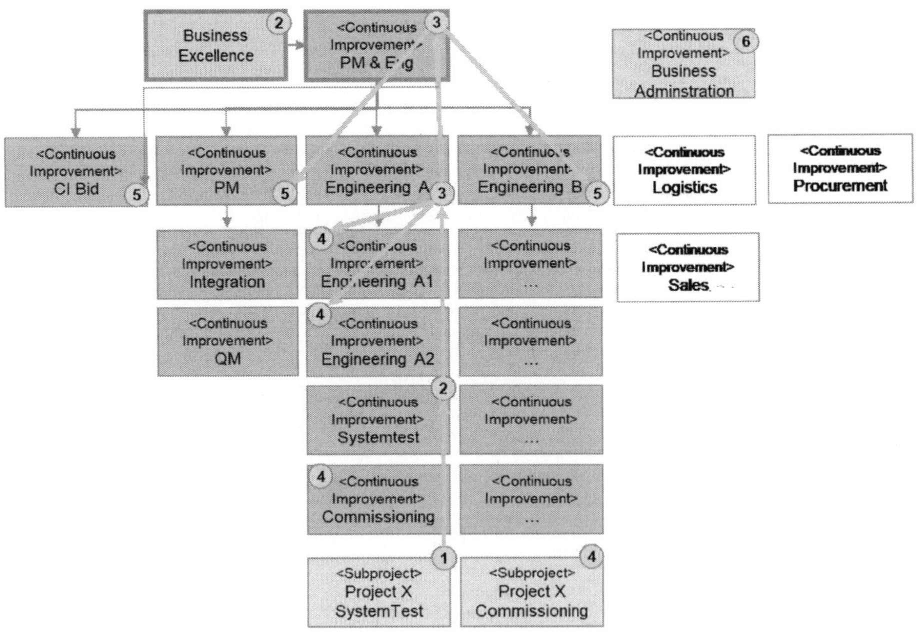

Figure 9-2: Case Study 1. Implementation started with System Test in a crisis project, before the CI-team for system test has been created. CI activities have then been extended to all PM and Engineering teams, other teams have been started with a lower intensity.

The sequence of implementation was:

1. Start with a critical subproject
2. Establish a CI team for the critical process, that is the reason for the critical subproject. Establish a business excellence team to train employees in CI methods.
3. Create a CI-team on a high level (End-to-End), to allow a fast escalation
4. Establish continuous improvement for all upstream and downstream teams and organize their know-how transfer to projects
5. Add neighbor departments that contribute to the same products
6. Establish Viable Business in other departments (like business administration)

We used three levels of continuous improvement teams to connect the (top) team with end-to-end-responsibility with the teams that are responsible for a defined set of processes. The end-to-end team was led by the head of product management, engineering and projects. The mid-level was led by department managers and the detail level was led by group managers. This design led to 1 + 5 + 20 teams.

During the next 12 months we sent 12 colleagues to a 15-days-lean-expert training and around 100 colleagues (incl. al team leaders) to a two-day lean training. With this base we extended the coverage of improvement teams to the end-to-end-process.

Results

This implementation was very successful in two perspectives

1. The critical project could be accelerated. We could significantly reduce the expected non-conformance cost by 20 million euros.
2. We established 20 improvement teams on the detail level with end-to-end-coverage from tender processes to final project acceptance.
3. We started agile project management in some customer projects.
4. We implemented A3 problem solving stories with a net benefit of 15 million euros.

Unfortunately, it was not possible to integrate those teams of the business which were not part of Engineering and Project Management. This excluded especially the sales teams and teams like logistics and procurement.

All the teams and their collaborations were stable until the carve out of the business and the exchange of the head of PM and engineering and the Viable Business coach. Several years later around half of the teams are still using the methods.

Observations

Learning Organization

This first implementation was the beginning of Viable Project Business for two reasons:

1. It moved the focus from critical projects to the learning organization
2. It showed the differences between project teams and continuous improvement teams and thus gave the idea of team types and groups of teams (cell types and tissues)
3. It showed that we need to understand groups of teams in order to develop standardized methods and collaborations.

End-to-End Coverage

It is difficult to get a full end-to-end-coverage, but it is important. If teams do not participate in continuous improvement, many problems cannot be solved. Additionally, it reduces the motivation of all other teams.

For a good end-to-end coverage, it needs a motivated senior manager who ensures that every department contributes. Ideally, the rollout starts with top management.

Usage of Predefined Team Types

During the case study we already used descriptions for every team type. This description contained a list of methods that were used and thus needed to be trained.

The descriptions separated four groups of methods:

1. Methods to make the team run: Agile sprints, Kanban boards, ...
   These methods are independent from the team type
2. Specific methods for the team type: e.g. Value Stream Analysis, A3 for CI-Teams
3. Collaborations with teams of the same type
4. Collaborations with other team types.

With group 1 we set a standard for all team types, with group 2 we described what a specific team type really has to do. This separation already introduced an object-oriented thinking with a common base class and several derived classes.

Interesting was especially group 3. The collaborations with other teams are only necessary, because the work cannot be done by one team alone. Doing the work with many teams, is the reason for these collaborations. These collaborations need to contain everything that is necessary to keep the group of teams together and to make them work.

Viable Business

In this implementation the first idea of an organic collaboration was born, but it was still unclear what the ideal rules for collaborations are. It was still unclear, which team types really exist.

During biological studies it turned out that the organ systems, tissues, cell types and collaborations can only be understood if the cell itself is understood. Secondly, it is clear that multicellular organisms are not created somewhere in the middle. Their ontogenesis starts by splitting an existing stem cell in a defined way.

The new focus on cells allowed us to understand, why cells are so successful on their own, and how they can collaborate in organisms. Thus, it widened the perspective from organism, cell types and tissue down to molecular biology and chemistry and up to design of multicellular organisms.

Conclusion

The organizations introduced Viable Business with the focus on Continuous Improvement Teams and Project Teams.

We have learned that Learning organizations should be understood as Viable Systems. They consist of teams like organisms consist of cells. It needs a much more detailed translation of cells to teams, to understand the higher-level structures.

Additionally, we have seen that it is helpful to implement the system with a top-down-approach starting with top management. Case study 2 shows the first trial.

9.3 Case Study 2

Background

The organization is responsible for project management, mechanical engineering and installation for large plants for the metal industries. It collaborates with the organization described in case study 1.

The implementation project started 09/13 and ended 09/14.

Implementation

Unlike the organization in case study 1, the top management decided to implement Viable Business on a volunteer base (Figure 9-3). Some of the departments were highly interested to introduce the concept. Top management did not want to establish regular meetings on a high management level, but they wanted to attend the improvement meetings regularly.

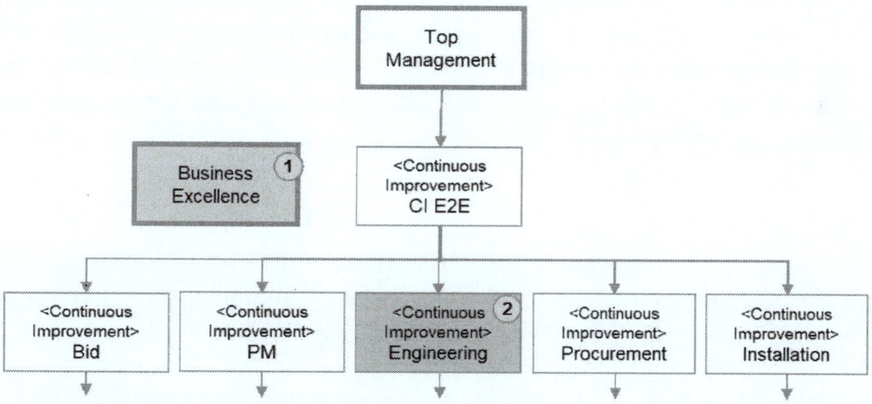

Figure 9-3: Case Study 2. Implementation started top down but on a volunteer base.

Implementation started 09/13 with nomination of the Viable Business coach, the selection of the teams and the installation of an Obeya. Like in case study one all team leads have been trained in a 2-day training. The lean coordinator attended a 15-day expert training.

The lean coordinator trained all the team members, coached the introduction of lean and agile in the selected departments and tried to convince the neighbor departments to start as well.

Results

The organization could implement lean and agile methods successfully in some selected departments, but they did not establish an end-to-end coverage. Cross-functional problem solving could not be systematically managed. However, during the project the organization could implement A3 problem solving stories with a net benefit of 15 million Euro.

The implementations of the volunteer departments are sustainable. Five years later, the continuous improvement activities in these departments are still ongoing.

Observations

Volunteer Implementation

Top management wanted to implement lean and agile but did not convince the next management level. Each department could start lean and agile on a volunteer base. Some of the engineering departments with around 100 employees started, others did not. Project management, procurement, sales and other engineering departments also did not participate.

As expected, this approach created no end-to-end coverage. As a consequence, it was difficult to solve all the cross-functional problems. The scope was limited to the participating departments.

Active Part of Top Management

In this case study the top management had no own activities regarding Viable Business. Thus, they were not part of the transformation. This way they acted as a role model and gave an example to the next management levels, which also did not use lean and agile methods.

A deeper reason for the missing management activities may be the strong focus on continuous improvement. Top management cannot contribute to the detailed process-related or technical discussions. In later case studies, the role of top management and leadership was defined much clearer. With the focus on resource management, people development, culture, and overall analysis these teams can give their contributions.

Conclusion

The organization successfully implemented lean and agile methods in several departments, but it did not reach an end-to-end-coverage. Thus, the implementation had not the architecture of a living organism.

We have learned that Viable Business should not be implemented on a volunteer base. Continuous Improvement needs the end-to-end coverage to be able to solve cross-functional problems. Top management has to ensure that all management levels are motivated and contribute to the implementation. Top management needs to have its own team and be a role model for the transformation.

9.4 Case Study 3

Background

The organization develops solutions for energy transmission solutions. In the previous years it produced high losses due to high non-conformance cost nearly every project.

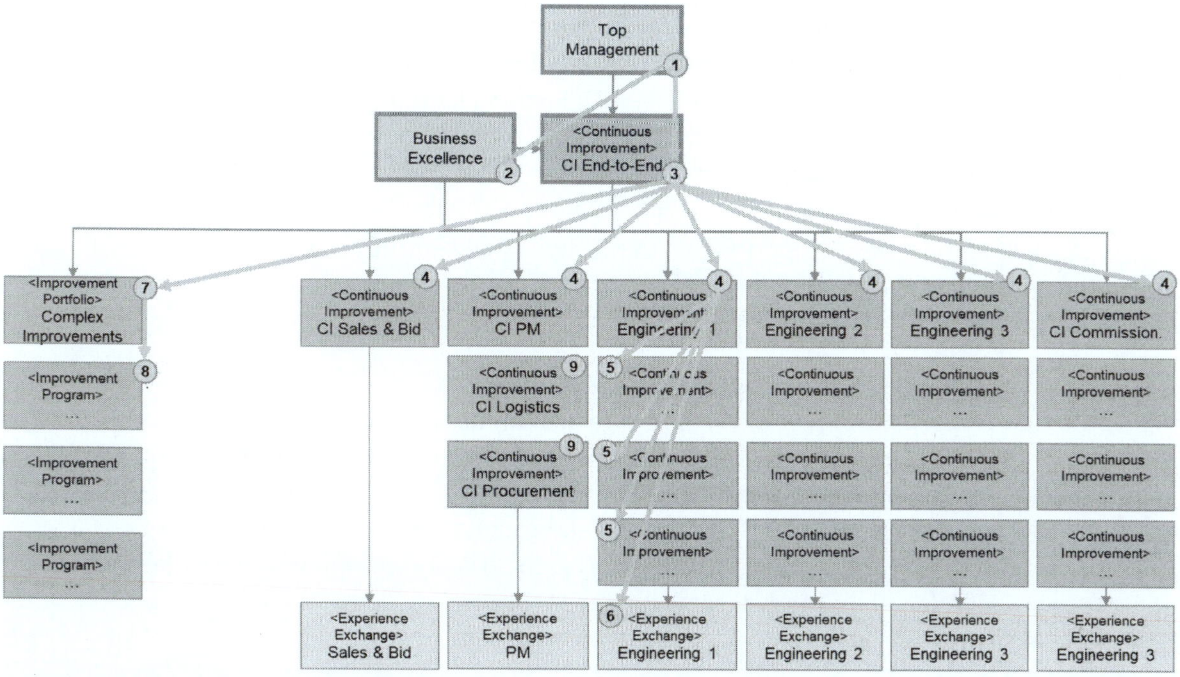

Figure 9-4: Case Study 3. Implementation started with top management using the division of teams. It creates a full end-to-end-coverage.

The organization had already a lean implementation in 2013, which introduced kanban boards, value streams and improvement methods for SW-engineering teams. However, this couldn't solve the main problems, which needed to cover the entire end-to-end process and a teamplay to solve the cross-functional problems efficiently.

The implementation project started 02/15 and ended 09/17. Viable Business was the main improvement activity of the organization between 2015 and 2017 and contained all other improvements.

Implementation

Directly at the beginning the management team decided to implement Viable Business with a top-down approach (see Figure 9-4)

1. Establish Top Management Team
    - Analyze targets and current situation
    - Define mission, vision, scorecard
    - Define future teamplay
2. Nominate the coordinator for the Viable Business implementation.
3. Establish a CI E2E team that meets weekly and serves as the anchor point for all other activities
    - Purpose: Design entire CI organization and provide best working conditions for projects
4. Establish one CI-team for each main department (6 teams)
    - Purpose: Provide best working conditions (resources, budget...) for CI teams on detail level

5. Establish one CI-team on detail level for each group (~25 teams)
    - Purpose: Provide best working conditions for projects
    - Methods: Standup, Visualization with Kanban, Kaizen, Value Stream, Work Instructions, A3
6. Setup an experience exchange team
    - Regular communication between heads of subprojects and CI teams
7. Create a team to manage the portfolio of complex improvements
8. Add all complex improvement programs.
    - Program management with Stakeholder Analysis, Benefit Maps, Alignment to KPI system
    - Connection of programs with CI teams
9. Add teams that are not part of the line organization but part of the value stream
    - E.g. Logistics and procurement teams were organizationally outside the business
10. Implement Hoshin Kanri / Target cascading
    - Definition of four future states / breakthrough targets
    - Discussion / detailing of future states with 30 teams on 4 management levels
    - Definition of quarterly targets for each team
11. Pilot agile methods in first project and subproject teams

This setup is strictly top-down. The first steps always follow the logic of cell division. Steps 3-5 took around three months for each split. This allowed to grow to 1-3-9-40-45-60 Viable Teams in around 18 months.

Results

The most important A3 problem solving stories showed a net benefit of 20′ €. The organization could additionally implement several hundred kaizen improvements.

This implementation helped to reduce non-conformance cost significantly. Most new projects are in time and budget or better. The organization is again market leader for this project type

Observations

Top-Down Approach

The top-down approach is very strong. It gives top management the opportunity to actively lead the transformation and it shows a strong commitment.

The exact sequence of the team splits is critical in practice. The introduction of improvement teams has to be at the same speed as the experience exchange teams. In this organization the agile customer projects were introduced too late. This leads to an unbalanced system, where the CI teams have no internal customers for a long time.

The implementation with three levels of CI teams might take too long for some customers. In such a case, it is possible to connect the CI E2E team directly with some frontrunning CI teams on the detail level and postpone the mid-level teams until they are really necessary.

Scorecard

At the beginning of the project the classical balanced scorecard with dimensions finance, customer, processes and people has been used. With the new knowledge of the functional overview diagram the KPIs have been restructured and linked to the functions. This creates a scorecard which has six areas, one for each stakeholder group, and one for the internal functions. This shows that KPIs are measurement points of the system dynamics, and links them to stakeholder value propositions and functions.

In the practical implementation the team did not accept the high number of performance indicators and focused more on cost, duration of project execution.

Main Key Performance Indicators

The introduction of a good KPI system in project business is also very difficult. Very important KPIs are cost and duration. It requires a very good modularization of plant products to compare the KPIs between projects; otherwise they are always unique like the project scope. On the other hand, the KPIs are needed early to create transparency for all improvement teams and to get quantitative success stories.

Improvement Programs

Improvement programs can be used, whenever an A3 is undersized and a dedicated team has to be founded. Their usage allows to manage all kind of improvements. Kaizen for small, A3 for medium programs for large improvements. The interaction with the CI teams that define standards and train and coach projects is always the same.

The introduction of improvement programs already requires stable continuous improvement teams, which are able to define and train standards. In this organization we started too early with the large improvements. If CI teams are not established, the programs have problems to establish a change of standard processes.

Most programs did not follow the recommendation to use small tranches and do the transition in steps. This leads to programs that are in the development phase for a long time and have no feedback cycles.

Culture

The entire introduction of Viable Business requires a culture change program. A parallel culture change program was done but it was not fully integrated in the other activities.

Conclusion

The organization had the best Viable Business implementation. The focus was on the setup of continuous improvement teams with an end-to-end coverage. The implementation had a strong contribution to the turnaround of the entire business.

We have seen that the top-down approach with split of teams was very successful. However, the exact sequence should be reconsidered. If we follow the ontogenesis of a multicellular organism, we can see that the organism has to be alive at every single point in time. Thus, it makes no sense for an organism to create the complete lung first before it starts with the heart. Instead everything has to be balanced.

Large suborganizations, like CI with 3 levels, should start with the top, create some detail teams and add the middle layer last. This allows them to fulfill their purpose as soon as possible and maintain their overall structure at the same time.

In project business the improvement organization and the projects have to learn and apply the methods at the same time. Since some organizations have more than 10 customer projects for each CI team, this is not possible, without delaying the setup of CI teams, and the start of improvement, significantly. A better solution is to start with the experience exchange teams in parallel with the CI teams and create a 1:1 mapping between them. This gives the CI teams a customer community. Projects and subproject can be added later.

9.5 Case Study 4

Background

The organization has three main locations in Germany and develops logistics products and different types of logistics plants.

The plant type with the highest revenue produces high losses due to high non-conformance cost in most of the projects. Another plant type has the challenge to reduce lead-time significantly to meet market requirements. Most of the reasons are not project-specific but reoccurring in all projects.

The implementation project started 04/18 and ended 02/19.

Implementation

First activities were the organizational analysis, the development of the team collaboration diagram and the first top management trainings.

The architecture is based on the one used in case study 3. The main difference is to separate leadership teams from continuous improvement teams (Figure 9-5). Leadership teams can focus more on resource management, people development and culture, and stakeholder relations. Continuous improvement teams are responsible for processes and know-how. They are led by experienced technical project managers and experts.

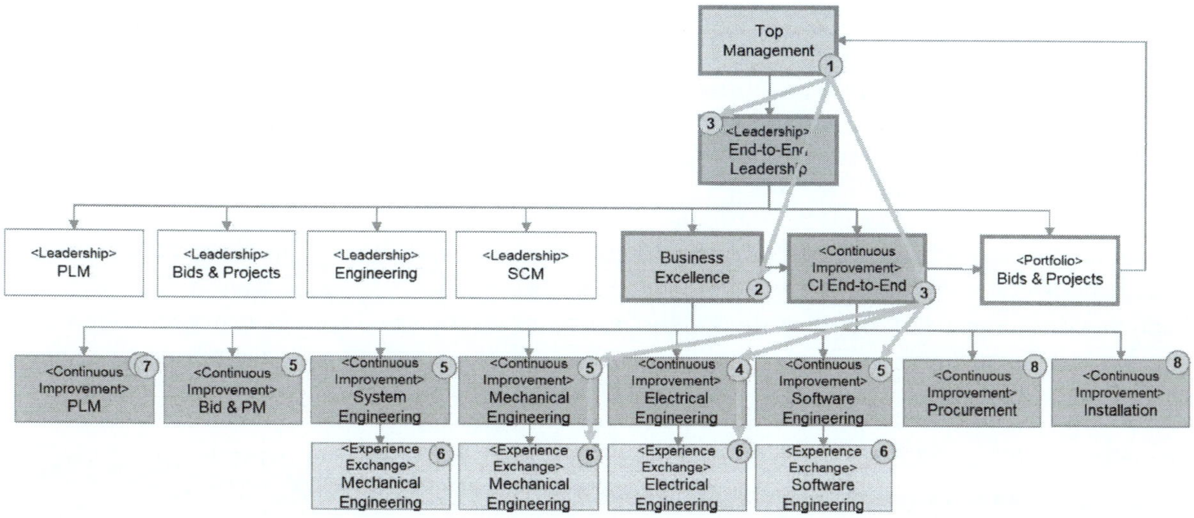

Figure 9-5: Case Study 4. CI teams care about processes and improvement; Leadership teams care about resources, people development and culture.

Due to the three different locations it was not possible to start with the top management team. Instead, the implementation started with three teams in parallel.

1. Leadership team (restricted to project management and engineering only)
2. Continuous improvement team E2E (restricted to project management and engineering)
3. Business excellence team

All three teams together were responsible to add 6 CI teams on a detail level. The leadership team was responsible for staffing, the business excellence team provided the methodic know-how and the CI E2E team was responsible for the alignment.

In 09/18 the scope was extended to the second location, where two CI teams for procurement and site have been added.

All CI teams together created the value stream analysis, described the most important processes with work instructions and created 46 A3, mainly cross-functional, problem solving stories.

The introduction of Viable Business has been steered directly by top management

Business Results

Continuous Improvement could establish transparency and a stream of A3 problem solving stories. During the first nine months the continuous improvement teams created 46 A3 improvements. The stream of A3s creates each month new improvements with a net benefit of 2 million euros.

The 46 improvements showed a typical time delay between the first idea and the realization of benefits. After some months there is a bottleneck in the improvement pipeline: There are many more ideas than resources, and many improvements remain for a longer time in the plan 3 phase and wait for resources.

Observations

Separation of Leadership Teams and Continuous Improvement Team

The separation of the leadership teams and the continuous improvement teams has some benefits and introduces some new challenges.

Having full-time employees in the CI E2E team allows to make fast progress on the top-level. The team produced a comprehensive value stream analysis with around 300 processes, defined the improvement targets and assigned them to the processes on the value stream. Cross-functional improvements could also be managed very professionally. The team created 46 A3 problem solving stories in 6 months.

The challenge is the linking of the leadership and continuous improvement teams. In case study three there was only one tree of teams. Teams were connected by the managers. Now the continuous improvement teams are managed by experts. This creates a disconnect between experts and managers.

Top Management Driven Implementation

If top management tries to drive continuous improvement, it bypasses the leadership teams. This weakens the authority of department and group heads and degrades them to observers.

Collaboration Between Several Locations

Case Study three had the luxury situation that it had only one location. Their projects were international, but project management, the engineering and procurement teams are at the same location.

In case study four the entire value stream was covered by teams at three different locations. In some cases (e.g. mechanical engineering) one process is done by employees on two locations.

Especially the leadership teams and the CI E2E team need digital collaboration tools. This includes tools for management of tasks, like digital kanban boards, tools for documenting best practices, like WIKI, and tools for specialized tasks like A3 or Value Streams.

Implementation in Tranches

It is not possible to train all competences of a team in a single step. This makes it difficult to report an implementation progress, because after the training all teams are in a long implementation phase.

Program management techniques show how to do it better. Introduction of Viable Business is a change of the organizational state and can be done in small tranches. Tranche one contains the setup of the team, the establishment of the teamplay, the orientation on (internal) customer value, and some basic methods. Tranche two contains some more advanced techniques. Especially methods like A3 problem solving cannot be trained in a single step. It makes sense to teach the basic first, and to add more details in a later tranche. Each tranche can be implemented the same way:

1. Prepare to ensure that all preconditions are given
2. Inform the team in a kickoff
3. Train team members in the new methods
4. Coach and support team members during implementation
5. Monitor the sustainable implementation and support the team

Thus, the know-how for every team type needs to be packaged in tranches. Every single team can be trained and coached step-by-step one tranche after the other. This allows to monitor progress more detailed than before.

Conclusion

Viable Business has been implemented successfully at one location. We have established very strong collaboration between all the continuous improvement teams and to the customer projects.

We have also learned that we need some improvements for the next Viable Business projects:

1. Be more consequent in staffing the team leads to ensure a better linkage of teams
2. Emphasize the role of leadership teams and inhibit that they are bypassed by top management
3. Provide digital solution for the improvement machine (incl. value stream, problem solving, kanban board, dashboards) to allow collaboration between several locations
4. Use program management techniques for the implementation of Viable Business more strictly: Split know-how of team types to tranches to create new competences

The case study showed no new biological questions.

9.6 Summary

Here we consolidate the answers to our eight case study questions and give recommendations:

Question 1: Did the organization establish continuous improvement teams with end-to-end coverage?

The organizations in case study 3 and 4 could create an end-to-end-coverage because they started top-down. Bottom-up approaches like in case study 1, or volunteer approaches like in case study 2 face the problem that important departments do not participate in continuous improvement.

A start with top management is best. It has to be ensured that also the management levels that have the resources are on board and provide the resources for the CI-teams. If this layer is missing (see case study 2), the organization gets either a problem to give resources to CI-teams, or to create the end-to-end coverage. In both cases this leads to weak improvement activities.

Question 2: Which resources were available for the continuous improvement teams?

If an organizational group has 20 persons, the CI-team should consist of the group manager and 4-5 experts of the group. All CI team members should spend at least one (better two) days per week for continuous improvement. This includes the definition of standards, the support of projects and the implementation of small improvements with kaizen. Larger improvements (e.g. implementation of A3) need additional resources.

Question 3: Which methods have been trained and coached?

Each team type uses different methods and needs a specialized training. CI teams need to be trained in the methods that have been explained in Chapter 8.

The setup of teams goes step by step. In each step new methods will be implemented. Every step should start with a kickoff, the preparation of all tools and information, and a training for the team members. All this should give best working conditions for the usage of the new methods in reality.

The training needs to be done by the business excellence team of the organization. This is necessary to continuously enhance the know-how in the organization. Thus, it is recommended to train the business excellence team first, and handover the trainings to them step by step.

Question 4: How were the meetings organized to ensure fast organization?

Case Study 1, 3 and 4 used a common schedule for continuous improvement teams, where the end-to-end team meets last and the detail teams meet directly before.

One example is case study 3, where the CI teams on detail level meet on Monday, the medium level meets on Tuesday morning, the End-to-End team meets Tuesday 13:00h and top management meets Tuesday 15:00h. This allows an escalation of 30 hours between 5 management level.

The meetings on detail level are scheduled inverse to the process chain. The team responsible for the last process meets first. If a problem is detected during a meeting, there is a chance to give to the right team which meets some hours later.

The communication between teams is very fast due to two reasons: The special scheduling of meetings and the fact that the heads of all the teams meets weekly and know each other.

Question 5: Did the teams have aligned targets?

Best results had case study 3 by implementing Hoshin Kanri. All CI teams have a quarterly review, retrospective and planning meeting, which are also cascaded.

At the beginning of the fiscal year the targets and future states are defined by top management and refined on each level of the CI teams. Continuous improvement is aligned to the targets. Major improvement programs are also aligned to the future states.

Other case studies have no target-cascading via Hoshin Kanri implemented, but they had defined the major improvement targets like reduction of cost overrun, cost reduction, reduction of lead-time or tender cost.

Question 6: How did CI teams document knowledge and transfer them to projects?

Case study 3 und 4 used Office-Tools like Excel or a Wiki to document standard processes.

The experience showed that it is difficult to convince engineers to describe processes, including presenting their unique know-how. It is important to use tools with a high usability and to avoid asking them for too many formalities like meta-data.

The best start would be work instructions with structures that are as simple as a cooking recipe. Here the expectations of the quality management department are normally much higher. However, the implementation needs to go step by step.

If the CI team members are the recognized experts in their field, the training and coaching is typically no problem. Unexperienced team members ask them anyway.

Question 7: How did CI teams ensure that projects follow the standards?

In case study 1 and 4 the CI team members were also subproject managers. This ensures that process changes are used in the projects. In case study 3 this was not possible due to the high number of projects. The experience exchange teams helped to transfer the knowledge.

Question 8: Which benefits could be realized?

Each of the organizations could realize more than 10′ € benefits in the first year by implementing A3 problem solving stories. All the improvements with kaizen cards are not counted here. Also, the improved communication cannot be counted.

In case study 3 the introduction of Viable Project Business contributed to the overall turnaround of the business.

Summary

The case studies show that Viable Project Business can be implemented. It helps to strengthen the continuous improvement organization by ensuring that resources and know-how is available. The centralized scheduling of the team meetings leads to fast escalation. The joint target setting of the continuous improvement teams lead to directed improvement towards the organization's objectives.

The benefits for the organization depend on the invested resources in continuous improvements. A systematic top-down implementation with sufficient resources is able to achieve a turnaround.

Thus, the answer to our initial research question 2: *"Can the system dynamics of Life be used to design a management system that helps us to significantly improve project business?"* can be answered with yes.

10 Summary and Outlook

10.1 Translation from Biology to Business

This work introduced a new perspective on organizations by providing a systematic translation from biology to business.

The translation of autopoiesis in this work is different from other translations. It is based on the current knowledge of biology and it is focused on organizations rather than social systems in general. The main differences to existing work are:

Whereas Luhmann constructs autopoietic organizations based on the communication of decisions[523], this work constructs them based on best working conditions for processes. Thus, it is about actions. This allows building the bridge to methods of lean management.

Whereas Beer and Malik focus on cybernetics as the management of control[524], this work also focuses on that what has to be controlled. It points out the purpose of control in cells, which is to optimize self-creation and self-protection, and translates it to business. This allows to detail the operations of system 1 in the Viable System Model.[525]

Whereas Padgett and Powell focus on the evolution of autocatalytic systems and their translation to micro-economy, this work describes autocatalysis in detail, without its evolution, and translates it to processes inside an organization.[526]

The translation in this work is not a pure functional translation. Instead, it analyses the relation of function and form, to be able to translate the requirements for the form as well. This solves the hen-egg problem that processes cannot be understood without components and vice versa. The resulting translation of autopoiesis is more specific and can be implemented with existing lean and agile methods.

10.2 Viable Business

This work introduces a description of a management system that is based on the system dynamics of life.

Viable Business Management System

The Viable Business Management System is a management system that utilizes system dynamics of life in order to achieve sustained profitable growth. It fulfills the definition of a management system as published in ISO 9000.[527]

---

[523] See (Luhmann, 2011)
[524] See (Beer, 1995), (Beer, 1990) and (Malik, 2016)
[525] See section 2.7 for the description of the 5 subsystems in the Viable System Model (VSM)
[526] See (Padgett, et al., 2012)
[527] See section 1

C. Dachs, *Viable Project Business*, Contributions to Management Science, https://doi.org/10.1007/978-3-030-62904-5_10

Life as the Best-of-Best Benchmark for Sustainably Growing Systems

The base for the Viable Business Management Systems are detail translations from biology to business. Biologists understand in detail, why cells can duplicate every 30 minutes, why cells are able to accelerate each chemical reaction by a factor of 100 billion and why billions of them can collaborate in the human body. All of that happens in nature, even under ever-changing conditions. The architecture is proven since 3.8 billion years. This work describes the concepts and translates them to a business environment.

Utilizing the System Dynamics of Life

By integrating the translations to one system, it is possible to see the system dynamics (Figure 10-1), with its reinforcing and balancing loops, rather than the single components.

These system dynamics are the reason for the growth of living systems. The inner driving force is autocatalysis: The cell creates catalysts that accelerate each of its chemical reactions. The target of the chemical reactions is to create even more catalysts. This leads to the main reinforcing loop between form (catalyst) and function (create catalyst).

All the chemical reactions need resources from outside. Sometimes the cell establishes a mutually beneficial symbiosis with other cells and exchanges resources. It is a give and take. These symbioses are also reinforcing loops that strengthen the cell.

The growth of the cell is limited at the bottleneck of this system. Thus, it has control mechanisms to identify and eliminate this bottleneck.

This picture leads to the system dynamics of the Viable Business Management System:

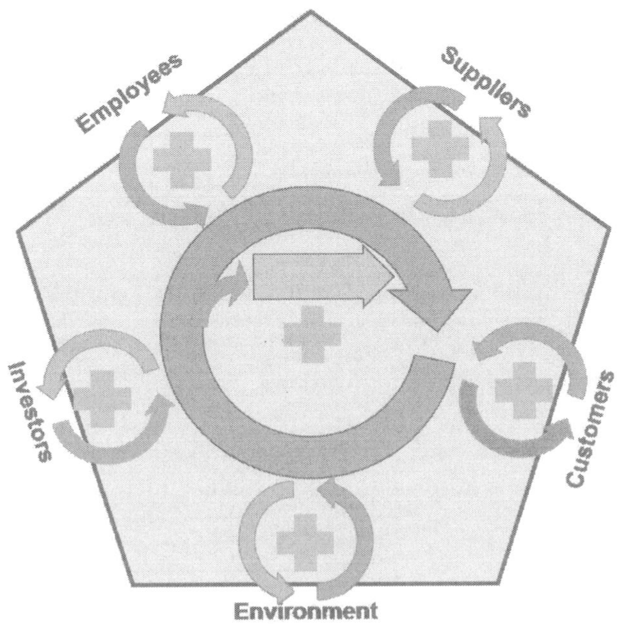

Figure 10-1: Reinforcing loops of Viable Business. The inner reinforcing loop represents the creation of best working conditions based on available resources. The outer reinforcing loops are translations of the symbiosis of lifeforms. They describe the mutually beneficial give and take of resources.

Functional Overview Diagram

The system dynamics can be detailed in a functional overview diagram of the Viable Business Management System (Figure 10-2). Each function contributes to the goal to create best working conditions (catalysts). All functions are connected circularly and have clear inputs and outputs. This allows connecting very different processes like supplier selection, knowledge management, people development and resource management in one holistic picture.

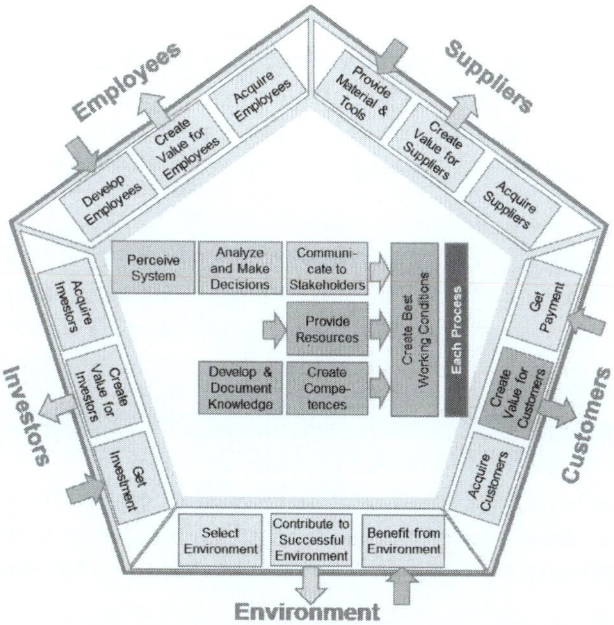

Figure 10-2: Functional Overview Diagram. The stakeholder-related functions are a detailing of the symbiosis with stakeholders. All functions contribute to the creation of best working conditions for all processes, even under changing conditions.

It is possible to detail each function. Function "Create Value for Customers", for example contains the entire End-to-End-value stream, that can be expressed by a value stream analysis.

Identifying and eliminating bottlenecks in this system means to understand the system dynamics and to identify those functions that are problematic. The responsibles can then analyze the function with a value stream analysis, design a future state and develop improvement measures. Management can provide resources to the team that improve the bottleneck.

It is worth the mention some translations of biology that help to understand this functional overview diagram.

Autocatalysis, Self-Reference and Metabolism

The creation of best working conditions is based on competences and available resources. Resources come from the stakeholders (e.g. Employees, Suppliers) and are further developed by the organization to create best working conditions for each process.

These functions are similar to the metabolic pathways in cells: cells import material from outside (maybe from symbionts), convert the material to amino-acids, and use the blueprint in the DNA, to create proteins. Some proteins are catalytic and are able to accelerate all these chemical reactions.[528]

System Boundary, Compartments

Cells have a cell membrane that allows to create high concentration of catalysts and reactants inside. They also have organelles, like the endoplasmic reticulum, which have the best environment for a set of chemical reactions. All this helps to create highest concentration of success factors.

The translation to business shows that many lean and agile methods like collocated teams, avoidance of split-heads, zoning, or the usage of meeting rules have the same target. All these concepts do not produce anything, but they create high concentration for processes and thus accelerate them.[529]

Adaptation

A cell is also able to gather inside and outside information. They can combine the information in signal pathways and influence their chemical reactions. The purpose of adaptation is to identify and eliminate the bottlenecks of growth – especially in an ever-changing environment.

In a management system this is the collection of numbers, data and facts about own resources and processes and about the outside world of stakeholders and markets. The information is used to identify current or future bottlenecks of growth and to make decisions to eliminate these bottlenecks. Finally, the organization influences business processes. In a VUCA world the organization has the same ever-changing environment and demand for adaptation like many living organisms.[530]

Symbiosis and Endosymbiosis

This model describes all stakeholders as symbionts of the organization.[531]

Humans are not part of the organization, even if the relationship is very strong. There is a give and take between employee and organization. Humans fulfill a role, invest motivation and knowledge, in return they expect a salary, and a good work environment.

Eukaryotic cells have the mitochondria as endosymbionts. According to the endosymbiosis theory, mitochondria were independent cells that were enclosed by a eukaryotic cell. The eukaryotic cell provides the entire environment to let the mitochondria grow, and in return the mitochondria provides ATP as the standardized energy carrier to the surrounding cell.

The relationship between organization and employee is not so close like in this biological example, but it has some similar characteristics. Humans are carrier of skills and energy. Both factors are

---

[528] See sections 5.1 for the translation of autocatalysis, 5.2 for DNA and RNA, and 5.3 for the metabolic network that converts all inputs to catalysts.
[529] See sections 5.4 and 5.8 for the translation of cell membrane, membrane proteins and organelles.
[530] See section 5.6 for the translation of adaptation mechanisms
[531] See section 5.7 for a description of symbiosis

where know-how of another organization is imported to improve the own processes. Future work about the evolution of the Viable Business could be based on works of Futuyma.[534]

3. Foundation

The entire translation is based on the ability to translate the foundation of space and time to a psychological and social world. This foundation is needed because it gives a stable inventory of terms that can be used to translate form and function.

The causal chains in biological systems can be analyzed because we understand that they are based on laws of nature. Molecular biology is always based on physics and chemistry. Every explanation, why an organization is more successful than another one, is finally based on laws of nature.

The same strong foundation is needed for social systems. Future works can strengthen this foundation. They can analyze space and time, energy, temperature, enthalpy, entropy and other concepts and translate them in more detail.

4. Model for Simulation

Future work could create a quantitative model for Viable Business that helps to simulate the behavior of such a system. Even if such a model is not calibrated, it can be used to teach managers the system dynamics of an organization.

Ideally, this software is like a game, which both teaches and entertains.

5. Detail translation of cells

Many mechanisms of the cell have not been translated yet.

One example is the exact understanding of biocatalysts. This work has not differentiated between the enzyme and its active site. It has also not discussed the co-enzymes. Post-translational modifications like special mechanisms of protein-folding, heat shocks, are also missing.

The annotations in each chapter describe ideas for these translations. The textbooks "Molecular biology of the cell" or "Life" contain more than enough material that can be translated.

6. Translation for other business types

This work is focused on continuous improvement in project business. Thus, it focused on only one organ system of one kind of animal.

Further research can adapt the models to software business, service business or product business.

7. Lean and Agile

The organizations in the case studies used existing lean and agile methods as building blocks. Future research could develop and test methods that follow the system dynamics of life more strictly.

---

[534] See (Futuyma, 2013)

One Examples are guiding principles. Today they are lists that have been developed based on experience. Every author has a different set of principles. However, if such a principle is true in general, then it expresses a logical dependency, that can be localized in the system dynamics. "We continuously improve our processes" omits the effect "to continuously increase efficiency and effectiveness" and "to free resources for other tasks". With the holistic approach Viable Business can be used to localize many of the principles in the functional overview diagram.

Another example is meeting rules. Viable Business compares meetings or other structures with organelles and emphasize that all rules help to increase concentration of success factors.

8. Business management textbooks

The translation work can also start from the other side. Business management textbooks cover a large number of topics, that are not integrated in Viable Business yet.

One example is the structure of the balance statement or income statement. How can this be mapped to the functional structure of Viable Business? What about non-physical assets like documented knowledge and competences? They correspond to DNA and RNA. Why does a cell not forget to create them? How is it possible to create an accounting that reflects the system dynamics?

9. Viable Consulting

If an organization is similar to an organism, then a business consultant is like a doctor. It makes sense to share the anatomic and physiologic know-how about organizations and collect experience about typical deviations (root cause and symptoms). The methods of physiology can be transferred to a business environment.

A shared picture of anatomy and physiology of the organizations allows many consultants to collaborate professionally.

10. Digitalization of Viable Business

If the system dynamics are clearly understood, it is possible to develop specific management support systems that are focused on sustained growth of the organization. They can start by managing daily improvement work, and stepwise extend their scope, like in autonomous driving, until a software is able to manage an entire organization.

Like an imaging system in healthcare, such a system can be shared by many organizations and thus learn faster than managers.

Appendices

Abbreviations

The following abbreviations are used in this work. New definitions in this work are in boldface type.

5S	Sort, Set-in-order, Shine, Standardize, Sustain
A3	Problem solving story in A3 paper format
ATP	Adenosine triphosphate
CI	Continuous Improvement
DNA	Deoxyribonucleic acid
E2E	End to End
EBIT	Earnings before interests and taxes
HR	Human Resources
IEC	International Electro-technical Commission
IPMA	International Project Management Institute
ISO	International Standardization Organization
KPI	Key performance indicator
LVO	Large Viable Organization
NCC	Non-conformance cost
P3O	Project, Program and Portfolio Office
PMI	Project Management Institute
RNA	Ribonucleic acid
	(mRNA = messenger RNA, rRNA = ribosomal RNA, tRNA = transfer RNA)
SOP	Standard Operating Procedure
TWI	Training Within Industry
VO	Viable Organization
VBMS	Viable Business Management System

C. Dachs, *Viable Project Business*, Contributions to Management Science, https://doi.org/10.1007/978-3-030-62904-5

VPBMS	Viable Project Business Management System
VSM	Viable System Model (Stafford Beer)
VT	Viable Team

Glossary

The glossary defines the terms that are used in this book. In some cases, a short explanation or some examples are added.

Since this book does not intend to explain biology or business in detail, the glossary is kept to the minimum that is necessary to allow reading of this book. Definitions of business terms are mainly taken from ISO standards, large organizations or textbooks. Definitions of biological terms are mainly taken from biological textbooks.

5S
A methodology consisting of five workplace practices – sorting, straightening, shining, standardizing, and sustaining – that are conducive to visual controls and lean production.
(Krajewski, et al., 2016 S. 633)

Amino acid
organic molecule containing both an amino group and a carboxyl group.
(Alberts, et al., 2015 p. G:2)

Amino acids are the components of proteins.

Adenosine-triphosphate (ATP)
Nucleoside triphosphate composed of adenine, ribose, and three phosphate groups.
The principal carrier of chemical energy in cells.
(Alberts, et al., 2015 p. G:3)

Auto-catalyst
system of objects and processes, where every process is accelerated by a catalyst that has been produced the system itself
(own definition)

Best working conditions
Sum of all production factors, including material, know-how and the right environment that lead to an effective and efficient process
(own definition)

Bottleneck
A capacity constraint resource whose available capacity limits the organization's ability to meet the product volume, product mix, or demand fluctuation required by the marketplace.
(Krajewski, et al., 2016 S. 630)

Viable Business focuses on the growth of the organization. The bottleneck is the resource that limits this growth.

Business Catalyst
All production factors that influence the speed of a process but are not consumed by the process.
(own definition)

This definition excludes all kind of material, but it includes the environment.

C. Dachs, *Viable Project Business*, Contributions to Management Science, https://doi.org/10.1007/978-3-030-62904-5

Catalyst
Substance that can lower the activation energy of a reaction (thus increasing its rate), without itself being consumed by the reaction.
(Alberts, et al., 2015 p. G:5)

Competence
Ability to apply knowledge and skills to achieve intended results
(ISO 9000, 2015 p. 53)

Continuous improvement
Recurring activity to enhance performance
(ISO 9000, 2015 p. 30)

Remark: Original term in ISO 9000 is "continual improvement" instead of "continuous improvement"

Deoxyribonucleic acid (DNA)
Polynucleotide formed from covalently linked deoxyribonucleotide units. The store of hereditary information with a cell and the carrier of this information from generation to generation.
(Alberts, et al., 2015 p. G:9)

Efficiency
Relationship between the result achieved and the resources used
(ISO 9000, 2015 p. 46)

Effectiveness
Extent to which planned activities are realized and planned results are achieved
(ISO 9000, 2015 p. 46)

Enzyme
Protein that catalyzes a specific chemical reaction
(Alberts, et al., 2015 p. G11)

Function
Intended or accomplished purpose or task
(IEC, 2009 S. 12)

Improvement
Activity to enhance performance
(ISO 9000, 2015 p. 30)

Interested Party
Person or *Organization* that can affect, be affected by, or perceive itself to be affected by a decision or activity
(ISO 9000, 2015 p. 28)

Kanban
A Japanese word meaning "card" or "visible record" that refers to cards used to control the flow of production through a factory.
(Krajewski, et al., 2016 S. 634)

Large Viable Organization
A Viable Organization consisting of other Viable Organizations
(own definition)

Large Viable Organizations are the translation of multicellular organism that consists of cells.

Management
Coordinated activities to direct and control an organization
(ISO 9000, 2015 p. 31)

Management System
Set of interrelated or interacting elements of an organization to establish policies and objectives, and processes to achieve those objectives
(ISO 9000, 2015 p. 36)

Metabolism
The sum total of the chemical processes that take place in living cells. All of catabolism and anabolism.
(Alberts, et al., 2015 p. G:19)

Mission
Organization's purpose for existing as expressed by top management
(ISO 9000, 2015 p. 38)

Objective
Results to be achieved
(ISO 9000, 2015 p. 42)

Organelle
Subcellular compartment or large macromolecular complex, often membrane-enclosed, that has a distinct structure, composition, and function. Examples are nucleus, nucleolus, mitochondrion, Golgi apparatus, and centrosomes.
(Alberts, et al., 2015 p. G:23)

Organization
Person or group of people that has its own functions with responsibilities, authorities and relationships to achieve its objectives
(ISO 9000, 2015 p. 27)

Performance
Measurable result
(ISO 9000, 2015 p. 45)

Policy
Intentions and direction of an organization as formally expressed by its top management
(ISO 9000, 2015 p. 38)

Process (technical view of IEC)
Set of interacting operations by which material, energy or information is transformed, transported or stored
(IEC, 2009 S. 12)

Process (quality management view of ISO 9001)
Set of interrelated of interacting activities that use inputs to deliver an intended result
(ISO 9000, 2015 p. 33)

Product
Output of an organization that can be produced without any transaction taking place between the organization and the customer
(ISO 9000, 2015 p. 44)

Product
Intended or accomplished result of labor, or of a natural or artificial process
(IEC, 2009 S. 12)

Protein
The major macromolecular constituent of cells. A linear polymer of amino acids linked together by peptide bonds in a specific sequence.
(Alberts, et al., 2015 p. G:26)

Program
Temporary, flexible organization created to coordinate, direct and oversee the implementation of a set of related projects and activities in order to deliver outcomes and benefits related to the organization's strategic objectives.
(Axelos, 2011 p. 5)

Project
Unique process, consisting of a set of coordinated and controlled activities with start and finish dates, undertaken to achieve an objective conforming to specific requirements, including the constraints of time, cost and resources.
(ISO 9000, 2015 p. 34)

Requirement
Need or expectation that is stated, generally implied or obligatory
(ISO 9000, 2015 p. 39)

Resource
An organization's physical or virtual entities (human or other) that are of limited availability and can be used to undertake operations or business change.
(Axelos, 2011 p. 132)

Ribosome
Particle composed of rRNAs and ribosomal proteins that catalyzes the synthesis of protein using information provided by mRNA.
(Alberts, et al., 2015 p. G:28)

Ribozyme
An RNA molecule with catalytic activity.
(Alberts, et al., 2015 p. G:28)

RNA (ribonucleic acid)
Polymer formed from covalently linked ribonucleotide monomers.
(Alberts, et al., 2015 p. G:28)

RNA world
Hypothesis that early life on Earth was based primarily on RNA molecules that both stored genetic information and catalyzed biochemical reactions.
(Alberts, et al., 2015 p. G:28)

Strategy
Plan to achieve a long-term or overall objective
(ISO 9000:2015)

Structure
Organization of relations among objects of a system describing constituency relations (consists of, is-a-part-of)
(IEC, 2009 S. 12)

Success
Achievement of an objective
(ISO 9000, 2015 p. 43)

Sustained success
Success over a period of time
(ISO 9000, 2015 p. 43)

System
Set of interrelated objects considered in a defined context as a whole and separated from the environment
(IEC, 2009 S. 11)

Top Management
Person or group of people who directs and controls an organization at the highest level
(ISO 9000, 2015 p. 35)

Transcription
Copying of one strand of DNA into a complementary RNA sequence by the enzyme RNA polymerase.
(Alberts, et al., 2015 p. G:32)

Translation
Process by which the sequence of nucleotides in an mRNA molecule directs the incorporation of amino acids into protein. Occurs on a ribosome.
(Alberts, et al., 2015 p. G:32)

Value Stream Mapping
A qualitative lean tool for eliminating waste or muda that involves a current state drawing, a future state drawing, and an implementation plan.
(Krajewski, et al., 2016 S. 641)

Viable Business Management System
Abstract management system of a Viable Organization, which can be used as a starting point for specific designs
(own definition)

Viable Organization
An organization that implements the system dynamics of life to achieve sustained success
(own definition)

Viable Organization is the translation of biological organism to business

Viable Project Business Management System
Abstract management system of a Large Viable Organization that can be a starting point for specific designs in project business
(own definition)

Viable Team
Viable Organization that has no contained Viable Organizations
(own definition)

Viable Team is the translation of a cell to business. It is the smallest entity.

Vision
Aspiration of what an organization would like to become as expressed by top management
(ISO 9000, 2015 p. 38)

Work Environment
Set of conditions under which work is performed
(ISO 9000, 2015 p. 36)

References

Accenture. 2018. *Living Business - Achieving Sustainable Growth through Hyper-Relevance.* s.l. : Accenture, 2018.

Achacoso B. Theodore, Yamamoto William S. 1991. *AY's Neuroanatomy of C. elegans for Computation.* Boca Raton : CRC Press, 1991.

Agile Manifesto. 2018. Agile Manifesto. [Online] 2018. [Cited: 02 04, 2018.] agilemanifesto.org.

Alberts, Bruce, et al. 2015. *Molecularbiology of the cell, 6thEdition.* Abingdon : Garland Science, 2015.

Allen, David. 2013. *Getting things done.* New York : Penguin Books, 2013.

Axelos. 2011b. *Management of Portfolios.* London : TSO, 2011b.

—. 2010b. *Management of Value.* London : TSO, 2010b.

—. 2011a. *Managing Successful programs.* London : TSO, 2011a.

Bartelmann, Matthias, et al. 2015. *Theoretische Physik.* Heidelberg : Springer Verlag, 2015.

Beer, Stafford. 1995. *Brain of the firm.* New York : Wiley, 1995.

—. 1990. *Diagnosing the system for Organisations.* Chichester : John Wiley & Sons, 1990.

Berg, Jeremy M., Tymoczko, John L., Gatto, Gregory J. Jr. and Stryer, Lubert. 2015. *Biochemistry, Eigth Edition.* New York : W.H. Freeman and Company, 2015.

Berghaus, Margot. 2011. *Luhmann leicht gemacht, 3. Auflage.* Köln : Böhlau Verlag, 2011.

Bicheno, John and Holweg, Matthias. 2009. *The Lean Toolbox.* Buckingham : PICSIE, 2009.

BioCyc. 2018. BioCyc Database Collection. [Online] 2018. [Cited: 02 04, 2018.] biocyc.org.

—. 2018. BioCyc Database Collection. [Online] 2018. [Cited: 02 04, 2018.] biocyc.org.

Brunner, Franz J. 2008. *Japanische Erfolgskonzepte.* München : Carl Hanser Verlag, 2008.

Butterhof, Luisa. 2017. *Bewältigung der aktuellen Herausforderungen von Projektmanagent in Infrastrukturprojekten durch agile MEthoden.* Erlangen : Masterarbeit an FAU Erlangen, 2017.

Covey, Stephen R. 1989. *7 Habits of Highly Effective People.* New York : Simon & Schuster, 1989.

Csikszentmihalyi, Mihaly. 2003. *Good Business. Leadership,Flow, and the Making of Meaning.* New York : Viking, 2003.

Deming, W. Edwards. *Out of the crisis.* Cambrige 2000 : MIT Press.

Dinero, Donald A. 2005. *Training within Industry - The Foundation of Lean.* Boca Raton ,FL, USA : CRC Press, 2005.

C. Dachs, *Viable Project Business*, Contributions to Management Science, https://doi.org/10.1007/978-3-030-62904-5

EcoCyc. 2018. EcoCyc E. coli database. [Online] 2018. [Cited: 02 04, 2018.] ecocyc.org.

Exner, Alexander, Exner, Hella and Hochreiter, Gerhard. 2009. *Selbststeuerung von Unternehmen.* Frankfurt : Campus Verlag, 2009.

Fuchs, Christoph and Golenhofen, Franziska. 2019. *Mastering Disruption and Innovation in Product Management.* Heidelberg : Springer, 2019.

Futuyma, Douglas J. 2013. *Evolution, Third Edition.* Sunderland : Sinauer Associates, 2013.

Gerberich, Claus W. 1998. *Benchmarking.* Planegg : Haufe Verlagsgruppe, 1998.

Goldratt, Eliyahu M. 1992. *The Goal. Excellence in Manufacturing.* Great Barrington, USA : North River Press, 1992.

—. 1990. *What is this thing called Theory Of Constraints and how should it be implemented.* Great Barrington : North River Press, 1990.

Goldspink, Chris and Kay, Robert. 2009. Autopoiesis and Organizations: A biological view of social system change and methods for its study. *Elsevier Science.* 2009.

Gorecki, Pawel and Pautsch, Peter. 2016. *Lean Management.* München : Hanser Verlag, 2016.

—. 2013. *Praxisbuch Lean Management.* München : Carl Hanser Verlag, 2013.

Graupp, Patrick and Wrona, Robert J. 2011. *Implementing TWI.* Boca Raton : CRC Press, 2011.

Hall, John E. 2016. *Guyton and Hall: Textbook of Medical Phsiology, 13th Edition.* Philadelphia : Elsevier, 2016.

Holacracy . 2018. Holacracy - Self-management practice for organizations. *Holacracy - Self-management practice for organizations.* [Online] 2018. [Cited: 02 05, 2018.] www.holacracy.org.

Hordijk, Wim and Mike, Steel. 2012. Autocatalytic sets extended: Dynamics, Inhibition and a Generalization. *Journal of Systems Chemistry.* 2012.

Hordijk, Wim. Autocatalytic sets: From the Origin of Life to the Economy. *Biosciencemag, Vol63, N.11.* pp. 877-881.

IEC. 2009. *IEC 81346.* Geneva, Switzerland : International Electrotechnical Commission, 2009.

IPMA. 2018. IPMA. *International Project Management Organization.* [Online] 2018. [Cited: 03 01, 2018.] www.ipma.world.

IPMA-ICB. 2018. ICB - Individual Competence Baseline. *ICB - Individual Competence Baseline.* [Online] 2018. [Cited: 03 01, 2018.] http://products.ipma.world/ipma-product/icb/.

IPMA-OCB. OCB - Organizational Competence Baseline. *OCB - Organizational Competence Baseline.* [Online] [Cited: 03 01, 2018.] http://products.ipma.world/ipma-product/ocb/.

IPMA-PEB. 2018. PEB - Project Excellence Baseline. *PEB - Project Excellence Baseline.* [Online] 2018. [Cited: 03 01, 2018.] http://products.ipma.world/ipma-product/peb/.

ISO 10004. 2018. *Quality Management - Customer satisfaction - Guidelines for monitoring and measuring.* Vernier, Geneva : ISO, 2018.

ISO 10013. 2001. *Guidelines for quality management system documentation.* Geneva : ISO, 2001.

ISO 9000. 2015. *ISO 9000.* Brüssel : CENELEC, 2015.

ISO 9001. 2015. *ISO 9001.* Brüssel : CENELEC, 2015.

ISO 9004. 2015. *ISO 9004.* Brüssel : CENELEC, 2015.

Jackson, Thomas L. 2006. *Hoshin Kanri for the Lean Enterprise.* Boca Raton, FL, USA : CRC Press, 2006.

Kaplan, Robert S., Norton, David P. 2001. *Die Strategiefokussierte Organisation.* Stuttgart : Schäffer-Poeschel Verlag, 2001.

Kieser, Alfred and Ebers, Mark. Stuttgart. *Organisationstheorien.* 2019 : Kohlhammer, Stuttgart.

Kirchhof, Robert. 2003. *Ganzheitliches Komplexitätsmanagement.* Wiesbaden : Deutscher Universitätsverlag GmbH, 2003.

Klipp, Edda, et al. 2012. *Systems Biology.* Weinheim : Wiley VCH Verlag, 2012.

Krajewski, Lee J., Malhotra, Manoj K. and Ritzman, Larry P. 2016. *Operations Management - Processes and Supply Chains - Eleventh Edition.* Harlow, UK : Pearson Education Limited, 2016.

Kuse, Miriam and Sandner, Franziska. 2009. *Allgemeine Pharmakologie.* München : Urban & Fischer Verlag, 2009.

Laloux, Frederic. 2017. *Reinventing Organizations.* München : Verlag Franz Vahlen, 2017.

Larman, Craig and Vodde, Bas. 2017. *Large-Scale Scrum - More with LeSS.* Heidelberg : dpunkt Verlag, 2017.

Lean Enterprise Institute. 2018. Lean Enterprise Institute. [Online] 2018. [Cited: 02 04, 2018.] www.lean.org.

Leech, Lawrence P. 2005. *Lean Project Management: Eight Principles for Success.* Advenced Projects : Boise, 2005.

Liker, Jeffrey K. and Hoseus, Michael. 2008. *Lean Culture.* New York : McGraw Hill, 2008.

Liker, Jeffrey K. and Meier, David P. 2007. *Praxisbuch - Der Toyota Weg.* München : Finanzbuch-Verlag, 2007.

—. 2008. *Toyota-Talent.* München : Finanzbuchverlag, 2008.

Liker, Jeffrey K. 2006. *Der Toyota-Weg.* München : Finanzbuch Verlag, 2006.

Lindner, Alexandra and Becker, Peter. 2010. *Wertstromdesign.* München : Carl Hanser Verlag, 2010.

Luhmann, Niklas and Baecker, Dirk. 2017. *Einführung in die Systemtheorie.* Heidelberg : Carl-Auer Systeme Verlag, 2017.

Luhmann, Niklas. 2011. *Organisation und Entscheidung.* Wiesbaden : VS-Verlag, 2011.

Malik, Fredmund. 2008. *Strategie des Managements komplexer Systeme - 10. Auflage.* Bern : Haupt Verlag, 2008.

—. 2016. *Strategy for managing complex systems.* Frankfurt : Campus Verlag GmbH, 2016.

Maturana, Humbert R. and Varela, Francisco J. 1980. *Autopoiesis and Cognition.* Kluwer : Dordrecht, 1980.

McKinsey & Co. 2018. *Organization: The five trademarks of agile organizations.* s.l. : McKinsey & Co, 2018.

Meadows, Donella H. 2008. *Thinking in Systems.* Chelsea : Chelsea Green Publishing, 2008.

Meyer, Bertrand. 2014. *Agile! The Good, The Hype and the Ugly.* Heidelberg : Springer International Publishing, 2014.

Nennstiel, Simon. 2013. *Allgemeine Pathologie.* München : Urban & Fischer Verlag, 2013.

Office of Government Commerce. 2008. *Portfolio, Programme and Project Offices.* London : TSO, 2008.

Ohno, Taiichi. 1988. *Toyota Production System.* Cambridge, MA, USA : Productivity Press, 1988.

Open Worm. 2018. OpenWorm. *Building the first digital life worm.* [Online] 2018. [Cited: 02 28, 2018.] www.openworm.org.

Padgett, John F. and Powell, Walter W. 2012. *The Emergence of Organizations and Markets.* Princeton : Princeton University Press, 2012.

PMI. 2018. Project Management Institute. [Online] 2018. [Cited: 02 05, 2018.] www.pmi.org.

Project Management Institute. 2008. *Project Management Body of Knowledge (PMBOK Guide) - Forth Edition.* Newton Square, Pennsylvania : Project Management Institute, 2008.

—. 2008. *The Standard for Program Management.* Newtown Square, Pennsylvania : Project Management Institute, 2008.

Pyzdek, Thomas and Keller, Paul. 2010. *The Six Sigma Handbook, Third Edition.* New York : McGraw Hill, 2010.

Radzicka, A. and Wolfenden, R. 1995. A proficient enzyme. *Science.* 1995, Vol. Vol. 267, Issue 5194.

Randers, Jorgen. 2012. *2052: A Global Forecast for the Next Forty Years.* Vermont, USA : Chelsea Gren Publishing, 2012.

Randolph, Justus. 2009. A Guide to Write the Dissertation Literature Review. *Practical Assessment, Research and Evaluation.* 2009, Vol. 14, p. 13.

Renz-Polster, Herbert and Krautzig, Steffen. 2012. *Basislehrbuch Innere Medizin.* München : Urban & Fischer Verlag, 2012.

Robbins, Stephen P. and Judge, Timothy A. 2017. *Organizational Behavior, Seventeeenth Edition.* Harlow, UK : Pearson Education Limited, 2017.

Robbins, Stephen P. 2001. *Organisation der Unternehmung, 9. Auflage.* München : Pearson Studium, 2001.

Robertson, Brian J. 2014. *Holacracy - The new management system for a rapidly changing world.* New York : Henry Holt & Company, 2014.

Rohen, Johannes W. and Lütjen-Drecoll, Elke. 2000. *Funktionelle Histologie.* Stutgart : Schattauer, 2000.

Rother, Mike and Shook, John. 2009. *Learning to see.* Cambridge, MA, USA, lean.org : Lean Enterprise Institut, 2009.

Rother, Mike. 2010. *Toyota KATA.* New York : McGrawHill, 2010.

Rüger, Stephan. How to write a good PhD thesis and survive the viva. [Online] [Cited: 02 05, 2018.] http://people.kmi.open.ac.uk/stefan/thesis-writing.pdf.

Sadava, David, et al. 2011. *Biologie, 9. Auflage.* Heidelberg : Spektrum Akademischer Verlag, 2011.

Schwaber, Ken. 2004. *Agile Project Management with SCRUM.* Redmond : Microsoft Press, 2004.

SEI. 2018. CMMI for Development, Version 1.3. [Online] 2018. [Cited: 02 05, 2018.] https://resources.sei.cmu.edu/asset_files/TechnicalReport/2010_005_001_15287.pdf.

Senge, Peter M. 2006. *The Fifth Discipline.* New York : Currency Doubleday, 2006.

Shook, John. 2008. *Managing to Learn.* Cambridge : Lean Enterprise Institute, 2008.

Sobek, Durward K. II and Smalley, Art. 2008. *Understanding A3 Thinking.* Boca Raton : CRC Press, 2008.

Speckmann, Erwin-Josef, Hescheler, Jürgen and Köhling, Rüdiger. 2013. *Physiologie, 6. Auflage.* München : Urban & Fischer, 2013.

Sterman, John D. 2010. *Busines Dynamics, Systems Thinking and Modeling for a complex world.* New Delhi : Tata McGraw Hill Education Private Limited, 2010.

Tagkapoulos, Ilias, Liu, Yir-Chung and Tavazoie, Saeed. 2008. Predective behavior within microbial genetic networks. *Science.* 2008, Vol. 320, 5881.

Taylor, Peter. 2016. *Leading Successful PMOs.* New York : Routledge, Taylor and Francis Group, 2016.

The Productivity Press Development Team. 1999. *Cellular Manufacturing, One Piece Flow for Workteams.* New York : Productivitiy Press, 1999.

—. 2002. *Identifying Waste on the Shopfloor.* New York : Productivity Press, 2002.

—. 2002a. *Pull Production for the Shopfloor.* New York : Produtivity Press, 2002a.

—. 2002. *Standard Work on the Shopfloor.* New York : Productivity Press, 2002.

Training within Industry Program. 2009a. *Job Instruction.* Bellingham : Enna Products, 2009a.

—. 2009b. *Job Methods.* Bellingham : Enna Products, 2009b.

—. 2009c. *Job Relations.* Bellingham : Enna Products, 2009c.

—. 2009. *Problem Solving.* Bellingham : Enna Products, 2009.

—. 2009. *Program Development.* Bellingham : Enna Products, 2009.

Tuckman, Bruce and Jensen, Marie Ann. 1977. Stages of Small-Group Development Revisited. *Group Organization Studies.* 2 1977, pp. 419-427.

Tuckman, Bruce. 1965. Developmental sequence in small groups. *Psychological Bulletin.* 1965.

VDMA. 2016. *Evolution?, Industry 4.0 in Industrial Plan Manufacturing: Revolution or Evolution.* Frankfurt : VDMA, 2016.

—. 2018. *Into the future - with innovation and the capacity to adapt.* Frankfurt : VDMA, 2018.

Vester, Heinz-Günther. 2009. *Kompendium der Soziologie I: Grundbegriffe.* Wiesbaden : VS Verlag für Sozialwissenschaften, 2009.

—. 2009. *Kompendium der Soziologie II: Die Klassiker.* Wiesbaden : VS Verlag für Sozialwissenschaften, 2009.

—. 2010. *Kompendium der Soziologie III: Neuere Soziologische Theorien.* Wiesbaden : VS Verlag für Sozialwissenschaften, 2010.

Weiler, Elmar and Nover, Lutz. 2008. *Allgemeine und molekulare Botanik.* Stuttgart : Georg Thieme Verlag, 2008.

Weilkiens, Tim. 2014. *Systems Engineering mit SysML, UML 3. Auflage.* Heidelberg : dpunkt Verlag, 2014.

Wiegand, Bodo and Franck, Philip. 2011. *Lean Administration I.* Mülheim an der Ruhr : Lean Management Institut, 2011.

Wiegand, Bodo and Frnck, Philip. 2009. *Lean Administration II.* Aachen : Lean Management Institut, 2009.

Wiener, Norbert. 1949. *Cybernetics: Or Control and Communicaion of Animal and Machine.* Paris : MIT Press, 1949.

Wiewiora, Anna, et al. 2009. Barriers to effective knowledge transfer in project-based organisations. *Proceedings of the 2009 International Conference on Global Innovation in Construction Proceedings.* 2009.

Wilkinson, Ian B., et al. 2017. *Oxford Handbook of Clinical Medicine, 10th Edition.* Oxford : Oxford University Press, 2017.

Womack, James P. and Jones, Daniel T. 1996. *Lean Thinking: Banish waste and create walth in your coporation.* New York : Free Press, 1996.

Womack, James P. and Jones, Daniel T., Roos Daniel. 2007. *The machine that changed the world.* London : Simon & Schuster, 2007.

Yin, Robert K. 2018. *Case Study Research and Applications - 6th edition.* Los Angeles : SAGE Publications, 2018.

The manufacturer's authorised representative in the EU is Springer Nature Customer Service Centre GmbH, Europaplatz 3, 69115 Heidelberg, Germany. If you have any concerns regarding our products, please contact ProductSafety@springernature.com

Printed and bound by CPI Group (UK) Ltd, Croydon, CR0 4YY
07/10/2026
02225517-0002